Wilhelm Stoll

Value Distribution Theory for Meromorphic Maps

Aspects of Mathematics

Aspekte der Mathematik

Editor: Klas Diederich

Vol. E1: G. Hector/U. Hirsch, Introduction to the Geometry of Foliations, Part A

Vol. E2: M. Knebusch/M. Kolster, Wittrings

Vol. E3: G. Hector/U. Hirsch, Introduction to the Geometry of Foliations, Part B

Vol. E4: M. Laska, Elliptic Curves over Number Fields with Prescribed Reduction Type

Vol. E5: P. Stiller, Automorphic Forms and the Picard Number of an Elliptic Surface

Vol. E6: G. Faltings, G. Wüstholz et al., Rational Points
(A Publication of the Max-Planck-Institut für Mathematik, Bonn)

Vol. E7: W. Stoll, Value Distribution Theory for Meromorphic Maps

---

Vol. D1: H. Kraft, Geometrische Methoden in der Invariantentheorie

The texts published in this series are intended for graduate students and all mathematicians who wish to broaden their research horizons or who simply want to get a better idea of what is going on in a given field. They are introductions to areas close to modern research at a high level and prepare the reader for a better understanding of research papers. Many of the books can also be used to supplement graduate course programs.
The series comprises two sub-series, one with English texts only and the other in German.

Wilhelm Stoll

Value Distribution Theory for Meromorphic Maps

V

Springer Fachmedien Wiesbaden GmbH

Prof. Dr. ***Wilhelm Stoll*** is Professor of Mathematics at the University of Notre Dame, Notre Dame, Indiana 46556, USA

AMS Subject Classification: 32 H 30, 32 A 22

ISBN 978-3-663-05294-4 ISBN 978-3-663-05292-0 (eBook)
DOI 10.1007/978-3-663-05292-0
1985

Originally published by Friedr. Vieweg & Sohn Verlagsgesellschaft in 1985.

Produced by Lengericher Handelsdruckerei, Lengerich

Dedicated to the memory of

Yozo Matsushima

CONTENTS

Preface

Value distribution theory studies the behavior of mermorphic maps. Let $f : M \longrightarrow N$ be a meromorphic map between complex manifolds. A target family $\mathfrak{A} = (E_a)_{a \in A}$ of analytic subsets E_a of N is given where A is a connected, compact complex manifold. The behavior of the inverse family $f^*(\mathfrak{A}) = (f^{-1}(E_a))_{a \in A}$ is investigated. A substantial theory has been created by many contributors. Usually the targets E_a stay fixed. However we can consider a finite set $\mathfrak{G}$ of meromorphic maps $g : M \longrightarrow A$ and study the incidence $f(z) \in E_{g(z)}$ for $z \in M$ and some $g \in \mathfrak{G}$.

Here we investigate this situation: M is a parabolic manifold of dimension m and $N = \mathbb{P}_n$ is the n-dimensional projective space. The family of hyperplanes in $\mathbb{P}_n$ is the target family parameterized by the dual projective space $\mathbb{P}_n^*$. We obtain a Nevanlinna theory consisting of several First Main Theorems, Second Main Theorems and Defect Relations and extend recent work by B. Shiffman and by S. Mori. We use the Ahlfors-Weyl theory modified by the curvature method of Cowen and Griffiths.

The Introduction consists of two parts. In Part A, we sketch the theory for fixed targets to provide background for those who are familar with complex analysis but are not acquainted with value distribution theory. In Part B, we outline the results of this monograph. A specialist can read Part B directly, and has only occasionally to look up some definitions, notations or facts in Part A.

The main part consists of 11 Sections, which can be read independently from the introduction. In Section 1, a new contraction operator is introduced, which is most helpful. After necessary preparations in Section 2, the First Main Theorem for an abstract operation is proved in Section 3. The general form of the First Main Theorem proves to be most helpful in its various and sometimes surprising applications. In Section 4, the definition of associated maps is recapitulated and convenient notations for the interaction of the associated maps of the given map $f : M \longrightarrow \mathbb{P}_n$

and the target map $g : M \longrightarrow \mathbb{P}_n^*$ are introduced. This interaction is studied in Section 5. Here the stress curvature formula together with the stress gradient formula is most likely the deepest result of these investigations. These identities should be of considerable interest to complex differential geometry quite independently from value distribution. They enable us to prove the Alfors Estimates for moving targets in Section 6. The treatment of general position in Section 7 is another highpoint in this research. The new contraction operator simplifies and clarifies the exposition. A gauge measure and the First Main Theorem for the exterior product permit us to solve the problem of general position for moving targets. The Second Main Theorem and the defect relation are established in Section 8. They are applied in Section 9 to create a value distribution theory over a function field of rank $m - 1$. As a special case and up to minor modifications in the assumptions, Shiffman's defect relation for functions is obtained. The particular choice of the holomorphic form B defining the differential operator in the associated maps is surprising and sheds new light upon this method which I introduced thirty years ago. An example is calculated in Section 10. It shows that deriving the Second Main Theorem is like walking a tight rope. In Section 11, the defect relation of Mori is improved and established on parabolic covering manifolds. References and an index conclude the monograph.

This research was carried out from January 1982 to July 1982 at the University of Notre Dame and completed at the Research Institute for Mathematical Sciences, Kyoto University, Kyoto, Japan where I was from May 13 to August 16 1983. In the Summer of 1984, Part A of the introduction was written on the request of the editor and Part B revised accordingly. Also Section 11 was added.

In March 1982, I was invited to the Centro de Investicacion y de estudios a vanzados del IPN, Mexico, for a week, where I lectured on a very preliminary version of the Second Main Theorem. In the Spring of 1982, I learned about Shiffman's result, which was available to me, when I developed the theory now written up in Section 9. The Research Institute for Mathematical Sciences, Kyoto, held a conference on value distribution theory July 4 to July 8, 1983 where I reported on the results of this monograph in a series of lectures. I learned about Mori's results at this conference.

This research was supported by the University of Notre Dame through a leave during the Academic Year 1981–82, by the National Science Foundation Grants MCS 82-01158 and DMS 84-04921, by the Japan Society for the Promotion of Science and by the Research Institute for Mathematical Science, Kyoto University. I thank all these agencies and institutions for their help and support without which this work would have been impossible.

Spring Dawn Reader designed the non-standard characters for this monograph and typeset the entire manuscript on the IBM Personal Computer using the Radio Shack TRS-80 DMP 2100 printer and a program designed for mathematical manuscripts written by R. J. Milgram of Stanford University. I thank here for her great energy, remarkable expertice and dedication to this project.

On April 9, 1983, Professor Yozo Matsushima died in Osaka. He was a great mathematician, a fine gentleman, a long-time colleague and a good friend. I dedicate these pages to him.

Wilhelm Stoll

Letters

Latin Capital	Latin Small	German Capital	German Small	Greek Capital	Greek Small
A	a	𝔄	𝔞	Α	α
B	b	𝔅	𝔟	Β	β
C	c	ℭ	𝔠	Γ	γ
D	d	𝔇	𝔡	Δ	δ
E	e	𝔈	𝔢	Ε	ε
F	f	𝔉	𝔣	Ζ	ζ
G	g	𝔊	𝔤	Η	η
H	h	ℌ	𝔥	Θ	θ
I	i	ℑ	𝔦	Ι	ι
J	j	𝔍	𝔧	Κ	κ
K	k	𝔎	𝔨	Λ	λ
L	l	𝔏	𝔩	Μ	μ
M	m	𝔐	𝔪	Ν	ν
N	n	𝔑	𝔫	Ξ	ξ
O	o	𝔒	𝔬	Ο	ο
P	p	𝔓	𝔭	Π	π
Q	q	𝔔	𝔮	Ρ	ρ
R	r	ℜ	𝔯	Σ	σ, ς
S	s	𝔖	ſ	Τ	τ
T	t	𝔗	𝔱	Υ	υ
U	u	𝔘	𝔲	Φ	φ
V	v	𝔙	𝔳	Χ	χ
W	w	𝔚	𝔴	Ψ	ψ
X	x	𝔛	𝔵	Ω	ω
Y	y	𝔜	𝔶		
Z	z	ℨ	𝔷		

Introdution

A. Value Distribution Theory for Fixed Targets

First we sketch some features of classical value distribution theory for fixed targets to provide the necessary background.

A1. Functions of one variable. The degree of a polynomial measures its growth, counts the number of points in its fibers and provides a valuation of the ring of polynomials. A polynomial splits into linear factors. Within 50 years, 1875–1925, strikingly similar features were discovered in the theory of transcendental entire functions. Weierstrass [116] provided the factorization. Picard [74] showed, that all values are assumed except perhaps one. Jensen's formula connects the number of zeros with the growth. The order of an entire function is an analog to the degree of a polynomial.

However, these results were difficult to extend to meromorphic functions since the maximum modulus could no longer serve as a growth measure. Within three years, 1922–1925, R. Nevanlinna created value distribution, which revolutionized the theory of entire functions and extended it to meromorphic functions solving the problem. Ever since, value distribution has remained Nevanlinna theory. We will outline his two main theorems.

A2. Classical value distribution of one variable. For $r > 0$ define

$$\mathbb{C}[r] = \{r \in \mathbb{C} \mid |z| \leqslant r\} \qquad \mathbb{C}(r) = \{z \in \mathbb{C} \mid |z| < r\} \tag{A.1}$$

$$\mathbb{C}\langle r\rangle = \{z \in \mathbb{C} \mid |z| = r\} \qquad \mathbb{C}_* = \mathbb{C} - \{0\}. \tag{A.2}$$

On any complex manifold, the exterior derivative splits $d = \partial + \bar{\partial}$ and twists to

(A.3) $$d^c = \frac{i}{4\pi}(\bar{\partial} - \partial) .$$

An exhaustion τ of $\mathbb{C}$ is defined by $\tau(z) = |z|^2$ such that $dd^c\tau$ is a Kaehler metric and such that $\log \tau$ is harmonic. Define $\sigma = d^c \log \tau$. Then $d\sigma = 0$. The form σ induces a rotation invariant measure on each circle $\mathbb{C}\langle r\rangle$ with

(A.4) $$\int_{\mathbb{C}\langle r\rangle} \sigma = 1 \qquad \text{for } 0 < r \in \mathbb{R} .$$

The **Riemann sphere** $\mathbb{P}_1$ is realized as an euclidean sphere of diameter 1 in $\mathbb{R}^3$. Let Ω be the rotation invariant surface measure on $\mathbb{P}_1$ with total measure 1. Let $\square\, w;a\, \square$ be the **chordal distance** between the points w and a in $\mathbb{P}_1$. If $a \in \mathbb{P}_1$ is fixed and if $w \in \mathbb{P}_1 - \{a\}$ is variable, we have

(A.5) $$\Omega(w) = -\, dd^c \log \square\, w;a\, \square^2 .$$

If $a = \infty$, and if $\mathbb{P}_1 = \mathbb{C} \cup \{\infty\}$ is identified with the compactified plane

(A.6) $$\Omega = dd^c \log(1 + \tau) = \frac{dd^c\tau}{(1+\tau)^2} \qquad \text{on } \mathbb{C} .$$

Let f be a non-constant, meromorphic function on $\mathbb{C}$. Then $f : \mathbb{C} \longrightarrow \mathbb{P}_1$ is an open, holomorphic map which spreads each disk $\mathbb{C}(t)$ over an open subset of $\mathbb{P}_1$. The spread volume is

(A.7) $$A_f(t) = \int_{\mathbb{C}(t)} f^*(\Omega) > 0 .$$

The **spherical image function** A_f increases and is continuous. For $0 \leq s < r$, the **Ahlfors–Shimizu Characteristic function** T_f is defined by

$$\text{(A.8)} \qquad T_f(r,s) = \int_s^r A_f(t) \frac{dt}{t} > 0 .$$

For fixed s, the characteristic serves as a growth measure for f, increases, is of class C^1 and is convex in $\log r$. If $\mu : \mathbb{P}_1 \longrightarrow \mathbb{P}_1$ is a holomorphic isometry, then $T_{\mu \circ f} = T_f$ in particular $T_{(1/f)} = T_f$.

Take $a \in \mathbb{P}_1$. The **compensation function** m_f of f for a is defined by

$$\text{(A.9)} \qquad m_f(r;a) = \int_{\mathbb{C}\langle r\rangle} \log \frac{1}{\Box\, f, a\, \Box}\, \sigma > 0 \qquad \text{for all } r > 0 .$$

A **divisor** is an integral valued function $\nu : \mathbb{C} \longrightarrow \mathbb{Z}$ whose support

$$\text{(A.10)} \qquad \operatorname{supp} \nu = \{z \in \mathbb{C} \mid \nu(z) \neq 0\}$$

is a closed set of isolated points. The **counting function** n_ν and the **valence function** N_ν of ν are defined by

$$\text{(A.10)} \qquad n_\nu(t) = \sum_{z \in \mathbb{C}[t]} \nu(z) \qquad \text{for } 0 \leq t \in \mathbb{R}$$

$$\text{(A.11)} \qquad N_\nu(r,s) = \int_s^r n_\nu(t) \frac{dt}{t} \qquad \text{for } 0 < s < r \in \mathbb{R} .$$

The divisors form a module ϑ, and n_ν and N_ν are additive in $\nu \in \vartheta$. If $\nu \geq 0$, then n_ν and N_ν, for fixed s, increase and are non-negative. For $z \in \mathbb{C}$, let $\mu_f^a(z)$ be the **a–multiplicity** of f at

a. Then μ_f^a is a non– negative divisor on $\mathbb{C}$, whose **counting function** and **valence function** are denoted by $n_f(t;a)$ and $N_f(r,s;a)$ respectively. The **argument principle**

(A.12)
$$n_f(t,0) - n_f(t,\infty) = \int_{\mathbb{C}\langle t\rangle} \frac{df}{f}$$

if $\mathbb{C}\langle t\rangle \cap \operatorname{supp}\left[\mu_f^0 + \mu_f^\infty\right] = \emptyset$ implies the **Jensen Formula** for $0 < s < r$:

(A.13)
$$N_f(r,s;0) - N_f(r,s;\infty) = \int_{M\langle r\rangle} \log |f| \sigma - \int_{M\langle s\rangle} \log |f| \sigma .$$

For $0 < s < r$, we have Nevanlinna's **First Main Theorem** (FMTH)

(A.14)
$$\boxed{T_f(r,s) = N_f(r,s;a) + m_f(r;a) - m_f(s;a)}$$

The theorem is a conservation principle. The measure for the fiber over a plus the compensation for the outflow over the boundaries of the annulus $\mathbb{C}(r) - \mathbb{C}[s]$ balances to a quantity independent of the fiber over a. All terms in (A.14) are non–negative permitting estimates:

(A.15)
$$\boxed{N_f(r,s;a) \leqq T_f(r,s) + m_f(s;a)}$$

Up to an additive constant, the growth measure T_f of f majorizes the fiber measure N_f of $f^{-1}(a)$. We easily calculate

(A.16)
$$\int_{a\in\mathbb{P}_1} \log \frac{1}{\square\, w,a\, \square}\, \Omega(a) = \frac{1}{2}$$

for all $w \in \mathbb{P}_1$. An exchange of integration yields

(A.17)
$$\int_{a\in\mathbb{P}_1} m_f(r;a)\Omega(a) = \frac{1}{2}\,.$$

Now (A.14) and differentiation yield the **Mean Value Theorem**

(A.18)
$$T_f(r,s) = \int_{a\in\mathbb{P}_1} N_f(r,s;a)\Omega(a)$$

(A.19)
$$A_f(t) = \int_{a\in\mathbb{P}_1} n_f(t;a)\Omega(a)\,.$$

For a boundary integral representation of the characteristic some preparations are needed. For $\mathfrak{z} = (z_1,z_2)$ and $\mathfrak{w} = (w_1,w_2)$ define the

(A.20) **inner product** $\langle\mathfrak{z},\mathfrak{w}\rangle = z_1w_1 + z_2w_2$

(A.21) **exterior product** $\mathfrak{z}\wedge\mathfrak{w} = z_1w_2 - z_2w_1$

(A.22) **hermitian product** $(\mathfrak{z}\,|\,\mathfrak{w}) = z_1\bar{w}_1 + z_2\bar{w}_2$

(A.23) **the norm** $\|\mathfrak{z}\| = \sqrt{(\mathfrak{z}\,|\,\mathfrak{z})} = \sqrt{|z_1|^2 + |z_2|^2}$

Put $\mathbb{C}^2_* = \mathbb{C}^2 - \{0\}$. Identify the Riemann sphere with the projective line: $\mathbb{P}_1 = \mathbb{C}^2_*/\mathbb{C}_*$. Let $\mathbb{P} : \mathbb{C}^2_* \longrightarrow \mathbb{P}_1$ be the residual map. If we identify $\mathbb{P}_1 = \mathbb{C} \cup \{\infty\}$, then $\mathbb{P}(z,w) = z/w$ and $\mathbb{P}(w,0) = \infty$ for $z \in \mathbb{C}$ and $w \in \mathbb{C}_*$.

A holomorphic vector function $\mathfrak{v} = (g,h) : \mathbb{C} \longrightarrow \mathbb{C}^2_*$ is said to be a **reduced representation** of f if $f = \mathbb{P}\circ\mathfrak{v} = g/h$. Stokes Theorem yields the **boundary integral representation** of the characteristic

(A.24)
$$T_f(r,s) = \int_{\mathbb{C}\langle r\rangle} \log \| \mathfrak{w} \| \sigma - \int_{\mathbb{C}\langle s\rangle} \log \| \mathfrak{w} \| \sigma$$

which is most helpful to gain estimates of the characteristic. Also (A.13) and (A.24) prove the FMTH (A.14) easily. If f_1 and f_2 are meromorphic functions on $\mathbb{C}$, then (A.24) yields

(A.25)
$$T_{f_1+f_2}(r,s) \leqslant T_{f_1}(r,s) + T_{f_2}(r,s) + \text{const.}$$

(A.26)
$$T_{f_1 f_2}(r,s) \leqslant T_{f_1}(r,s) + T_{f_2}(r,s) + \text{const.}$$

which corresponds to the valuation property of the degree of polynomials.

If f is constant, the definitions (A.7) and (A.8) still hold with $A_f = T_f = 0$. If f is not constant, then $T_f(r,s) \longrightarrow \infty$ and even

(A.27)
$$T_f(r,s)/\log r \longrightarrow A_f(\infty) = \lim_{t\to\infty} A_f(t) \qquad \text{for } r \longrightarrow \infty$$

where $A_f < \infty$ if and only if f is rational.

If f is an entire function, the **maximum modulus**

(A.28)
$$M_f(r) = \operatorname{Max}\{\, |f(z)| \mid z \in \mathbb{C}[r] \}$$

is a natural growth measure of f. If we use the reduced representation $\mathfrak{w} = (1,f)$ in (A.24), we obtain

(A.29)
$$T_f(r,s) \leqslant \log M_f(r) + \tfrac{1}{2} \log 2 \ .$$

Take $0 < \theta < 1$ and $0 < s < r$. Since $\log \| \mathfrak{w} \| > 0$ is subharmonic, the Poisson formula implies

(A.30)
$$\log M_f(\theta r) \leqslant \frac{1+\theta}{1-\theta} \left(T_f(r,s) + m_f(s,\infty)\right)$$

Again, T_f is justified as a growth measure.

In the FMTH, the valence function usually dominates, while the compensation function remains small. The proper measure is the **defect**

(A.31)

$$0 \leqslant \delta_f(a) = \liminf_{r\to\infty} \frac{m_f(a)}{T_f(r,s)} = 1 - \limsup_{r\to\infty} \frac{N_f(r,s;a)}{T_f(r,s)} \leqslant 1$$

which does not depend on s. If f does not assume the value a, then $\delta_f(a) = 1$. If f is transcendental, and if f assumes a only finitely many times, then $\delta_f(a) = 1$. Fatou's Lemma and the Mean Value Theorem easily show that $\delta_f(a) = 0$ for almost all $a \in \mathbb{P}_1$.

Nevanlinna's Second Main Theorem (SMTH) improves this result tremendously: Let f be a non-constant meromorphic function on $\mathbb{C}$. Let $\mathfrak{w} : \mathbb{C} \longrightarrow \mathbb{C}^2_*$ be a reduced representation of f. The ramification (or Wronski) divisor ρ of $\mathfrak{w} \wedge \mathfrak{w}'$ does not depend on the choice of $\mathfrak{w}$. Take a finite subset $S \neq \emptyset$ of $\mathbb{P}_1$. Take $0 < s \in \mathbb{R}$. Then there is a constant $c_s > 0$ and a subset E_s of finite measure in $\mathbb{R}^+$ such that

(A.32)

$$\boxed{N_\rho(r,s) + \sum_{a\in S} m_f(r,a) \leqslant 2T_f(r,s) + c_s \log(r\, T_f(r,s))}$$

for all $r \in \mathbb{R} - E_s$ with $r > s$. This **Second Main Theorem** implies the **defect relation**

(A.33)

$$\boxed{\sum_{a\in S} \delta_f(a) \leqslant 2}$$

A value $a \in \mathbb{P}_1$ is **deficient** for f if $\delta_f(a) > 0$. Thus there are at most countably many deficient points and we can write

(A.34)
$$\boxed{\sum_{a\in\mathbb{P}_1} \delta_f(a) \leqslant 2}$$

Picard's Theorem follows, but the defect relation is stronger and deeper.

The deficiencies of meromorphic functions have been studied extensively. We cannot outline the vast theory here, but mention these points: Baerenstein's spread relation [3] has helped greatly with functions of finite lower order. Most strikingly, Drasin [24] solved Nevanlinna's Umkehr problem: "Given a non–negative function δ on $\mathbb{P}_1$ positive at at most a countable number of points such that the sum of all function values of δ does not exceed 2, there exists a meromorphic function f on $\mathbb{C}$ with $\delta_f = \delta$." A meromorphic function f of finite lower order has additional restrictions on its deficiencies, since

(A.35)
$$\sum_{a\in\mathbb{P}_1} \delta_f(a)^{\alpha} < \infty$$

converges for $\alpha \geqslant \frac{1}{3}$ (Hayman [39] for $\alpha > \frac{1}{3}$, Weitsman [117] for $\alpha = \frac{1}{3}$), where $\alpha = \frac{1}{3}$ is optimal.

Ahlfors created a value distribution theory of covering surfaces with differentiated versions of the two Main Theorems. Wittich and his school studied the value distribution of meromorphic solutions of complex differential equations. Value distribution was essential in the composition of meromorphic functions, in product representation of functions, the study of function classes, the existence of boundary values and many other topics. The theory was lifted to Riemann surfaces from the study of algebroid functions to the study of holomorphic maps between arbitrary Riemann surfaces.

The reader can find a present day view in the memorial volume 7 for Rolf Nevanlinna, Annales Academiae Scientiarum Fennicae Series AI Mathematics 1982, in particular the contributions of O. Lehto [48], W. H. J. Fuchs [28], W. K Hayman [40], S. Rickman [76] and H. Wittich [121]. Also we recommend the books by G. Valiron [113], R. Nevanlinna [66], [67], M. Tsuij [111], W. K. Hayman [39], B. Ja. Levin [53], L. Sario and K. Noshiro [78], H. Wittich [121], F. Gross [37] and A. Dinghas [23].

Value distribution has been extended outside the district of one dimensional complex analysis, to subharmonic functions in $\mathbb{P}^n$, see W. K. Hayman and P. B. Kennedy [41], to quasi-regular maps from $\mathbb{R}^n$ into itself, see Rickman [77] and of course to several complex variables which will be the topic of the rest of this monograph.

A3. <u>The beginning of value distribution in several variables</u>. Naturally, we hope for a good Nevanlinna theory for holomorphic maps $f : \mathbb{C}^m \longrightarrow \mathbb{P}_m$ of maximal rank. Most fibers are discrete and easy to count. Yet in 1922, Bieberbach [6], extending a result of Fatou [27] constructed a biholomorphic map $f : \mathbb{C}^m \longrightarrow D$, where D is open but not dense in $\mathbb{C}^m$. Our hopes are shattered.

For meromorphic functions on $\mathbb{C}^m$, Picard's Theorem is trivial; yet the extension of Nevanlinna theory is not obvious. In 1938, H. Kneser [46] achieved the break through. He restricted the given meromorphic function f to the complex lines through the origin and took the integral average of the value distribution functions over the parameter space $\mathbb{P}_{n-1}$ of these lines. A FMTH like (A.14) emerges. Had Kneser stopped here, his result would be of little consequence. However, he expressed the terms in the FMTH in geometrically meaningful ways, which today remain basic to value distribution of functions and maps of several, independent complex variables. Therefore we shall formulate his result here.

Abbreviate $\mathbb{C}^m = M$ and $M_* = M - \{0\}$. An exhaustion τ of M is defined by $\tau(\mathfrak{z}) = \|\mathfrak{z}\|^2$. The standard Kaehler metric on M is given by $dd^c\tau = \upsilon > 0$. On M_* define $\omega = dd^c \log \tau \geqq 0$ and $\sigma = d^c \log \tau \wedge \omega^{m-1}$. Then $d\sigma = 0$. Thus τ is strictly parabolic. For any $S \subseteq M$, let $S[r]$, $S(r)$, $S\langle r\rangle$ be the intersection of S with, respectively, the closed ball, the open ball, the sphere of radius $r > 0$ centered at $0 \in M$. Then σ induces the rotation invariant measure of total measure 1 on each sphere $M\langle r\rangle$. Let f be a non-constant, meromorphic function on M. The **spherical image** A_f of f is defined for $t > 0$ by

(A.36) $$A_f(t) = t^{2-2m} \int_{M[t]} f^*(\Omega) \wedge \upsilon^{m-1}$$

$$= \int_{M[t]} f^*(\Omega) \wedge \omega^{m-1} + A_f(0)$$

where $A_f(0) = \lim_{t\to 0} A_f(t)$ is an integer. If f is determined at 0, which Kneser assumed, then $A_f(0) = 0$. The **characteristic** of f is defined for $0 < s < r$ by

(A.37) $$T_f(r,s) = \int_s^r A_f(t)\,\frac{dt}{t}$$

where $T_f(r,0)$ exists if and only if f is determined at a.

Take $a \in \mathbb{P}_1$. Let $\mu_f^a(\mathfrak{z})$ be the a-multiplicity of f at $\mathfrak{z} \in M$. Let $F_a = \{\mathfrak{z} \in M \mid \mu_f^a(\mathfrak{z}) > 0\}$ be the support of μ_f^a which is empty or a pure $(m-1)$-dimensional analytic subset of M. The **counting function** of μ_f^a is defined for $t > 0$ by

(A.38) $$n_f(t;a) = t^{2-2m} \int_{F_a[t]} \mu_f^a \upsilon^{m-1} = \int_{F_a[t]} \mu_f^a \omega^{m-1} + \mu_f^a(0)$$

where υ^{m-1} is the euclidean volume element of the variety F_a. The **valence function** of μ_f^a is defined for $0 < s < r$ by

(A.39) $$N_f(r,s;a) = \int_s^r n_f(t,a)\,\frac{dt}{t}$$

The **compensation function** of f for a is defined for $r > 0$ by

$$m_f(r;a) = \int_{M\langle r\rangle} \log \frac{1}{\Box\, f\, ;\, a\, \Box}\, \sigma \tag{A.40}$$

As said before, the functions are the integral averages over $\mathbb{P}_{n-1}$ over the corresponding functions on the lines through the origin which gives the **First Main Theorem**

$$T_f(r,s) = N_f(r,s;a) + m_f(r;a) - m_f(s;a)\ . \tag{A.41}$$

Kneser adds two important theorems: 1) "If L is a linear subspace of M, then $T_{f\,|\,L}$ majorizes T_f." He proves this theorem for holomorphic functions, but it is true for meromorphic functions. 2) "Let f be an entire function of finite order on M with $f(0) = 1$. Take $s > 0$ maximal with $f(\mathfrak{z}) \neq 0$ for all $\mathfrak{z} \in M(s)$. Then $f = he^g$ where g is a polynomial and h is an entire function whose logarithm can be expressed in M(s) as an integral over the zero set F_0 of f." This is the analogon to the Weierstrass product theorem. Later Stoll [91] showed that such an integral over the given zero set F_0 in fact constructs an entire function h whose growth is controlled by the growth of F_0 . P. Lelong [49], [51] constructed the same canonical function h independently using a different integral representation. Later, Ronkin [77] gave still another integral representation. A theory of functions of finite order on M was created. In [92], Stoll used the canonical function to construct theta functions to given 2m-periodic divisors, the first outside application of value distribution theory. All these results were surveyed by Stoll [101].

The technical difficulties at the time should not be under estimated. Several complex variables was in its infant state. Many technical devices were not available. For instance, Kneser used integrals and the Stokes Theorem for bordered domains on analytic sets. This was established only later by Lelong [50] and Tung [112].

Kneser did not prove the Second Main Theorem, which was proved by Stoll [93] in a much more general setting, but contains the above setting, see Satz 24.2.

There is another corner stone of value distribution in several complex variables. In 1897, E. Borel [7] reformulated and proved Picard's Theorem in the form: "If F and G are entire functions on $\mathbb{C}$ without zeros such that $F + G = 1$, then F and G are constant." More generally he shows, if $F_1, \dots, F_n$ are zero free, entire functions with $F_1 + \dots + F_n = 1$, than at least one F_j is constant. We would expect, that the beautiful result of Borel would be mentioned in most standard textbooks on basic complex analysis in conjunction with Picard's Theorem. This is not the case. Thus we should not be surprised by D. J. Newman's question in the 1984 Mathematical Intelligencer [68]: "Do there exist three non-constant, non-vanishing, entire functions F, G, H on $\mathbb{C}$ such that $F + G + H = 1$?" Borel's Theorem does not extend to infinitely many non-constant, zero free functions:

(A.42)
$$\sum_{p=0}^{\infty} \frac{1}{p!} e^{pz-e^z} = 1 .$$

In modern, geometrical language, Borel's Theorem reads: "A holomorphic map $f : \mathbb{C} \longrightarrow \mathbb{P}_n$ which misses $n + 2$ hyperplanes in general position is linearly degenerated." A local Borel Theorem in several variables was proved by M. Green [33], [34].

In 1932, H. Cartan [12] proved a <u>Second Main Theorem</u> for a holomorphic map $f : \mathbb{C} \longrightarrow \mathbb{P}_n$. Let $\mathfrak{v} = (v_0, \dots, v_n) : \mathbb{C} \longrightarrow \mathbb{C}_*^{n+1}$ be a holomorphic vector function with $f = \mathbb{P} \circ \mathfrak{v}$ where $\mathbb{P} : \mathbb{C}_*^{n+1} \longrightarrow \mathbb{P}_n$ is the residual map. Then the <u>characteristic</u>

(A.43)
$$T_f(r) = \int_{\mathbb{C}\langle r\rangle} \log \|\mathfrak{v}\|\sigma - \log\|\mathfrak{v}(0)\|$$

is an increasing function, convex in $\log r$, which does not depend on the choice of $\mathfrak{v}$. Assume that $v_j(0) \neq 0$ for $j = 0, \dots, n$ and that

$v_0, \dots, v_n$ are linearly independent over $\mathbb{C}$. Take $q > n + 1$. Take $\mathfrak{a}_j = (a_{jn}, \dots, a_{jn}) \neq 0$ in $\mathbb{C}^{n+1}$ for $j = 1, \dots, q$. Assume that the matrix $(\mathfrak{a}_1, \dots, \mathfrak{a}_q)$ has rank $n + 1$. Define $F_j = a_{j0}v_0 + \dots + a_{jn}v_n$. Then $F_j(0) \neq 0$. Let $\mu^0_{F_j}$ be the zero divisor of F_j. Define μ_j by $\mu_j(z) = \mathrm{Min}(n, \mu^0_{F_j}(z))$ for $z \in \mathbb{C}$. Then there exists a constant $c > 0$ and a set E of finite measure in $\mathbb{R}^+$ such that

(A.44) $$(q - n - 1)T_f(r) \leq \sum_{j=1}^{q} N_{\mu_j}(r,0) + c \log(r\, T_f(r))$$

for all $r \in \mathbb{R}^+ - E$.

We kept close to Cartan's formulation, which does not contain much geometry. He does not introduce a compensation function and the FMTH appears only in the form of an estimate (A.15). He uses the maximum norm instead of the euclidean norm to define T_f. He observes that his SMTH implies Borel's Theorem. His proof rests upon the Lemma of the logarithmic derivative, which much later was extended to meromorphic functions on $\mathbb{C}^m$ by Vitter [115], who proved the Second Main Theorem for meromorphic maps $f : \mathbb{C}^m \longrightarrow \mathbb{P}_n$ by Cartan's method. Also see Biancofiore and Stoll [5] and Stoll [107].

H. Weyl and J. Weyl [118], [119] and L. Ahlfors [1] pioneered the theory of value distribution of holomorphic curves. They established the two Main theorems and the Defect relation for the intersection of a holomorphic map $f : M \longrightarrow \mathbb{P}_n$ with the hyperplanes in $\mathbb{P}_n$, where M is a Riemann surface. In a stunning performance, Ahlfors obtained a defect relation for associated maps. H. Wu [126] reorganized this theory in a modern fashion. In 1953–54 Stoll [93] united the Ahlfors–Weyl Theory and the Kneser Theory into a value distribution theory of meromorphic maps $f : M \longrightarrow \mathbb{P}_n$, where M is a non–compact Kaehler manifold. The two Main Theorems and the Defect relation were established for the intersection of the image of f with the hyperplanes in $\mathbb{P}_n$. We will sketch this theory now.

In subsequent years many different extension and versions of value distribution theory were given. They cannot be sketched here. For an account see Stoll [102].

A4. The First Main Theorem for meromorphic maps.

a) Hermitian geometry. We introduce some concepts which are also needed in Section B. Let V be a complex vector space of dimension $n+1$. Define $V_* = V - \{0\}$. Let V^* be the dual vector space. Let $\langle \mathfrak{w}, \mathfrak{v} \rangle$ be the inner product between $\mathfrak{w} \in V$ and $\mathfrak{v} \in V^*$. Let $\bigwedge_p V$ be the exterior product space. For $\mathfrak{w} \in \bigwedge_p V$ and $\mathfrak{v} \in \bigwedge_q V$ the exterior product $\mathfrak{w} \wedge \mathfrak{v}$ is defined. If $1 \leqslant q \leqslant p \leqslant n+1$, the interior product $\mathfrak{w} \llcorner \mathfrak{v} \in \bigwedge_{p-q} V$ is defined for $\mathfrak{w} \in \bigwedge_p V$ and $\mathfrak{v} \in \bigwedge_q V^*$. Since $V = V^{**}$, the roles of V and V^* can be interchanged.

Let $\mathbb{P}(V) = V_*/\mathbb{C}_*$ be the projective space associated to V. Let $\mathbb{P} : V_* \longrightarrow \mathbb{P}(V)$ be the residual map. For $A \subseteq V$, define $\mathbb{P}(A) = \mathbb{P}(A \cap V_*)$. The Grassmann cone of order $p \in \mathbb{Z}[0,n]$

$$\text{(A.45)} \qquad \widetilde{G}_p(V) = \{ \mathfrak{w}_0 \wedge \ldots \wedge \mathfrak{w}_p \mid \mathfrak{w}_j \in V \}$$

and the Grassmann manifold $G_p(V) = \mathbb{P}(\widetilde{G}_p(V))$ of order p are defined. If $x = \mathbb{P}(\mathfrak{x}) \in G_p(V)$, the $(p+1)$-dimensional linear subspace

$$\text{(A.46)} \qquad E(x) = \{ \mathfrak{w} \in V \mid \mathfrak{x} \wedge \mathfrak{w} = 0 \}$$

is assigned, where $\ddot{E}(x) = \mathbb{P}(E(x))$ is a projective plane of dimension p in $\mathbb{P}(V)$. Put $q = n - p - 1$. For $a = \mathbb{P}(\mathfrak{a}) \in G_q(V^*)$ the $(p+1)$-dimensional linear subspace

$$\text{(A.47)} \qquad E[a] = \{\mathfrak{w} \in V \mid \mathfrak{a} \llcorner \mathfrak{w} = 0\}$$

is assigned, where $\ddot{E}[a] = \mathbb{P}(E[a])$ is a projective plane of dimension p in $\mathbb{P}(V)$.

A positive definite hermitian form $(\;|\;)$ on V is called a **hermitian product** (or metric) on V, and V together with $(\;|\;)$ is said to be a **hermitian vector space**. On V a norm $\|\mathfrak{x}\| = \sqrt{(\mathfrak{x}\,|\,\mathfrak{x})}$ is defined. Natural hermitian products are induced on V^*, $\bigwedge_p V$, $\bigotimes_p V$ (**tensor product**), $\circledast_p V$ (**symmetric tensor product**). If V and W are hermitian vector spaces, so are the direct sum $V \oplus W$ and the tensor product $V \otimes W$.

Define $\tau_p : \bigwedge_{p+1} V \longrightarrow \mathbb{R}_+$ by $\tau_p(\mathfrak{x}) = \|\mathfrak{x}\|^2$. The **Fubini–Study form** on $\mathbb{P}(\bigwedge_{p+1} V)$ is the unique positive form Ω_p of bidegree $(1,1)$ with $\mathbb{P}^*(\Omega_p) = dd^c \log \tau_p$. Also write $\Omega = \Omega_0$.

Take $x = \mathbb{P}(\mathfrak{x}) \in \mathbb{P}(\bigwedge_{p+1} V)$ and $a = \mathbb{P}(\mathfrak{a}) \in \mathbb{P}(\bigwedge_{q+1} V^*)$ with $0 \leqslant q \leqslant p \leqslant n$. If $\mathfrak{x} \llcorner \mathfrak{a} \neq 0$, then (x,a) is said to be **free** for $\llcorner$ and $x \llcorner a = \mathbb{P}(\mathfrak{x} \llcorner \mathfrak{a})$ exists. In any case the **projective distance** from x to a is defined by

$$\text{(A.48)} \qquad \square\, x \dot{\llcorner} a \,\square = \frac{\|\mathfrak{x} \llcorner \mathfrak{a}\|}{\|\mathfrak{x}\|\;\|\mathfrak{a}\|}.$$

If $x \in G_p(V)$ and $a \in G_q(V^*)$, then $0 \leqslant \square\, x \dot{\llcorner} a \,\square \leqslant 1$. If $p = q$, then $\mathfrak{x} \llcorner \mathfrak{a} = \langle \mathfrak{x}, \mathfrak{a} \rangle \in \mathbb{C}$. We write

$$\text{(A.49)} \qquad 0 \leqslant \square\, x;a \,\square = \square\, x \dot{\llcorner} a \,\square = \frac{|\langle \mathfrak{x}, \mathfrak{a} \rangle|}{\|\mathfrak{x}\|\;\|\mathfrak{a}\|} \leqslant 1 .$$

If $p = q = 0$, if $a \in \mathbb{P}(V^*)$ is fixed and if $w \in \mathbb{P}(V) - \ddot{E}[a]$ is variable,

(A.50) $$\Omega(w) = - dd^c \log \Box\, w;a\, \Box^2 .$$

If $x \in \mathbb{P}(V)$, we have (Weyl [119])

(A.51) $$\int_{a \in \mathbb{P}(V^*)} \log \frac{1}{\Box\, x,a\, \Box}\, \Omega^n(a) = \frac{1}{2} \sum_{\mu=1}^{n} \frac{1}{\mu} .$$

b) Hermitian vector bundles. The operations $\oplus$, $\otimes$, $\circledast$, L, Λ extend to vector bundles. The concept of a hermitian metric along the fibers of a vector bundle is defined. A holomorphic vector bundle together with a hermitian metric along its fibers is called a hermitian vector bundle. Let $\pi : W \longrightarrow N$ be a holomorphic vector bundle over a pure k-dimensional complex manifold N. The vector space of holomorphic sections of W over N is denoted by $\Gamma(N,W)$. For $s \in \Gamma(N,W)$, let $Z(s)$ be the zero set. If $\dim_x Z(s) \leqslant k - 2$, then s is said to be reduced at $x \in N$. If $\dim Z(s) \leqslant k - 2$, then s is reduced.

Let M be a complex manifold. A holomorphic map $f : M \longrightarrow N$ pulls back W to the bundle $f^*(W)$ over M, the relative product (π,f). The standard model is defined by

(A.52) $$f^*(W) = \{(x,w) \in M \times W \mid f(x) = \pi(w)\}$$

where $\tilde{\pi} : f^*(W) \longrightarrow M$ with $\tilde{\pi}(x,w) = x$ is a vector bundle and where $\tilde{f} : f^*(W) \longrightarrow W$ with $\tilde{f}(x,w) = w$ is a bundle map over f. If $s \in \Gamma(N,W)$ a lifted section $\hat{f}(s) \in \Gamma(M,f^*(W))$ is defined by $\tilde{f}(s)(x) = (x,s(f(x)))$. If s is reduced, $\hat{f}(s)$ may not be reduced.

Let Y be another holomorphic vector bundle over N. Let $\eta : W \longrightarrow Y$ be a surjective bundle homomorphism with kernel X. Let κ be a hermitian metric along the fibers of W and let $X^\perp$ be the differentiable subbundle of W perpendicular to X. Then one and only one hermitian metric along the fibers of Y, also denoted by κ and called the quotient metric, exists such that $\eta : X^\perp \longrightarrow Y$ is an isometry.

c) **Hermitian line bundles.** Let L be a holomorphic line bundle over N with hermitian metric κ along its fibers. Let $c(L,\kappa)$ be the associated **Chern form**. If $U \neq \emptyset$ is open in N and if $s \in \Gamma(U,L)$ with $Z(s) = \emptyset$, then

(A.53) $$c(L,\kappa)\,|\,U = -\, dd^c \log \|s\|_\kappa^2 .$$

Let V^* be a linear subspace of $\Gamma(N,L)$ with $0 < \dim V^* = n + 1 < \infty$. Put $V = V^{**}$. A bundle map $\eta : N \times V^* \longrightarrow L$, called the **evaluation**, is defined by $\eta(x,\alpha) = \alpha(x) \in L_x$ for all $x \in N$ and $\alpha \in V^*$. By definition V^* **spans** L if and only if η is surjective.

Take $a = \mathbb{P}(\alpha) \in \mathbb{P}(V^*)$, then $E_L[a] = Z(\alpha)$ only depends on a. Let ℓ be a hermitian metric on V^*. The **distance** from $x \in N$ to $E_L[a]$ is defined by

(A.54) $$\Box\, x,a\, \Box_\kappa = \Box\, x,a\, \Box_{\kappa,\ell} = \frac{\|\alpha(x)\|_\kappa}{\|\alpha\|_\ell}$$

where (κ,ℓ) is said to be **distinguished** if $0 \leqslant \Box\, x,a\, \Box_\kappa \leqslant 1$ for all $x \in N$ and $a \in \mathbb{P}(V^*)$. If N is compact, κ or ℓ can be multiplied by a constant to make the pair distinguished.

Assume that V^* spans L. Then we can take for κ the induced ℓ. We have $\Box\, x,a\, \Box_{\ell,\ell}$. For $x \in N$, the map $\eta_x : V^* \longrightarrow L_x$ defined by $\eta_x(\alpha) = \alpha(x)$ is linear with kernel $E[\varphi(x)]$ where $\varphi(x) \in \mathbb{P}(V)$ is uniquely defined. This **dual classification map** $\varphi : N \longrightarrow \mathbb{P}(V)$ is holomorphic with

(A.55) $$0 \leqslant \Box\, x,a\, \Box_{\ell,\ell} = \Box\, \varphi(x);a\, \Box \leqslant 1 .$$

Thus (ℓ,ℓ) is distinguished. The name is explained as follows. The **tautological bundle**

(A.56) $$S_0(V) = \{(x,\mathfrak{z}) \in \mathbb{P}(V) \times V \,|\, \mathfrak{z} \in E(x)\}$$

is a subbundle of the trivial bundle $\mathbb{P}(V) \times V$ and defined the **quotient bundle** $Q_0(V)$ by the exact sequence

$$0 \longrightarrow S_0(V) \longrightarrow \mathbb{P}(V) \times V \longrightarrow Q_0(V) \longrightarrow 0, \tag{A.57}$$

whose dual sequence is given by

$$0 \longrightarrow Q_0(V)^* \longrightarrow \mathbb{P}(V) \times V^* \xrightarrow[\eta]{} S_0(V)^* \longrightarrow 0, \tag{A.58}$$

where $H = \mathcal{O}(1) = S_0(V)^*$ is the **hyperplane section bundle** on $\mathbb{P}(V)$ with $V^* = \Gamma(\mathbb{P}(V),H)$. If $x \in \mathbb{P}(V)$, then $H_x = E(x)^*$ is the dual of $E(x)$. If $\alpha \in V^*$, then $\eta(x,\alpha) = \alpha | E(x) \in E(x)^*$. The hermitian metric ℓ on V^* induces a hermitian metric along the fibers of $\mathbb{P}(V) \times V^*$ which defined a quotient metric ℓ along the fibers of H where the Fubini–Study form coincides with the Chern form: $\Omega = c(L,\ell)$. Then $L = \varphi^*(H)$ is the pullback of H under φ such that

$$c(L,\ell) = \varphi^*(c(H,\ell)) = \varphi^*(\Omega) . \tag{A.59}$$

d) **Meromorphic maps**. Let M and N be connected, complex manifolds with $\dim M = m$. Let $A \neq \emptyset$ be an open subset of M such that $M - A$ is analytic. Then A is dense in M. Let $f : A \longrightarrow M$ be a holomorphic map. The closure $\Gamma(f)$ of the graph $\{(x,f(x)) | x \in A\}$ in $M \times N$ is said to be the **closed graph** of f. The map f is said to be **meromorphic** on M if $\Gamma(f)$ is analytic in $M \times N$ and if the projection $\pi : \Gamma(f) \longrightarrow M$ is proper. The **indeterminacy** $I_f = \{x \in M | \#\pi^{-1}(x) > 1\}$ is analytic in M with $\dim I_f \leq m - 2$. Obviously $I_f \cap A = \emptyset$. The map f continues to a holomorphic map $f : M - I_f \longrightarrow N$. A meromorphic function is a meromorphic map into $\mathbb{P}_1$ not identic to ∞.

Let V be a complex vector space of dimension $n + 1 > 1$. Take $N = \mathbb{P}(V)$ and take A as above. Let $f : A \longrightarrow \mathbb{P}(V)$ be a holomorphic map. Let $U \neq \emptyset$ be an open, connected subset of M. A holomorphic map $\mathfrak{v} : U \longrightarrow V$ is said to be a **representation** of f if $\mathfrak{v} \not\equiv 0$ and if $f(x) = \mathbb{P}(\mathfrak{v}(x))$ for all $x \in U \cap A$ with $\mathfrak{v}(x) \neq 0$. If $a \in U$, then $\mathfrak{w}$ is a **representation** at a. The representation is **reduced** if $\dim \mathfrak{w}^{-1}(0) \leq m - 2$. The map f is meromorphic if and only if there is a representation at every point of M. A meromorphic map admits a reduced representation on every Cousin II domain in M. If $\mathfrak{w} : U \longrightarrow V$ is a reduced representation, then $U \cap I_f = \mathfrak{w}^{-1}(0)$.

Let $V_M = M \times V$ be the trivial bundle. Up to isomorphism, there exists a holomorphic line bundle L_f and a holomorphic section $F = F_f$ of $V_M \otimes L_f$ over M uniquely determined by the property:

"Let $\mathfrak{w} : U \longrightarrow V$ be a reduced representation of f. Define $\tilde{\mathfrak{w}} \in \Gamma(U, V_M)$ by $\tilde{\mathfrak{w}}(x) = (x, \mathfrak{w}(x))$. Then $\mathfrak{w}^{\Delta} \in \Gamma(U, L_f)$ exists with $Z(\mathfrak{w}^{\Delta}) = 0$ such that $F | U = \tilde{\mathfrak{w}} \otimes \mathfrak{w}^{\Delta}$."

Mostly we write $\tilde{\mathfrak{w}} = \mathfrak{w}$. Let H be the hyperplane section bundle on $\mathbb{P}(V)$. Then $L_f | (M - I_f)$ is isomorphic to $f^*(H)$ on $M - I_f$. Therefore we call L_f the **hyperplane section bundle** of f and F_f the **representation section** of f. (See Proposition 2.1). The section F_f overcomes the possible absense of global reduced representations.

e) **Divisors**. Since M is a manifold, we can identify a divisor with its multiplicity function. Let $f \not\equiv 0$ be a holomorphic function on M. Take $x \in M$. Let U be an open neighborhood of x and let $\alpha : U \longrightarrow \mathbb{C}^m(1)$ be a biholomorphic map with $\alpha(x) = 0$. For each integer $q \geq 0$ there exists uniquely a homogeneous polynomial P_q of degree q on $\mathbb{C}^m$ such that

$$f | U = \sum_{q=0}^{\infty} P_q \circ \alpha \tag{A.60}$$

converges uniformly on every compact subset of U. Since $f \not\equiv 0$, there is a number $\mu = \mu_f^0(x)$, called the **zero multiplicity of f at x** and depending on f and x only, such that

$P_0 \equiv P_1 \equiv \dots \equiv P_{\mu-1} \equiv 0 \not\equiv P_\mu$. The function $\mu_f^0 : M \longrightarrow \mathbb{Z}$ is called the <u>zero divisor of f</u>. An integral valued function $\nu : M \longrightarrow \mathbb{Z}$ is said to be a <u>divisor</u> on M if for every $x \in M$ there is an open, connected neighborhood U of x and holomorphic functions $g \not\equiv 0$ and $h \not\equiv 0$ such that $\nu | U = \mu_g^0 - \mu_h^0$. The <u>support</u> of a divisor

(A.61) $$\operatorname{supp} \nu = \overline{\{x \in M \mid \nu(x) \neq 0\}}$$

is either empty or an analytic set of pure dimension $m - 1$. A divisor ν is said to be non-negative if $\nu \geqslant 0$. The set of all divisors on M is a module ϑ_M. Now we shall introduce specific divisors which will be needed.

α) <u>Simple divisors</u>. Let S be an analytic subset of pure dimension $m - 1$ of M. Then there exists one and only one divisor ν_S with $S = \operatorname{supp} \nu_S$ and $\nu_S(x) = 1$ for every regular point x of S. Then $\nu_S \geqslant 0$. The divisor ν_S is called <u>simple</u>. If S is irreducible, ν_S is called <u>prime</u>.

β) <u>Decomposition</u>. If $\nu \not\equiv 0$ is a divisor, there is a unique, locally finite set $\mathfrak{B}$ of irreducible, $(m-1)$-dimensional analytic subsets B of M and a family $(k_B)_{B \in \mathfrak{B}}$ of non-zero integers such that

(A.62) $$\nu = \sum_{B \in \mathfrak{B}} k_B \nu_B$$

where $\nu \geqslant 0$ if and only if $k_B > 0$ for all $B \in \mathfrak{B}$. Given such a set $\mathfrak{B}$ and family $(k_B)_{B \in \mathfrak{B}}$, then (A.62) defines a divisor. Given $p \in \mathbb{Z}$, the p-<u>truncated divisor</u> $\nu^{(p)}$ of ν is defined by

(A.63) $$\nu^{(p)} = \sum_{B \in \mathfrak{B}} \operatorname{Min}(p, k_B) \nu_B .$$

γ) <u>Divisors of sections in vector bundles</u>. Let W be a holomorphic vector bundle over M and let $s \not\equiv 0$ be a holomorphic section in W

over M. Then one and only one (zero)–**divisor** μ_s of s is defined by the property:

"Let $U \neq \emptyset$ be an open, connected subset of M. Let $t \in \Gamma(U,W)$ be a reduced section. Let $h : U \longrightarrow \mathbb{C}$ be a holomorphic function such that $s = ht$ on U, then $\mu_s | U = \mu_h^0$."

If U is a Cousin II domain, t and h exist. Obviously, $\mu_s \geqslant 0$ and $\operatorname{supp} \mu_s \subseteq Z(s)$. For line bundles we have equality. If $\nu \geqslant 0$ is a divisor, a holomorphic line bundle L and $s \in \Gamma(M,L)$ exist with $\nu = \mu_s$.

δ) Divisor of a vector function. Let V be a complex vector space. Let $\mathfrak{w} : M \longrightarrow V$ be a holomorphic vector function with $\mathfrak{w} \not\equiv 0$. A section $\tilde{\mathfrak{w}} \in \Gamma(M, M \times V)$ is defined by $\tilde{\mathfrak{w}}(x) = (x, \mathfrak{w}(x))$. thus the divisor $\mu_{\mathfrak{w}} = \mu_{\tilde{\mathfrak{w}}}$ of $\mathfrak{w}$ is defined.

ε) Pullback divisor. Let $\nu : M \longrightarrow \mathbb{Z}$ be a divisor. Let $f : M \longrightarrow N$ be a holomorphic map with $f(M) \not\subseteq \operatorname{supp} \nu$. The pullback divisor $f^*(\nu)$ on M is defined by the property:

"Let U be an open, connected subset of N with $\tilde{U} = f^{-1}(U) \neq \emptyset$. Let $g \not\equiv 0$ and $h \not\equiv 0$ be holomorphic functions on U with $\nu | U = \mu_g^0 - \mu_h^0$. Then $f^*(\nu) | \tilde{U} = \mu_{g \circ f}^0 - \mu_{h \circ f}^0$."

If $f : M \longrightarrow N$ is meromorphic, then $f : M - I_f \longrightarrow N$ is holomorphic and $f^*(\nu)$ is defined on $M - I_f$. Since $\dim I_f \leqslant m - 2$, the divisor continues uniquely to a divisor $f^*(\nu)$ on M.

Let W be a holomorphic vector bundle on N. Take $0 \not\equiv s \in \Gamma(N,W)$ with $f(M) \not\subseteq \operatorname{supp} \mu_s$, then $f^*(\mu_s)$ exists. Also a section $\hat{f}(s) \in \Gamma(M, f^*(W))$ exists with $\hat{f}(s) \not\equiv 0$ if $f(M) \not\subseteq Z(s)$. then $\mu_{\hat{f}(s)} \geqslant f^*(\mu_s)$, where equality holds for line bundles. Abbreviate $\mu_f^s = \mu_{\hat{f}(s)}$.

If V^* is a finite dimensional linear subspace of $\Gamma(N,W)$, if $a = \mathbb{P}(\alpha) \in \mathbb{P}(V^*)$, then $\mu_f^a = \mu_f^\alpha$ depends on a and f only.

If $\mathfrak{w} \not\equiv 0$ is a holomorphic vector function on N with $\mathfrak{w} \circ f \not\equiv 0$, then $\mu_{\mathfrak{w} \circ f} = \mu_f^{\tilde{\mathfrak{w}}} \geqq f^*(\mu_{\mathfrak{w}})$.

ζ) The a-divisor of a meromorphic function. A meromorphic function f can be regarded as a meromorphic map $f : M \longrightarrow \mathbb{P}_1$. Take $a \in \mathbb{P}_1$. Then $\nu_a = \nu_{\{a\}}$ is a prime divisor on $\mathbb{P}_1$. If $f \not\equiv a$, the a-divisor $\mu_f^a = f^*(\nu_a)$ is defined by ϵ). If $U \neq \emptyset$ is an open, connected subset of M, if $g \not\equiv 0$ and $h \not\equiv 0$ are co-prime holomorphic functions on U with $hf = g$, then $\mu_f^a | U = \mu_{g-ah}^0$ if $a \in \mathbb{C}$ and $\mu_f^\infty | U = \mu_h^0$. If $f \not\equiv 0$, then $\mu_f = \mu_f^0 - \mu_f^\infty$ is called the divisor of f. If f_1 and f_2 are meromorphic functions on M then $\mu_{f_1 f_2} = \mu_{f_1} + \mu_{f_2}$.

η) The intersection divisor. Let V be a complex vector space of dimension $n + 1$. Let $f : M \longrightarrow \mathbb{P}(V)$ be a meromorphic map. Take $a \in \mathbb{P}(V^*)$ with $f(M) \not\subseteq \ddot{E}[a]$. Using the hyperplane section bundle H on $\mathbb{P}(V)$, the intersection divisor μ_f^a is defined by ϵ). Also $\mu_f^a = f^*(\nu_{\ddot{E}[a]})$. If $U \neq \emptyset$ is an open, connected subset of M, if $\mathfrak{w} : U \longrightarrow V$ is a reduced representaton of f, if $a = \mathbb{P}(\alpha)$ with $\alpha \in V_*^*$, then $\mu_f^a | U = \mu^0_{\langle \mathfrak{w}, \alpha \rangle}$.

e) The basic integral theorems. Let M and N be connected complex manifolds of dimension m and n respectively. Let W be a holomorphic vector bundle over N. Take $s \in \Gamma(N,W)$. Let $f : M \longrightarrow N$ be a meromorphic map with $f(M) \not\subseteq Z(s)$. Define $S = \operatorname{supp} \mu_f^s$. Let κ be a hermitian metric along the fibers of W. Abbreviate

(A.64) $$\Lambda = \log\|s\|_\kappa^2$$

Let $G \neq \emptyset$ be an open, relative compact subset of M. Assume that the boundary $\partial G = \bar{G} - G$ is either empty or a pure $(2m-1)$-dimensional, real, class C^∞ submanifold of M oriented to the exterior of G. Let ξ be a form of class C^1 and degree $2m-1$ on M. Suppose $(\Lambda \circ f)\xi$ is integrable over ∂G. Then we have the **Singular Stokes Theorem**

(A.65) $$\int_G (\Lambda \circ f)d\xi + \int_G d(\Lambda \circ f) \wedge \xi = \int_{\partial G} (\Lambda \circ f)\xi \, .$$

Let χ be a form of class C^∞ and bidegree $(m-1,m-1)$ on M. Assume that $S \cap \mathrm{supp}(\chi \mid \partial G)$ has measure zero on S. Assume that $\chi \wedge d^c(\Lambda \circ f)$ is integrable over ∂G. Then the **Residue Theorem** holds

(A.66)
$$\int_G d\chi \wedge d^c(\Lambda \circ f) + \int_G \chi \wedge dd^c(\Lambda \circ f) + \int_{S \cap G} \mu_f^s \chi = \int_{\partial G} \chi \wedge d^c(\Lambda \circ f)$$

If we take $\xi = d^c\chi$, then $d\chi \wedge d^c\Lambda \circ f = d\Lambda \circ f \wedge d^c\chi$. We obtain the **Green–Residue Theorem**

$$\int_G \chi \wedge dd^c(\Lambda \circ f) - \int_G (\Lambda \circ f)dd^c\chi + \int_{S \cap G} \mu_f^s \chi$$

(A.67)
$$= \int_{\partial G} \chi \wedge d^c(\Lambda \circ f) - \int_{\partial G} (\Lambda \circ f)d^c\chi \, .$$

Proofs for even more general situations can be found in Stoll [103]. Also see Stoll [93], [96], [97], [98], [108] and Griffiths and King [35]. Before we proceed a number of important applications shall be listed.

1. <u>**APPLICATION**</u>: <u>**Theorem of Poincaré–Lelong**</u> [51]. Take $N = \mathbb{C}$, $W = \mathbb{C} \times \mathbb{C}$, $s \in \Gamma(\mathbb{C},W)$ with $s(z) = (z,z)$. Let $f \not\equiv 0$ be a holomorphic function on M. Then $\mu_f^s = \mu_f^0$ is the zero–divisor of f. We have $dd^c\Lambda \circ f \equiv 0$. Assume that χ has compact support which is contained in G for a good choice of G. Then

$$\text{(A.68)} \qquad \int_M \log |f|^2 \, dd^c\chi = \int_S \mu_f^0 \chi \, .$$

Thus the current $dd^c[\log|f|^2]$ is the current of integration over the zero divisor of f.

2. <u>**APPLICATION**</u>: <u>**The Argument Principle**</u> (Wirtinger [120], Stoll [93]). Take N, W, s and f as in the 1. Application. Assume $d\chi = 0$. Then $d^c\chi = 0$ and

$$\text{(A.69)} \qquad \int_{S\cap G} \mu_f^0 \chi = \int_{\partial G} \chi \wedge \frac{df}{f} = \int_{\partial G} \chi \wedge d^c \log |f|^2 \, .$$

3. <u>**APPLICATION**</u>. <u>**The Jensen Formula**</u> (Stoll [93]). Take N, W, s and f as in the 1. Application. Let $g \neq \emptyset$ be open with $\bar{g} \subset G$. Assume that the boundary $\partial g = \bar{g} - g$ is a pure $(2m-1)$–dimensional, real, class C^∞ submanifold of M oriented to the exterior of g. Assume that $\chi > 0$ on $\bar{G} - g$. Take $0 < R \in \mathbb{R}$. Let $0 \leqslant \psi \leqslant R$ be the unique continuous function on $\bar{G}$ which is of class C^∞ on $\bar{G} - g$ which solves the Dirichlet problem

$$\text{(A.70)} \qquad dd^c(\psi\chi) = 0 \text{ on } G - \bar{g} \qquad \psi|\bar{g} = 0 \qquad \psi|\partial G = R.$$

A function $\tilde{\psi}$ of class C^∞ exists on M such that $\tilde{\psi}|(\bar{G} - g) = \psi$. The Green–Residue Theorem implies

$$\int_{S\cap G} \mu_f^0(R - \tilde{\psi})\chi = \int_{\partial G} (\log |f|^2) d^c\psi \wedge \chi - \int_{g} (\log |f|^2) dd^c(\tilde{\psi}\chi)$$

$$\int_{S\cap g} \mu_f^0 \tilde{\psi}\chi = - \int_{\partial g} (\log |f|^2) d^c\psi \wedge \chi + \int_{g} (\log |f|^2)(dd^c\tilde{\psi}\chi)$$

which adds up to the <u>Jensen formula</u>

(A.71)

$$\int_{S\cap G} \mu_f^0(R - \psi)\chi = \int_{\partial G} (\log |f|^2) d^c\psi \wedge \chi - \int_{\partial g} (\log |f|^2) d^c\psi \wedge \chi$$

where $d^c\psi \wedge \chi$ defines a non-negative measure on ∂G and ∂g.

4. <u>**APPLICATION**</u>. <u>**The unintegrated First Main Theorem**</u> (Levine [54], Stoll [108]). Let $W = L$ be a line bundle. Then $dd^c\Lambda = -c(L_1, \kappa)$. Assume $d\chi = 0$. Then $d^c\chi = 0$. The Green–Residue Theorem becomes the <u>**unintegrated First Main Theorem**</u>

(A.72) $$\int_G f^*(c(L,\kappa)) \wedge \chi = \int_{S\cap G} \mu_f^s \chi - \int_{\partial G} \chi \wedge d^c \log \|s \circ f\|_\kappa^2 .$$

This is a balancing statement, the integral over G does not depend on s, and is non-negative if $\chi \geqslant 0$ and $c(L,\kappa) \geqslant 0$. The integral over $S \cap G$ measures the intersection of f with the zeros of the section s and the difference of these two integrals is compensated by a boundary integral. Unfortunately nothing can be said about the sign of the boundary integral, which makes estimates impossible. The existence of the integral has even to be assumed. To correct those short comings (A.72) will be "integrated."

5. <u>**APPLICATION**</u>. <u>**Poincaré dual**</u>. In addition to the assumptions in the 4. Application we require that $M = G = N$ is compact and that f is the identity. Then

(A.73) $$\int_M c(L,\kappa) \wedge \chi = \int_S \mu_s \chi$$

for every closed form χ of bidegree $(m-1,m-1)$ and class C^∞. thus the Chern form $c(L,\kappa)$ is the Poincaré-dual to the zero divisor of any $s \in \Gamma(M,L)$. If M is a compact Riemann surface, we take $\chi = 1$. Then (A.73) means, that the degree of the divisor of a holomorphic section in a line bundle equals the Chern number of the line bundle.

f) <u>The First Main Theorem</u>. Let M and N be connected complex manifolds of dimension m and n respectively. Let L be a holomorphic line bundle over N. Let κ be a hermitian metric along the fibers of L. Let $\chi \geqslant 0$ be a form of class C^∞ and bidegree $(m-1,m-1)$ on M with $d\chi = 0$. Let $f : M \longrightarrow N$ be a meromorphic map. Let (G,g,ψ) be a <u>condensor</u> on M. This means: g and $G \neq \emptyset$ are open, relative compact subsets of M with $\bar{g} \subset G$. The boundaries $\partial G = \bar{G} - G$ and $\partial g = \bar{g} - g$ are either empty or are pure $(2m-1)$-dimesional, real, submanifolds of class C^∞ of M oriented to the exterior of G respectively g. The function $\psi : \bar{G} \longrightarrow \mathbb{R}$ is continuous on $\bar{G}$ and of class C^2 on $\bar{G} - g$ and $\psi | \bar{g} = 0$ and $\psi | \partial G = R > 0$ are constant. On $\bar{G}$ we have $0 \leqslant \psi \leqslant R$. For $t \geqslant 0$ define $G_t = \{x \in G \mid \psi(x) < t\}$.

Let V^* be a linear subspace of $\Gamma(N,L)$ with $0 < \dim V^* < \infty$. Let ℓ be a hermitian metric on V^*. Assume that (κ,ℓ) is distinguished. Take $a \in \mathbb{P}(V^*)$ with $f(M) \not\subseteq E_L[a]$. Abbreviate $F_a = \operatorname{supp} \mu_f^a$. <u>the spherical image</u> A_f, <u>the characteristic</u> T_f, the <u>counting function</u> $n_f \geqslant 0$, the <u>valence function</u> $N_f \geqslant 0$, the <u>compensation functions</u> $m_f \geqslant 0$ and the <u>deficit</u> D_f are defined by

(A.74) $$A_f(G,L,\kappa) = \int_G f^*(c(L,\kappa) \wedge \chi$$

(A.75) $$T_f(G,L,\kappa) = \int_G (R - \psi) f^*(c(L,\kappa)) \wedge \chi = \int_0^R A_f(G_t,L,\kappa)dt$$

(A.76) $$n_f(G,a,L) = \int_{G\cap F_a} \mu_f^a \chi$$

(A.77) $$N_f(G,a,L) = \int_{G\cap F_a} \mu_f^a(R - \psi)\chi = \int_0^R n_f(G_t,a,L)dt$$

(A.78) $$m_f(\partial G,a,L,\kappa) = \int_{\partial G} \log \frac{1}{\Box\, f,a\, \Box_{\kappa,\ell}}\, 2d^c\psi \wedge \chi$$

(A.79) $$m_f(\partial g,a,L,\kappa) = \int_{\partial g} \log \frac{1}{\Box\, f,a\, \Box_{\kappa,\ell}}\, 2d^c\psi \wedge \chi$$

(A.80) $$D_f(\partial G,a,L,\kappa) = \int_{G} \log \frac{1}{\Box\, f,a\, \Box_{\kappa,\ell}}\, 2dd^c\psi \wedge \chi$$

If the line bundle L is non-negative, that is if $c(L,\kappa) \geqslant 0$, then $A_f(G,L,\kappa) \geqslant 0$ and $T_f(G,L,\kappa) \geqslant 0$. If we apply the same $\tilde{\psi}$ trick as in the 3rd application we obtain the **First Main Theorem of line bundles** and condensors.

(A.81)

$$\begin{aligned} &T_f(G,L,\kappa) \\ &= N_f(G,L,a) + m_f(\partial G,a,L,\kappa) - m_f(\partial g,a,L,\kappa) - D_f(G,a,L,\kappa) \end{aligned}$$

We have a balancing statement. T_f does not depend on a. A new term $D_f(G,a,L,\kappa)$ appears called the deficit. The characteristic depends nicely on the line bundles. If (L_1,κ_1) and (L_2,κ_2) are hermitian line bundles then

(A.82)
$$T_f(G,L_1,\kappa_1) + T_f(G,L_2,\kappa_2) = T_f(G,L_1 \otimes L_2,\kappa_1 \otimes \kappa_2)$$

The characteristic introduced by Nevanlinna turns out to be the integrated version of the characteristic of a line bundle introduced by Chern, a relation which holds in much more general situations.

If we take $N = \mathbb{P}(V)$ and $L = H = \mathcal{O}(1)$ as the hyperplane section bundle on $\mathbb{P}(V^*)$, then $V^* = \Gamma(\mathbb{P}(V),H)$ and $c(H,\ell) = \Omega$ is the Fubini Study form. Now $\square\, f,a \,\square_{\ell,\ell} = \square\, f,a\, \square$ is the distance from $f(x)$ to $\ddot{E}[a] = E_L[a]$ and μ_f^a is the intersection divisor with the hyperplane defined by a. We substitute all this in the definitions (A.74) and appreviate $A_f(G) = A_f(G,H,\ell)$, $T_f(G) = T_f(G,H,\ell)$, $n_f(G;a) = n_f(G;a,H)$, $N_f(G,a) = N_f(G,a,H)$, $m_f(\partial G,a) = m_f(\partial G,a,H,\ell)$, $m_f(\partial g,a) = m_f(\partial g,a,H,\ell)$ and $D_f(G;a) = D_f(G,a,H,L)$. We have the First Main Theorem

(A.83)
$$\boxed{T_f(G) = N_f(G,a) + m_f(\partial G,a) - m_f(\partial g,a) - D_f(G,a)}$$

Exchange of integration and (A.51) imply

(A.84)
$$\int_{a\in\mathbb{P}(V^*)} m_f(\partial G,a)\Omega^n(a) = \sum_{\mu=1}^{n} \frac{1}{\mu} \int_{\partial G} d^c\psi \wedge \chi$$

(A.85)
$$\int_{a\in\mathbb{P}(V^*)} m_f(\partial g,a)\Omega^n(a) = \sum_{\mu=1}^{n} \frac{1}{\mu} \int_{\partial g} d^c\psi \wedge \chi$$

(A.86)
$$\int_{a\in\mathbb{P}(V^*)} D_f(G,a)\Omega^n(a) = \sum_{\mu=1}^{n} \frac{1}{\mu} \int_{G-\bar{g}} dd^c\psi \wedge \chi$$

while Stokes Theorem implies

$$\text{(A.87)}\qquad \int_{G-\bar{g}} dd^c\psi \wedge \chi = \int_{\partial G} d^c\psi \wedge \chi - \int_{\partial g} d^c\psi \wedge \chi \,.$$

We obtain the <u>**Mean Value Theroem**</u>

$$\text{(A.88)}\qquad T_f(G) = \int_{a\in \mathbb{P}(V^*)} N_f(G,a)\Omega^n(a)$$

which does not hold in general for $T_f(G,L,\kappa)$.

Let $h \not\equiv 0$ be a meromorphic function on M. Put $N_h(G) = N_h(G,0) - N_h(G,\infty)$. Since $|h| = □\, h;0\, □/□\, h;\infty\, □$, the FMTH implies the <u>**General Jensen Formula**</u>

$$\text{(A.89)}\qquad N_{\mu_h}(G) = \int_{\partial G} \log |h|\, d^c\psi \wedge \chi - \int_{\partial g} \log |h|\, d^c\psi \wedge \chi - \int_{G-\bar{g}} \log |h|\, dd^c\psi \wedge \chi \,.$$

If there is a reduced representation $\mathfrak{w} : M \to V$ of f, take $\mathfrak{a} \in V^*$ such that $\langle \mathfrak{w}, \mathfrak{a}\rangle \not\equiv 0$. Define $a = \mathbb{P}(\mathfrak{a})$. Then $\|\mathfrak{w}\|\, \|\mathfrak{a}\|\, □\, f;a\, □ = \langle \mathfrak{w}, \mathfrak{a}\rangle$. If we apply (A.89) to $h = \langle \mathfrak{w}, \mathfrak{a}\rangle$, the FMTH implies

$$\text{(A.90)}\qquad \begin{aligned} T_f(G) &= \int_{\partial G} (\log \|\mathfrak{w}\|)2d^c\psi \wedge \chi - \int_{\partial g} (\log \|\mathfrak{w}\|)2d^c\psi \wedge \chi \\ &\quad - \int_{G-\bar{g}} (\log \|\mathfrak{w}\|)dd^c\psi \wedge \chi \,. \end{aligned}$$

g) <u>**Remarks to the First Main Theorem**</u>. We rediscover many useful features of the classical FMTH as well as the close connection to the Chern forms. However there are substantial disadvantages. We have a new deficit term and our set up depends on the arbitrary choices of ψ, R, χ. These problems can be resolved. We assume $\chi > 0$ with $d\chi = 0$ is given. If M is a Kaehler manifold with exterior form υ, we can choose $\chi = \upsilon^{m-1} > 0$. For ψ we take the solution of the Dirichlet problem $dd^c\psi \wedge \chi = 0$ on $G - \bar{g}$ with boundary conditions $\psi | \bar{g} = 0$ and $\psi | \partial G = R(G) > 0$ where the constant $R(G) > 0$ is chosen such that

$$\text{(A.91)} \qquad \int_{\partial G} 2d^c\psi \wedge \chi = \int_{\partial g} 2d^c\psi \wedge \chi = 1$$

The deficit is eliminated. We say that (G,g,ψ) is a harmonic condensor for χ.

If M is a non-compact Riemann surface, Weyl [119] took $\chi = 1$ and succeeded to establish a Second Main Theorem and Defect relation by exhausting M with harmonic condensors. Decisive for the type of defect relation is the question if the capacity $\mathfrak{C} = \inf(1/R(G))$ is positive or zero. Wu [126] recast and enlarged this theory. Stoll [93] extended the theory to $\dim M > 1$ by introducing the form χ. Thus Nevanlinna theory was fully established for meromorphic maps from Kaehler manifolds into projective space with the family of hyperplanes as targets. The exhaustion by condensors remains cumbersome, and we will see how this can be rectified.

h) **Higher Codimension**. Could value distribution be extended to other target families? The breakthrough came with the papers of Levine [54] and Chern [14] where the family of p-ℓplanes in projective space as the target family. In general we have this situation: Let M and N be connected complex manifolds of dimension m and n respectively. Let $f : M \longrightarrow N$ be a meromorphic map. A target family $\mathfrak{A} = (E_a)_{a \in A}$ of pure (n−p)–dimensional analytic subsets is given, where $m \geq p$ and where A is a compact complex manifold. Define $F_a = f^{-1}(E_a)$ for $a \in A$. We assume (respectively construct) a closed form $\Omega \geq 0$ of bidegree (p,p) and class C^∞ on N and for each $a \in A$ a form $\Lambda_a \geq 0$ of bidegree $(p-1,p-1)$ and class C^∞ on $N - E_a$ such that Λ_a has residue 1 on E_a and such that $dd^c\Lambda_a = \Omega$ on $N - E_a$. We take a closed form $\chi \geq 0$ of class C^∞ and bidegree $(m-p,m-p)$ and a condensor (G,g,ψ) as before, define the value distribution functions and obtain a FMTH containing a deficit term $D_f(G,a;\alpha)$. Now, however, it is impossible to eliminate the deficit, since $dd^c\psi \wedge \chi \equiv 0$ is an over determined system and in general the Dirichlet problem cannot be solved. The existence of the deficit accounts for the occurance of Bieberbach maps and other phenomena as the transcendental Bezout problem of Ph. Griffiths. Still we can consider the pseudoconcave case where $dd^c\psi \leq 0$ and the pseudoconvex case where $dd^c\psi \geq 0$. Integral averaging and capacity methods still allow Casorati–Weierstrass type theorems under reasonable assumptions. For these problems and the case of higher codimensions in general the reader is referred to: Hirschfelder [45]. Wu [125] Bott–Chern [8], Stoll [96], [97], [98], [99], [104], [105], Tung [112], Cowen [20], Shiffman [79], Molzon [56], Molzon–Shiffman–Sibony [59], Dektjarev [22], Griffiths–King [36] and of course the papers of Levine and Chern.

i) <u>Application of the First Main Theorem</u>. Internal applications of the value distribution functions and the FMTH are common and shall not be listed here. There are some interesting applications outside value distribution theory.

α) <u>The Theorem of Fatou</u>. Let (G,g,ψ) be a condensor on M with $d\psi \neq 0$ on ∂G. Let $\chi > 0$ be a form of class C^∞ and of bidegree $(m-1,m-1)$ on M. We do not require that χ is closed. Such ψ and χ always exist. If $f : G \longrightarrow \mathbb{P}(V)$ is a meromorphic map, the characteristic $T_f(G)$ still can be defined, but may be infinite. The map f is said to belong to the Nevanlinna class if $T_f(G) < \infty$. The condition does not depend on the choice of χ and ψ. Naturally, a bounded holomorphic function is in the Nevanlinna class. In extension of a result of L. Lempert [52] in $\mathbb{C}^n$, Patrizio [73] proved: "A meromorphic map $f : G \longrightarrow \mathbb{P}(V)$ in the Nevanlinna class admits admissible boundary values in the sense of Korany–Stein at almost every point of ∂G. If $f(M)$ is not contained in any hyperplane in $\mathbb{P}(V)$, if $a \in \mathbb{P}(V^*)$, then the set of all $x \in \partial G$ where f has an admissible boundary value at x belonging to $\ddot{E}[a]$ has measure zero in ∂G." This is a beautiful extension of the classical theorem of Fatou–Nevanlinna for the unit disc. Without the apparatus of value distribution thoery, Lempert's theorem could not have been reformulated, localized and extended to manifolds.

β) <u>Blaschke products</u>. Take ψ and χ as in α). A divisor $\nu \geq 0$ on G satisfies the Blaschke condition if $N_\nu(G) < \infty$. The condition does not depend on the choice of ψ and χ. The zero divisor of a holomorphic function in the Nevanlinna class satisfies the Blaschke condition by the FMTH. On the unit disc a divisor, which satisfies the Blaschke condition, is the zero divisor of a bounded holomorphic function, which is not true on the unit ball in $\mathbb{C}^m$ with $m > 1$, but Skoda [86], and Henkin [43] proved independently: A principal divisor satisfying the Blaschke condition is the zero divisor of a holomorphic function in the Nevanlinna class. Thus a difficult problem was solved. The classification of the zero divisors of functions in H^p is still unresolved.

γ) <u>Affine algebraic varieties</u>. Let A be a pure p–dimensional analytic subset in $\mathbb{C}^m$. Adopt the notations of A.3 around (A.36) and (A.37). The counting function n_A of A is defined by (Kneser [46], Lelong [50])

(A.92)
$$n_A(r) = \frac{1}{r^{2p}} \int_{A[r]} \upsilon^{2p} - \int_{A[r]} \omega^{2p} + n_A(0)$$

where the Lelong number $n_A(0)$ is an integer (Thie [110]).

Stoll [94] extended the previous theory to the analytic set, in applying the FMTH to the map $\mathbb{P} : A \longrightarrow \mathbb{P}_{n-1}$ and showed that A is affine algebraic if and only if $n_A(r)$ is bounded. For $p = m - 1$ this was shown earlier and is much easier.

δ) <u>Other applications</u>. The construction of Theta functions to prescribed (2m)–periodic divisor (Stoll [92]) was mentioned before. Later we will observe that Henson and Rubel [44] applied Nevanlinna theory, in particular the Lemma of the Logarithmic derivative to solve a problem in mathematical logic. Of course both applications involve the FMTH.

A5. Parabolic manifolds. The exhaustion by condensors is difficult to handle. Hence exhaustions have been used in recent years. Here the parabolic exhaustions are opportune since they possess a rather easy harmonic condensor system.

Let M be a connected complex manifold of dimension m. Let $\tau \geqslant 0$ be a non-negative, unbounded function of class C^∞ on M. For $0 \leqslant r \in \mathbb{R}$ and $A \subseteq M$ define

(A.93) $$A[r] = \{x \in A \mid \tau(x) \leqslant r^2\} \qquad A(r) = \{x \in A \mid \tau(x) < r^2\}$$

(A.94) $$A\langle r\rangle = \{x \in A \mid \tau(x) = r^2\} \qquad A_* = \{x \in A \mid \tau(x) > 0\}$$

(A.95) $$\upsilon = dd^c\tau \qquad \omega = dd^c \log \tau \qquad \sigma = d^c \log \tau \wedge \omega^{m-1} .$$

If $M[r]$ is compact for each $r > 0$, the function τ is said to be an <u>exhaustion</u> of M. The function τ is said to be <u>parabolic</u> if

(A.96) $$\omega \geqslant 0 \qquad \omega^m \equiv 0 \not\equiv \upsilon^m$$

on M_*. Then $\upsilon \geqslant 0$ on M. If τ is a parabolic exhaustion, (M,τ) is said to be a <u>parabolic manifold</u>. Define

(A.97) $$\hat{\mathcal{E}}_\tau = \{r \in \mathbb{R}^+ \mid d\tau(x) \neq 0 \quad \forall\, x \in M\langle r\rangle\}.$$

Then $\mathbb{R}^+ - \hat{\mathcal{E}}_\tau$ has measure zero. If $r \in \hat{\mathcal{E}}_\tau$, the boundary $\partial M(r) = M\langle r\rangle$ is a compact, real, $(2m-1)$-dimensional submanifold of class C^∞ of M, oriented to the exterior of $M\langle r\rangle$, and

(A.98) $$\varsigma = \int_{M\langle r\rangle} \sigma > 0$$

does not depend on $r \in \hat{\mathfrak{L}}_\tau$. If $r \geq 0$, then

$$\int_{M[r]} \upsilon^m = \varsigma r^{2m} . \tag{A.99}$$

The cartesian product of parabolic manifolds is parabolic. A non-compact Riemann surface is parabolic if and only if each subharmonic function bounded above is constant. $(\mathbb{C}^m,\tau_0)$ is parabolic where $\tau_0(\mathfrak{z}) = \|\mathfrak{z}\|^2$. Let $\pi : M \longrightarrow \mathbb{C}^m$ be a surjective, proper, holomorphic map. Then $\tau = \tau_0 \circ \pi$ is a parabolic exhaustion of M and (M,τ) is called a <u>parabolic covering manifold</u> of $(\mathbb{C}^m,\tau_0)$. Its generic sheet number is ς. Every m-dimensional connected, affine algebraic manifold is a parabolic covering manifold of $(\mathbb{C}^m,\tau_0)$. For more details see Stoll [103]. A parabolic manifold is said to be <u>strict</u> if $\upsilon > 0$.

THEOREM. (Stoll [105], [106].) If (M,τ) is a strict parabolic manifold, there is a biholomorphic map $h : M \longrightarrow \mathbb{C}^m$ with $\tau = \tau_0 \circ h$. Thus τ is an isometry.

Other proofs were given by D. Burns [9] and P. Wong [104].

A6. <u>The First Main Theorem on parabolic manifolds</u>. For $0 < s < r$ define

$$R(r,s) = \frac{1}{2m-2}\left[\frac{1}{s^{2m-2}} - \frac{1}{r^{2m-2}}\right] \quad \text{if } m > 1 \tag{A.100}$$

$$R(r,s) = \log \frac{r}{s} \quad \text{if } m = 1 \tag{A.101}$$

$$\psi_s = \mathrm{Max}(R(s,\sqrt{\tau}),0). \tag{A.102}$$

Then, except for the smoothness condition, $(M(r), M(s), \psi_s)$ is a condensor with

(A.103) $$2d^c\psi_s \wedge \upsilon^{m-1} = \sigma \qquad 2dd^c\psi_s \wedge \upsilon^{m-1} = 0$$

on $M[r]$. Thus the theory of section A4 applies and can be translated. Strictly speaking, the smoothness condition is assured only if r and s belong to $\hat{\mathcal{L}}_\tau$, but the Dirichlet problem is solved for all $0 < s < r$. Thus we have to require r and s in $\hat{\mathcal{L}}_\tau$ only if boundary integrals are involved.

Now, we rewrite the theory in the new terms. The <u>counting function</u> n_ν and the <u>valence function</u> N_ν of a divisor $\nu : M \longrightarrow \mathbb{Z}$ with $S = \operatorname{supp} \nu$ are given for $t > 0$ and $0 < s < r$ by

(A.104) $$n_\nu(t) = t^{2-2m} \int_{S[t]} \nu\upsilon^{m-1} = \int_{S_*[t]} \nu\omega^{m-1} + n_\nu(0)$$

(A.105) $$N_\nu(r,s) = \int_s^r n_\nu(t)\, \frac{dt}{t}$$

where $n_\nu(0) = \lim_{t \to 0} n_\nu(t)$. Since $M[0] \not\equiv \emptyset$ may consist of more than one point, we cannot expect $n_\nu(0) = \nu(0)$ as in (A.38), but for strictly parabolic manifolds these are Kneser's identities. If f is a meromorphic function on M, if $a \in \mathbb{P}_1$, the counting function and the valence function of the a-divisor μ_f^a of f are denoted by $n_f(t;a)$ and $N_f(r,s;a)$ and these functions for the divisor $\mu_f = \mu_f^0 - \mu_f^\infty$ of $f \not\equiv 0$ are denoted by $n_f(t)$ and $N_f(r,s)$. If $0 < s < r \in \hat{\mathcal{L}}_\tau$ and $s \in \hat{\mathcal{L}}_\tau$, then (A.89) translates to the <u>Jensen Formula</u>

(A.106) $$N_f(r,s) = \int_{M\langle r\rangle} \log |f|\sigma - \int_{M\langle s\rangle} \log |f|\sigma$$

which implies **Liouville's Theorem** on M.

Let L be a hermitian line bundle with metric κ on a connected complex manifold N. Let V^* be a linear subspace of $\Gamma(N,L)$ with $0 < \dim V^* < \infty$. Let ℓ be a hermitian metric on V^*. Assume that (κ,ℓ) is distinguished. Let $f : M \longrightarrow N$ be a meromorphic map. If $a \in \mathbb{P}(V^*)$ and $f(M) \subsetneq E_L[a]$, denote the **counting function** and **valence function** of the intersection divisor μ_f^a by $n_f(t;a,L)$ and $N_f(r,s;a,L)$ respectively where $t > 0$ and $0 < s < r$. The **compensation function** of f for a is given for $r \in \hat{\mathfrak{L}}_\tau$ by

(A.107) $$m_f(r;a,L,\kappa) = \int_{M\langle r\rangle} \log \frac{1}{\Box\, f, a\, \Box_{\kappa,\ell}}\, \sigma \geqq 0.$$

The **spherical image** and the **characteristic** of f for L are defined for $t > 0$ and $o < s < r$ by

(A.108) $$A_f(t,L,\kappa) = t^{2-2m} \int_{M[t]} f^*(c(L,\kappa)) \wedge \upsilon^{m-1}$$

(A.109) $$T_f(r,s;L,\kappa) = \int_s^r A_f(t,L,\kappa)\, \frac{dt}{t}\,.$$

For $0 < s < r \in \hat{\mathfrak{L}}_\tau$ and $s \in \hat{\mathfrak{L}}_\tau$ we obtain the **First Main Theorem**

(A.110)

$$\boxed{T_f(r,s;L,\kappa) = N_f(r,s;a,L) + m_f(r;a,L,\kappa) - m_f(s;a,L,\kappa)}$$

which extends m_f to all $r > 0$ as a continuous function such that (A.110) holds for all $0 < s < r$.

If $c(L,\kappa) \geqq 0$, then A_f and T_f are non-negative and increase. Hence

$$\text{(A.111)} \qquad A_f(0,L,\kappa) = \lim_{t\to 0} A_f(t,L,\kappa) \qquad A_f(\infty,L,\kappa) = \lim_{t\to 0} A_f(t,L,\kappa)$$

exist and we have

$$\text{(A.112)} \qquad A_f(t,L,\kappa) = \int_{M_*[t]} f^*(c(L,\kappa)) \wedge \omega^{m-1} + A_f(0,L,\kappa) .$$

If $M = N$ and if f is the identity, we drop the index f in the formulas (A.107) – (A.112).

Let V be a hermitian vector space of dimension $n + 1 > 1$. Take $N = \mathbb{P}(V)$ and $L = H = \mathcal{O}(1)$ as the hyperplane section bundle, then we obtain

(A.113)

$$A_f(t) = t^{2-2m} \int_{M[t]} f^*(\Omega) \wedge \upsilon^{m-1} = \int_{M_*[t]} f^*(\Omega) \wedge \omega^{m-1} + A_f(0)$$

$$\text{(A.114)} \qquad T_f(r,s) = \int_s^r A_f(t) \frac{dt}{t} \qquad \text{if } 0 < s < r$$

$$\text{(A.115)} \qquad m_f(r,a) = \int_{M\langle r\rangle} \log \frac{1}{\Box f, a \Box} \sigma \qquad \text{if } r \in \hat{\ell}_\tau$$

$$\text{(A.116)} \qquad A_f(0) = \lim_{t\to 0} A_f(t) \qquad A_f(\infty) = \lim_{t\to\infty} A_f(t)$$

where $A_f \geqslant 0$ and $T_f \geqslant 0$ increase and $m_f \geqslant 0$ extends to a continuous function on $\mathbb{R}^+$. Then we have the **First Main Theorem** for $0 < s < r$:

(A.117)

$$\boxed{T_f(r,s) = N_f(r,s,a) + m_f(r,a) - m_f(s,a)}$$

If $0 < s < r \in \hat{\ell}_\tau$ and $s \in \hat{\ell}_\tau$ and if $\mathfrak{v} : M \longrightarrow V$ is a reduced representation, we have

(A.118) $$T_f(r,s) = \int_{M\langle r\rangle} \log \|\mathfrak{v}\| \sigma - \int_{M\langle s\rangle} \log \|\mathfrak{v}\| \sigma .$$

A global representation of f may not exist. There is a substitute, which uses the representation section F_f in $V_M \times L_f$ where L_f is the hyperplane section bundle of f. Let κ be a hermitian metric along the fibers of f and $\lambda = \ell \otimes \kappa$ is a hermitian metric along the fibers of $V_M \otimes L_f$. By Proposition 3.2 we have the substitute

(A.119) $$T_f(r,s) - T_f(r,s,L_f,\kappa) = \int_{M\langle r\rangle} \log \|F_f\|_\lambda \sigma - \int_{M\langle s\rangle} \log \|F_f\|_\lambda \sigma .$$

The exhaustion of M is easy to handle now. We have to consider the limit $r \to \infty$ only. For instance

(A.120) $$\lim_{r\to\infty} \frac{T_f(r,s)}{\log r} = A_f(\infty) = \lim_{t\to\infty} A_f(t)$$

where $A_f(\infty) > 0$ if and only if f is not constant. If (M,τ) is strictly parabolic, that is, if $(M,\tau) = (\mathbb{C}^m,\tau_0)$, then f is rational if and only if $A_f(\infty) < \infty$ (Stoll [93]). Hence f is said to have **transcendental** growth if and only if $A_f(\infty) = \infty$.

The meromorphic map $f : M \longrightarrow \mathbb{P}(V)$ is said to be **linearly non-degenerate** if $f(M) \not\subseteq \ddot{E}[a]$ for all $a \in \mathbb{P}(V^*)$. If f is linearly non-degenerate, we have

(A.121) $$\int_{a\in\mathbb{P}(V^*)} m_f(r,a)\Omega^n(a) = \frac{1}{2}\left[\sum_{\mu=1}^{n} \frac{1}{\mu}\right] \varsigma$$

(A.122) $$T_f(r,s) = \int_{a\in \mathbb{P}(V^*)} N_f(r,s;a)\Omega^n(a).$$

For each $r > 0$, the incidence set $B_f(r) = \{a \in \mathbb{P}(V^*) \mid f(M(r)) \cap \ddot{E}[a] \neq \emptyset\}$ is measurable with

$$0 \leqslant b_f(r) = \int_{B_f(r)} \Omega^n \leqslant 1.$$

Since $N_f(r,s;a) = 0$ if $a \in \mathbb{P}(V^*) - B_f(r)$, we can replace $\mathbb{P}(V^*)$ by $B_f(r)$ in (A.22). If we integrate the FMTH over $B_f(r)$, we obtain

$$T_f(r,s) \leqslant b_f(r)T_f(r,s) + \int_{B_f(r)} m_f(r,a)\Omega^n(a) \leqslant b_f(r)T_f(r,s) + \frac{C_1}{2}\sum_{\mu=1}^{n}\frac{1}{\mu}$$

which implies

(A.123)

$$\boxed{0 \leqslant 1 - b_f(r) \leqslant \frac{C_1}{2T_f(r,s)}\sum_{\mu=1}^{n}\frac{1}{\mu}}$$

Thus if $n = 6$ and $C_1 = 1$, then $T_f(r,s) \geqslant 123$ assures us, that 99% of the hyperplanes are intersected by $f(M(r))$. Since $T_f(r,s) \longrightarrow \infty$ for $r \to \infty$, we obtain $b_f(r) \longrightarrow 1$ for $r \to \infty$. hence **f(M)** **intersects almost all hyperplanes**, which is a **Theorem of Casorati–Weierstrass**.

A7. The Ricci function. a) Notations. We will give a most useful example of the characteristic function. First some notations have to be introduced which will be needed later also for other purposes. For $0 \leqslant q \in \mathbb{Z}$ define

(A.124) $$i_q = \left[\frac{i}{2\pi}\right]^q q!(-1)^{\frac{q(q-1)}{2}}$$

Let M be a connected complex manifold of dimension m. A biholomorphic map

(A.125) $$\mathfrak{z} = (z^1, \dots, z^m) : U \longrightarrow U'$$

of an open subset U of M onto an open subset U' of $\mathbb{C}^m$ is called a <u>chart</u> of M (at x if $x \in U$). On U define

(A.126) $$\chi = \frac{i}{2\pi} \sum_{\mu=1}^{m} dz^\mu \wedge d\bar{z}^\mu$$

(A.127) $$\mathfrak{r} = dz^1 \wedge \dots \wedge dz^m$$

where $\mathfrak{r}$ is a holomorphic frame of the <u>canonical</u> <u>bundle</u> κ of M over U with

(A.128) $$\chi^m = i_m \mathfrak{r} \wedge \bar{\mathfrak{r}}$$

Dependence on $\mathfrak{z}$ is expressed by $U_\mathfrak{z}$, $U'_\mathfrak{z}$, $\mathfrak{r}_\mathfrak{z}$, $\chi_\mathfrak{z}$.

Let $\mathfrak{U} = \{U_\lambda\}_{\lambda \in \Lambda}$ be any open covering of M. For $\lambda = (\lambda_0, \dots, \lambda_p) \in \Lambda^{p+1}$ define

(A.129) $$U_\lambda = U_{\lambda_0 \dots \lambda_p} = U_{\lambda_0} \cap U_{\lambda_1} \cap \dots \cap U_{\lambda_p}$$

(A.130) $$\Lambda[p] = \Lambda[p,\mathfrak{U}] = \{\lambda \in \Lambda^{p+1} \mid U_\lambda \neq \emptyset\} .$$

A chart atlas $\{\mathfrak{z}_\lambda\}_{\lambda \in \Lambda}$ is a family of charts $\mathfrak{z}_\lambda : U_\lambda \longrightarrow U'_\lambda$ such that $\mathfrak{U} = \{U_\lambda\}_{\lambda \in \Lambda}$ is an open, so called associated, covering of M. Abbreviate $\mathfrak{r}_\lambda = \mathfrak{r}_{\mathfrak{z}_\lambda}$ and $\chi_\lambda = \chi_{\mathfrak{z}_\lambda}$. If $(\lambda,\mu) \in \Lambda[1]$, one and only one zero free holomorphic function $\Delta_{\lambda\mu}$ exists on $U_{\lambda\mu}$ such that

(A.131) $$\mathfrak{r}_\lambda = \Delta_{\lambda\mu} \mathfrak{r}_\mu \qquad \Delta_{\lambda\mu}\Delta_{\mu\lambda} = 1 \qquad \text{on } U_{\lambda\mu}$$

Then $\{\Delta_{\lambda\mu}\}_{(\lambda,\mu)\in\Lambda[1]}$ is a basic cocycle of the canonical bundle.

b) <u>The Ricci form</u>. Let $\Psi > 0$ be a positive form of class C^∞ and degree $2m$ on M. Let $\{\mathfrak{a}_\lambda\}_{\lambda\in\Lambda}$ be a chart atlas on M. For each $\lambda \in \Lambda$ there is a positive function Ψ_λ of class C^∞ on M such that $\psi | U_\lambda = \psi_\lambda \chi_\lambda^m$. If $(\lambda,\mu) \in \Lambda[1]$ then $\psi_\mu = \Psi_\lambda |\Delta_{\lambda\mu}|^2$ on $U_{\lambda\mu}$. Since $dd^c \log |\Delta_{\lambda\mu}|^2 \equiv 0$, one and only one form $\operatorname{Ric}\Psi$ of class C^∞ and bidegree $(1,1)$ exists on M such that $\operatorname{Ric}\Psi | U_\lambda = dd^c \log \psi_\lambda$. The form $\operatorname{Ric}\psi$ is called the <u>Ricci form</u> of Ψ. It does not depend on the selection of the chart altas.

One and only one hermitian metric κ_ψ is defined along the fibers of the canonical line bundle K such that $i_m \varphi \wedge \bar{\xi} = \kappa_\psi(\varphi,\xi)\psi$ for all forms φ and ξ of bidegree $(m,0)$ on M. Then $\operatorname{Ric}\psi$ is the Chern form

$$\operatorname{Ric}\psi = c(K,\kappa_\Psi) \,. \tag{A.132}$$

c) <u>The Ricci function</u>. Let (M,τ) be a parabolic manifold of dimension m. For $0 < s < r$ <u>the Ricci function</u> of Ψ is defined by

$$\operatorname{Ric}(r,s,\Psi) = \int_s^r \int_{M[t]} (\operatorname{Ric}\Psi) \wedge \upsilon^{m-1}\, t^{1-2m}\, dt. \tag{A.133}$$

From (A.132) we see that $\operatorname{Ric}(r,s,\Psi) = T(r,s,K,\kappa_\Psi)$ is the characteristic function of the canonical bundle K for the metric κ_Ψ.

Define $M^+ = \{x \in M \mid \upsilon(x) > 0\}$. Then $\operatorname{Ric}\upsilon^m$ is defined on M^+, but in order to define $\operatorname{Ric}(r,s,\upsilon^m)$ we need $M^+ = M$ which means that (M,τ) is isometric to $(\mathbb{C}^m,\tau_0)$ and that $\operatorname{Ric}\upsilon^m \equiv 0$. Therefore we have to use an indirect method to define the Ricci function of υ^m.

Let $\Psi > 0$ be any form of degree 2m and class C^∞ on M. A function $v \geqq 0$ of class C^∞ is defined by $\upsilon^m = v\Psi$. The set

(A.134) $$\mathcal{L}_\tau^0 = \{r \in \hat{\mathcal{L}}_\tau \mid (\log v)\sigma \text{ integrable over } M\langle r\rangle\}$$

does not depend on the choice of Ψ and $\mathbb{R}^+ - \mathcal{L}_\tau^0$ has measure zero. For $0 < s < r \in \mathcal{L}_\tau^0$ with $s \in \mathcal{L}_\tau^0$ the <u>Ricci function</u> of τ is defined by

(A.135) $$\mathrm{Ric}_\tau(r,s) = \frac{1}{2}\int_{M\langle r\rangle} (\log v)\sigma - \frac{1}{2}\int_{M\langle s\rangle} (\log v)\sigma + \mathrm{Ric}(r,s,\Psi)$$

and does not depend on the choice of the form Ψ. Is Ric_τ non-negative?

If (M,τ) is a parabolic covering manifold of $(\mathbb{C}^m,\tau_0)$, there is a proper, surjective, holomorphic map $\pi = (\pi_1, \ldots, \pi_m) : M \longrightarrow \mathbb{C}^m$ such that $\tau = \tau_0 \circ \pi$. The divisor ρ of $d\pi_1 \wedge \ldots \wedge d\pi_m$ is the branching divisor of τ. Then

(A.136) $$\mathrm{Ric}_\tau(r,s) = N_\rho(r,s) \geqq 0 .$$

Proofs are given in Stoll [103].

A8. <u>**Associated maps**</u>. **a)** <u>**General remarks**</u>. There are several methods to prove a <u>**Second Main Theorem**</u>. Cartan's method [12] was extended to several variables by Vitter [115] and Stoll [107] but is restricted to meromorphic maps $f : \mathbb{C}^m \longrightarrow \mathbb{P}(V)$. The method of Carlson–Griffiths–King [11], [35] works for meromorphic maps $f : M \longrightarrow N$ where M is a parabolic and N is projective algebraic but is restricted to $\dim M \geqq \operatorname{rank} f = \dim N$. Noguchi's method [71], [72] works in the same situation without the restriction, but the defect bound is not explicitly calculated. The method of Ahlfors–Weyl [1], [109] works for meromorphic maps $f : M \longrightarrow \mathbb{P}(V)$ where M is a parabolic manifold (Stoll [108]). The original integral averaging method has been replaced by a curvature method introduced by Cowen and Griffiths [21] if $M = \mathbb{C}$ and modified by Pit–Mann Wong [123] and Stoll [108] for parabolic manifolds M. We will outline this modified Ahlfors–Weyl method, since we can extend it to moving targets. Consciously or not, all the methods use the jet bundle, which shows up in our case in the concept of associated maps.

b) <u>The definition of associated maps on Riemann surfaces</u>. Weyl [118] and Ahlfors [1] first considered holomorphic curves $f : \mathbb{C} \longrightarrow \mathbb{P}(V)$. Let $\mathfrak{w} : \mathbb{C} \longrightarrow V_*$ be a reduced representation. For $0 \leqslant p \leqslant n$ holomorphic maps

$$\text{(A.137)} \qquad \mathfrak{w}_{\underline{p}} = \mathfrak{w} \wedge \mathfrak{w}' \wedge \ldots \wedge \mathfrak{w}^{(p)} : \mathbb{C} \longrightarrow \widetilde{G}_p(V)$$

are defined, where $\mathfrak{w}_p \not\equiv 0$ for $0 \leqslant p \leqslant n$, if and only if f is linearly non–degenerated. If so, the pth <u>associated map</u> $f_p : \mathbb{C} \longrightarrow G_p(V)$ is defined. For a holomorphic map $f : M \longrightarrow \mathbb{P}(V)$ on a Riemann surface M, the definition (A.137) still works for a reduced representation $\mathfrak{w} : U \longrightarrow V$ on a chart $z : U \longrightarrow U'$. If we switch to another reduced representation on another chart, $\mathfrak{w}_{\underline{p}}$ gains only a zero free holomorphic factor. Hence the pth associated map $f_p : M \longrightarrow G_p(V)$ is well defined by $f_p | U = \mathbb{P} \circ \mathfrak{w}_{\underline{p}}$ if f is linearly non–degenerate. See Weyl [119].

c) <u>The definition of associated maps on manifolds</u>. A distinguished differential operator is missing on a connected, complex manifold M of dimension $m > 1$. Therefore, Stoll [93] assume that a holomorphic form B of bidegree $(m - 1,0)$ is given on M. Let $\mathfrak{w} : U \longrightarrow V$ be a holomorphic vector function on a chart $\mathfrak{z} : U \longrightarrow U'$ of M. Define $\mathfrak{k}$ by (A.127). One and only one holomorphic vector function $\mathfrak{w}' : U \longrightarrow V$ is defined by

$$\text{(A.138)} \qquad \mathfrak{w}'\mathfrak{k} = d\mathfrak{w} \wedge B.$$

The operation can be iterated: $\mathfrak{w}^{(p)} = (\mathfrak{w}^{(p-1)})'$. Put $\mathfrak{w}^{(0)} = \mathfrak{w}$. Abbreviate

$$\text{(A.139)} \qquad \mathfrak{w}_{\underline{p}} = \mathfrak{w} \wedge \mathfrak{w}' \wedge \ldots \wedge \mathfrak{w}^{(p)} : U \longrightarrow \tilde{G}_p(V)$$

Dependence on the chart $\mathfrak{z}$ is indicated by $\mathfrak{w}'_{\mathfrak{z}}$, $\mathfrak{w}^{(p)}_{\mathfrak{z}}$ and $\mathfrak{w}_{\mathfrak{z}\underline{p}}$.

Let $f : M \longrightarrow \mathbb{P}(V)$ be a meromorphic map. If $\mathfrak{w}_{\underline{p}} \not\equiv 0$, for one choice of a reduced representation $\mathfrak{w} : U \longrightarrow V$ on a chart $\mathfrak{z} : U \longrightarrow U'$, then $\mathfrak{w}_{\underline{p}} \not\equiv 0$ for all possible choices and f is said to be __general of order p for $\underline{B}$__. If so, the pth associated map $f_p : M \longrightarrow G_p(V)$ is well defined as a meromorphic map by $f_p | U = \mathbb{P} \circ \mathfrak{w}_{\underline{p}}$ for all possible choices of $\mathfrak{w}$ and $\mathfrak{z}$.

A global definition can be given. Let $V_M = M \times V$ be the trivial bundle

$$\text{(A.140)} \qquad L_f[p] = (\textstyle\bigwedge_{p+1} V_M) \otimes (L_f)^{p+1} \otimes K^{\frac{p(p+1)}{2}}$$

admits a holomorphic section $F_p = F_{pf}$, called the $\underline{p}^{th}$ __representation section__ of f, which is uniquely defined by the property:

Let $\mathfrak{w} : U \longrightarrow V$ be a reduced representation on the chart $\mathfrak{z} : U \longrightarrow U'$. Define $\tilde{\mathfrak{w}}_{\underline{p}} \in \Gamma(Y, \bigwedge_{p+1} V_M)$ by $\tilde{\mathfrak{w}}_p(x) = (x, \mathfrak{w}_{\underline{p}}(x))$. Define $\mathfrak{k}$ by (A.127). Then

$$\text{(A.141)} \qquad F_p | U = \tilde{\mathfrak{w}}_{\underline{p}} \otimes (\mathfrak{w}^{\Delta})^{p+1} \otimes \mathfrak{k}^{\frac{p(p+1)}{2}}$$

Usually we write $\tilde{\mathfrak{w}}_{\underline{p}} = \mathfrak{w}_{\underline{p}}$. Then $F_p \not\equiv 0$, if and only if f is general of order p for B. Beware: The representation $\mathfrak{w}_{\underline{p}}$ of f may not be reduced. Thus F_{pf} and F_{f_p} cannot be identified.

There is a number $\ell_f = \ell_f(B) \in \mathbb{Z}[0,n]$, called the __generality index__ of f for B such that $F_p \not\equiv 0$ for $0 \leqslant p \leqslant \ell_f$ but $F_p \equiv 0$ for $p > \ell_f$. The map f is said to be __general for B__ if $\ell_f = n = \dim \mathbb{P}(V)$. If so, f is linearly non-degenerate. If M admits m analytically independent holomorphic functions, and if f is linearly non-degenerate, then there exists a holomorphic form B of bidegree $(m-1,0)$ on M such that f is general for B (Stoll [100]). Moreover if (M,τ) is a parabolic covering manifold of $(\mathbb{C}^m,\tau_0)$, and if $f : M \longrightarrow \mathbb{P}(V)$ is linearly non-degenerate, a holomorphic form B of bidegree $(m-1,0)$ exists on M such that f is general for B and such that

$$\text{(A.142)} \qquad m i_{m-1} B \wedge \bar{B} \leqslant (1+\tau)^{m-1} \upsilon^{m-1} \qquad \text{on } M.$$

A9. __The Plücker Difference Formula__. If $0 \leqslant p \leqslant \ell_f$, then $F_p \not\equiv 0$ and a divisor $\mu_{F_p} \geqslant 0$ is defined. The pth __stationary index__ divisor

$$\text{(A.143)} \qquad \iota_{f_p} = \mu_{F_{p-1}} - 2\mu_{F_p} + \mu_{F_{p+1}} \geqslant 0$$

is non-negative if $0 \leqslant p < \ell_f$. Here $\mu_{F_{-1}} \equiv 0$ by definition and $\mu_{F_0} = \mu_F \equiv 0$ by nature. Let Ω_p be the Fubini-Study form on $\mathbb{P}(\bigwedge_{p+1} V)$ and $G_p(V)$. Then

$$\text{(A.144)} \qquad \mathbb{H}_p = m i_{m-1} f_p^*(\Omega_p) \wedge B \wedge \bar{B} \geqslant 0.$$

If $p = \ell_f$, then $\mathbb{H}_{\ell_f} \equiv 0$. If $0 \leqslant p < \ell_f$, then $\mathbb{H}_p > 0$ outside a thin analytic subset of M where we can compute the Ricci form of $\mathbb{H}_p$:

(A.145) $$\mathrm{Ric}\, \mathbb{H}_p = f_{p-1}^*(\Omega_{p-1}) - 2f_p(\Omega_p) + f_{p+1}^*(\Omega_{p+1})$$

(A.146) $$m i_{m-1}\, \mathrm{Ric}\, \mathbb{H}_p \wedge B \wedge \bar{B} = \mathbb{H}_{p-1} - 2\mathbb{H}_p + \mathbb{H}_{p+1}\, .$$

For $0 \leq p \leq \ell_f$, define $h_p \geq 0$ on M^+ by $\mathbb{H}_p = h_p^2 \upsilon^m$ and put $h_p = 0$ on $M - M^+$. If $0 \leq p < \ell_f$ and $r \in \mathfrak{L}_\tau^0$, the integral

(A.47) $$S_{f_p}(r) = \int_{M\langle r\rangle} (\log h_p)\sigma$$

exists. If $0 \leq p < \ell_f$ and if $0 < s < r \in \mathfrak{L}_\tau^0$ with $s \in \mathfrak{L}_\tau^0$ we have the <u>Plücker Difference Formula</u>

(A.148)

$$\boxed{\begin{aligned} &N_{\ell_{f_p}}(r,s) + T_{f_{p-1}}(r,s) + T_{f_{p+1}}(r,s) \\ &= S_{f_p}(r) - S_{f_p}(s) + \mathrm{Ric}_\tau(r,s) \end{aligned}}$$

H. Weyl and J. Weyl [118], [119] established this formula first in the case $m = 1$. They liked the formula so much, that they called it the Second Main Theorem. We will reserve the name for the analogon to Nevanlinna's Second Main Theorem.

Hopefully, $S_{f_p}(r)$ and $\mathrm{Ric}_\tau(r,s)$ are "small" such that all the characteristics T_{f_p} have about the same growth. Then by adding up the second difference the SMTH follows, but there is still much to do. If we try to copy Ahlfors orignal proof, we are forced to require the identity $m i_{m-1} B \wedge \bar{B} = \upsilon^{m-1}$ which is impossible if $m > 1$. A closer inspection reveals that only $m i_{m-1} B \wedge \bar{B} \leq \upsilon^{m-1}$ is required. With this condition,

Stoll [93] obtained the SMTH. Later (Stoll [100]) it became clear that a weaker codition suffices, which we will discuss now.

A10. **Majorization.** The exhaustion τ is said to **majorize** B, if for every $r > 0$ there is a constant $c \geqslant 1$ such that

(A.149) $$0 \leqslant m i_{m-1} B \wedge \bar{B} \leqslant c\upsilon^{m-1} \quad \text{on} \quad M[r] .$$

The infimum of all these constants is called $Y_0(r)$. Then Y_0 increases. Define

(A.150) $$Y(r) = \lim_{r<t\to r} Y_0(t) .$$

Then $Y(r) \geqslant Y_0(r) \geqslant 1$ and Y increases. Also (A.149) holds with $c = Y(r)$. The function Y is called the **majorant** associated to τ and B.

Let (M,τ) be a parabolic covering manifold of $(\mathbb{C}^m,\tau_0)$. Let V be a complex vector space of dimension $n + 1 > 1$. Let $f : M \longrightarrow \mathbb{P}(V)$ be a linearly non-degenerate holomorphic map. Then there exists a holomorphic form B of bidegree $(m - 1,0)$ on M such that f is general for B and such that τ majorize B with a majorant Y satisfying

(A.151) $$Y(r) \leqslant (1 + r)^{2n-2} \quad \text{for all} \quad r > 0 .$$

See Stoll [100].

Let (M,τ) be a parabolic manifold of dimension m. Let V be a complex vector space of dimension $n + 1 > 1$. Let $f : M \longrightarrow \mathbb{P}(V)$ be a linearly non-degenerate map, which is general for the holomorphic form B of bidegree $(m - 1,0)$ on M. Assume that τ majorizes B with a majorant Y. Take $\varepsilon > 0$. Abbreviate

(A.152) $$Q_\varepsilon(r,s) = \mathfrak{S} \log Y(r) + \mathrm{Ric}_\tau(r,s) + \varepsilon \mathfrak{S} \log r$$

for $0 < s < r$. The inequality sign $\underset{\cdot}{\leq}$ means that the inequality holds except for a set of finite measure in the variable $r > 0$. Then we have the estimate (Stoll [108], Proposition 10.9)

$$T_{f_p}(r,s) \underset{\cdot}{\leq} 3^p T_f(r,s) + \frac{1}{2}(3^p - 1)Q_\Sigma(r,s) \tag{A.153}$$

A.11 <u>Projection</u>. Take $a = \mathbb{P}(\alpha) \in \mathbb{P}(V^*)$ and $p \in \mathbb{Z}[0,n]$. Let F_p be the pth representation section of f. Then $F_p \llcorner \alpha \not\equiv 0$. A non-negative divisor $\mu^a_{f_p} = \mu_{F_p \llcorner \alpha} - \mu_{F_p} \geq 0$ is defined, whose counting function and valence function are denoted by $n_{f_p}(t,a)$ and $N_{f_p}(r,s;a)$ respectively. For $r \in \hat{\ell}_\tau$ the <u>compensation function</u>

$$m_{f_p}(r;a) = \int_{M\langle r \rangle} \log \frac{1}{\square\, \dot{f}_p \llcorner a\, \square}\, \sigma \geq 0 \tag{A.154}$$

exists. A meromorphic map $f_p \llcorner a = \mathbb{P}(F_p \llcorner \alpha) : M \longrightarrow \mathbb{P}(\bigwedge_p V)$ is defined. The <u>First Main Theorem for projections</u> holds for $0 < s < r \in \hat{\ell}^0_\tau$ with $s \in \hat{\ell}_\tau$

$$T_{f_p}(r,s) = N_{f_p}(r,s;a) + m_{f_p}(r;a) - m_{f_p}(s;a) + T_{f_p \llcorner a}(r,s)\,. \tag{A.155}$$

Usually, (A.155) is derived from the FMTH's for f_p and $f_p \llcorner a$; compare Stoll [108] pages 152–153. In Section 3 of this paper, we will prove a general First Main Theorem which contains (A.155) as a very special case.

If $p = 0$, then $f = f_0$. Also $f \llcorner a$ is constant and $T_{f \llcorner a} \equiv 0$. We obtain the usual FMTH (A.117). If $p = n$, then f_n and $f_n \llcorner a$ are constant, $\mu^a_{f_n} \equiv 0$ and $\square\, \dot{f}_n \llcorner a\, \square = 1$. Thus (A.155) is trivial.

A.12 <u>The Ahlfors Estimates</u>. Take $p \in \mathbb{Z}[0,n]$ and $a \in \mathbb{P}(V^*)$. Abbreviate

(A.157) $$\Phi_p(a) = \square\, f_p ; a \,\square^2 \not\equiv 0\ .$$

Then

(A.158) $$0 \leqslant \Phi_{p-1}(a) \leqslant \Phi_p(a) \leqslant \Phi_n(a) \leqslant 1$$

(A.159) $$m i_{m-1}\, dd^c \Phi_p(a) \wedge B \wedge \bar{B} = (\Phi_{p+1}(a) - 2\Phi_p(a) + \Phi_{p-1}(a)) \mathbb{H}_p$$

(A.160)

$$m i_{m-1}\, d\Phi_p(a) \wedge d^c \Phi_p(a) \wedge B \wedge \bar{B} = (\Phi_{p+1}(a) - \Phi_p(a))(\Phi_p(a) - \Phi_{p-1}(a)) \mathbb{H}_p$$

For $0 < \beta \leqslant 1$, we obtain the fundamental estimate

(A.161)
$$\frac{\beta^2}{4} \frac{\Phi_{p+1}(a)}{\Phi_p(a)^{1-\beta}} \mathbb{H}_p$$
$$\leqslant m i_{m-1} dd^c \log\,(1 + \Phi_p(a)^{\beta}) \wedge B \wedge \bar{B} + \beta(\beta + 1) \mathbb{H}_p$$

For $0 < s < r$, the inequality yields the <u>**Ahlfors**</u> <u>**Estimate**</u> (Stoll [108])

(A.162)

$$\boxed{\int_s^r \int_{M[t]} \frac{\Phi_{p+1}(a)}{\Phi_p(a)^{1-\beta}} \mathbb{H}_p \frac{dt}{t^{2m-1}} \leqslant \frac{8}{\beta^2} (T_{f_p}(r,s) + c_0)}$$

Substituting (A.153), T_{f_p} can be replace by T_f. The Ahlfors estimate is the cornerstone of the proof of the SMTH. Ahlfors [1] obtained the first version by integral average methods. Cowen and Griffiths [21] introduced the curvature method for holomorphic maps $f : \mathbb{C} \longrightarrow \mathbb{P}(V)$. Pit-Mann Wong [123] modified the method for use on parabolic manifolds. This was necessitated by the more complicated singularity sets and intersection divisors of meromorphic maps.

Define $\beta(r,a)$ by $1/\beta(r,a) = 1 + T_{f_p}(r,s) + m_{f_p}(s;a)$ if $r \geqslant s$ and $\beta(r,a) = \beta(s,a)$ if $0 \leqslant r < s$. Then $0 < \beta(r,s) < 1$. For almost all $r > 0$ the integral

$$\text{(A.163)} \qquad F(r,a) = \int_{M\langle r\rangle} \frac{\Phi_{p+1}(a)h_p^2}{\Phi_p(a)^{1-\beta(r,a)}}\,\sigma$$

exists. Take $\varepsilon \in \mathbb{R}(0,1)$. By a method of Nevanlinna, the Ahlfors Estimate gives

$$\text{(A.164)} \qquad \begin{aligned} &\log^+ F(r,a) \\ &\overset{\cdot}{\leqslant} 3(1+\varepsilon)(\log T_f(r,s) + \log Y(r) + \log^+ \mathrm{Ric}_\tau(r,s)) + \varepsilon \log r \,. \end{aligned}$$

A.13 __**The Second Main Theorem**__. A finite subset A of $\mathbb{P}(V^*)$ is said to be in __**general position**__ if any subset S of A with $\#S = p + 1 \leqslant n + 1$ spans a projective plane of dimension p, in which case the family $\{\ddot{E}[a]\}_{a\in A}$ of hyperplanes is said to be in __**general position**__.

Let A be a finite subset of $\mathbb{P}(V^*)$ in general position with $k = \#A \geqslant n + 1$. Take $p \in \mathbb{Z}[0,n]$. Then there is a constant $c_p(A)$ such that we have the __**Product to Sum Estimate**__: Take any $x \in G_p(V)$. For each $a \in A$ take any $\beta(a) \in \mathbb{R}(0,1)$ and $P(a) \in \mathbb{R}[0,1]$. Then we have

$$\text{(A.165)} \qquad \prod_{a\in A} \frac{P(a)}{\square\, x\dot{\llcorner} a\, \square^{2-2\beta(a)}} \leqslant c_p(A) \left[\sum_{a\in A} \frac{P(a)}{\square\, x\dot{\llcorner} a\, \square^{2-2\beta(a)}}\right]^{n-p}$$

which implies

$$\text{(A.166)} \qquad \begin{aligned} &h_p^{2(n-p)} \left[\prod_{a\in A} \frac{\Phi_{p+1}(a)}{\Phi_p(a)}\right] \left[\prod_{a\in A} \Phi_p(a)^{\beta(r,a)}\right] \\ &\leqslant c_p(A) \left[\sum_{a\in A} \frac{\Phi_{p+1}(a)h_p^2}{\Phi_p(a)^{1-\beta(a)}}\right]^{n-p} . \end{aligned}$$

Put $c = \log c_p(A)$. With (A.155), integration over $M\langle r\rangle$ in (A.166) implies

(A.167)

$$(n-p)S_{f_p}(r) + \sum_{a\in A}(m_{f_p}(r,a) - m_{f_p}(s,a))$$

$$\leqslant (n-p)\int_{M\langle r\rangle} \log\left[\sum_{a\in A} \frac{\Phi_{p+1}(a)h_p^2}{\Phi_p(a)^{1-\beta(r,a)}}\right]\sigma$$

$$+ \sum_{a\in A} \beta(r,a)m_{f_p}(r,a) + c$$

$$\leqslant (n-p)\varsigma \log\left[\frac{1}{\varsigma}\sum_{a\in A} F(r,a)\right] + 1 + c$$

$$\leqslant (n-p)\varsigma \sum_{a\in A} \log^+ F(r,a) + \text{const.}$$

$$\underset{\bullet}{\leqslant} 3(1+\varepsilon)(n-p)k\varsigma(\log T_f(r,s) + \log Y(r) + \log^+ \mathrm{Ric}_\tau(r,s)) + \varepsilon \log r$$

for every $\varepsilon > 0$. Let $\theta = \mu_{F_n}$ be the Wronsky divisor. If we sum the Plücker Difference formula (A.148), we obtain

(A.168)
$$\sum_{p=0}^{n-1}(n-p)(S_{f_p}(r) - S_{f_p}(s)) + \frac{1}{2}n(n+1)\mathrm{Ric}_\tau(r,s)$$

$$= N_\theta(r,s) - (n+1)T_f(r,s)$$

Abbreviate

(A.169)

$$D(r,s) = \frac{3}{2}n(n+1)k\varsigma(1+\varepsilon)(\log T_f(r,s) + \log Y(r) + \log^+ \mathrm{Ric}_\tau(r,s))$$

$$+ \varepsilon \log r .$$

Then we have the <u>**Second Main Theorem**</u> (Stoll [108])

(A.170)

$$N_\theta(r,s) + \sum_{a\in A} m_f(r;a) \leq (n+1)T_f(r,s) + \frac{1}{2}n(n+1)\mathrm{Ric}_\tau(r,s) + D(r,s)$$

The Ricci defect R_f, the majorization defect Y_f, the ramification defect Θ_f and the Nevanlinna defect $\delta_f(a)$ are defined by

(A.171) $$R_f = \limsup_{r\to\infty} \frac{\mathrm{Ric}_\tau(r,s)}{T_f(r,s)}$$

(A.172) $$Y_f = \limsup_{r\to\infty} \frac{\log Y(r)}{T_f(r,s)} \geq 0$$

(A.173) $$\Theta_f = \liminf_{r\to\infty} \frac{N_\theta(r,s)}{T_f(r,s)} \geq 0$$

(A.174) $$\delta_f(a) = \liminf_{r\to\infty} \frac{m_f(r,s)}{T_f(r,s)} = 1 - \limsup_{r\to\infty} \frac{N_f(r,s)}{T_f(r,s)}$$

where $0 \leq \delta_f(a) \leq 1$. The defects R_f and Y_f are obstruction defects and we assume $R_f < \infty$ and $Y_f < \infty$ at least. The Second Main Theorem yields the Defect Relation

(A.175)

$$\Theta_f + \sum_{a\in A} m_f(r;a) \leq n + 1 + \frac{1}{2}n(n+1)(R_f + 3kC_{\sigma}Y_f)$$

Here Θ_f and Y_f depend on B, but m_f and R_f are independent of B.

If (M,τ) is a parabolic covering space of $(\mathbb{C}^m,\tau_0)$, if f has transcendental growth and if, as before, f is linearly non-degenerate, we can choose a holomorphic form B of bidegree $(m-1,0)$ on M such that f is general for B and such that τ majorizes B with $Y_f = 0$. Thus we obtain the <u>Defect Relation</u>

(A.176)
$$\boxed{\sum_{a\in A} \delta_f(a) \leq n + 1 + \frac{1}{2} n(n+1)R_f}$$

which is independent of B. If (M,τ) is affine algebraic, then $R_f = 0$ and

(A.177)
$$\boxed{\sum_{a\in A} \delta_f(a) \leq n + 1}$$

If $\upsilon > 0$ on M, Then (M,τ) is isometric to $(\mathbb{C}^m, \tau_0)$ and $R_f = 0$. In the situation (A.176), $R_f < \infty$ if f separates the fibers of π by a theorem of Noguchi [70]. If the meromorphic function h on M separates the fibers of π and if $T_h(r,s)/T_f(r,s) \longrightarrow 0$ for $r \longrightarrow \infty$, then $R_f = 0$ by the same theorem of Noguchi.

In (A.176), the defect Θ_f was dropped since it depends on B. Yet the term N_θ can be disected to obtain better estimates. Let $(\mu_f^a)^{(n)}$ be the intersection divisor μ_f^a truncated at level n (A.63). Let $N_f^{(n)}(r,s,a)$ be the valence function of $(\mu_f^a)^{(n)}$. The disection yields the <u>Second Main Theorem</u>

(A.178)
$$(k - n - 1)T_f(r,s) \leq \sum_{a\in A} N_f^{(n)}(r,s;a) + \frac{n(n+1)}{2} \mathrm{Ric}_\tau(r,s) + D(r,s)$$

See Smiley [87], [88] and Stoll [108], (13.21) with $p = 1$. Define the <u>truncated Nevanlinna defect</u>

(A.179)
$$\delta_f^{(n)}(a) = 1 - \limsup_{r\to\infty} \frac{N_f^{(n)}(r,s,a)}{T_f(r,s)}$$

with $0 \leq \delta_f(a) \leq \delta_f^{(n)}(a) \leq 1$. Now $\delta_f(a)$ can be replaced by $\delta_f^{(n)}(a)$ in (A.175) – (A.177).

A14. **Comments to the defect relation.** The behavior of the defects of a meromorphic map has been investigated in a number of papers. They cannot be reviewed here. Good introductions to the subject matter are: Griffiths [35], Stoll [107], Stoll [101], Stoll [108], Shiffman [82], Stoll [103], Wu [125], Weyl [119]. Now, we will list a few open problems:

a) **The Griffiths conjecture.** Let (M,τ) be a parabolic manifold of dimension m. Let N be a compact, connected, complex manifold of dimension n. Let L be a positive line bundle on N which is spanned by its sections. Let $f : M \longrightarrow N$ be a meromorphic map of rank $f = n$. Then $n \leq m$. Then Griffiths and King [36] proved a defect relation for the target family $\{E_L(a)\}_{a\in\mathbb{P}(\Gamma(N,L))}$. Does the same defect relation still hold if we assume only that $f(M)$ is not contained in any thin analytic subset of N. This is the **big Griffiths conjecture.** The problem is not easy. The conjecture becomes wrong if we only assume that $f(M) \not\subseteq E_L(a)$ for all $a \in \mathbb{P}(\Gamma(N,L))$ (Biancofiore [4]).

If V is a complex vector space of dimension $n + 1$, if $N = \mathbb{P}(V)$ and if $L = H^p$ is the $p > 1$ power of the hyperplane section bundle the conjectured defect sum is $(n + 1)/p$ and we speak of the **small Griffiths conjecture**.

b) **Higher codimension.** Various types of Casorati–Weierstrass theorems have been proved for target families of higher codimension, but no defect relation has been found. Perhaps there is none.

c) **Behavior of the defects.** The set of deficient value $a \in \mathbb{P}(V^*)$ with $\delta_f(a) > 0$ can be a continuum. How does this set look like? If $0 < \eta \leq 1$ what can we say about the set $\{a \in \mathbb{P}(V^*) \mid \delta_f(a) \geq \eta\}$.

d) **The Umkehr problem.** Given a countable subset A of $\mathbb{P}(V^*)$ such that every finite subset is in general position, given a function $\delta : A \longrightarrow \mathbb{R}(0,1]$ with $\sum_{a\in A} \delta(a) \leq n + 1$ is there a linearly non-degenerate meromorphic map $f : \mathbb{C}^m \longrightarrow \mathbb{P}(V)$ such that $\delta_f(a) = \delta(a)$ for all $a \in A$. In view of c) we do not perscribe the defects on the total space $\mathbb{P}(V^*)$.

A15. __Applications__. There are applications of value distribution theory some of which were already mentioned: The construction of Theta fucntions to periodic divisors Stoll [92], normal families of non-degenerate divisors Stoll [95], the characterization of affine algebraic varieties Stoll [94], analytic cycles of affine algebraic varieties Cornalba-Griffiths [18], the transcendental Bezout theorem, Cornalba-Shiffman [19], Stoll [100] and others, Tarski's High School Algebra Problem in Mathematical Logic Henson and Rubel [44].

The last application is the most unexpected. Rubel asked Stoll to prove the Lemma of the Logarithmic Derivative for meromorphic functions on polydiscs. In the paper, Stoll [109], so commissioned, value distribution on a polydisc was recast and the Lemma of the Logarithmic Derivative proved. The value distribution functions depend on a vector $\mathfrak{r} = (r_1, \ldots, r_m)$ and exceptional sets are taken on a p-dimensional subvariety of vectors $\mathfrak{r}$ where $0 < p \leqq m$. Later Spellecy [89] proved the defect relation for polydiscs.

Are there more applications? Perhaps the well-developed theory of value distribution of ordinary differential equations in the complex domain can be extended to partial differential equations.

In the next chapter, B, we will outline the theory of value distribution of movable targets developed in this monograph.

B. Value Distribution Theory for Moving Targets

B1. <u>**Initial**</u> <u>**remarks**</u>**.** Nevanlinna [66] conjectured that his defect relation remains valid, if the target points $a_1, \dots, a_q$ are replaced by mutually distinct, meromorphic target functions $g_1, \dots, g_q$ on $\mathbb{C}$, which grow slower than the given meromorphic function f on $\mathbb{C}$, that is

$$\text{(B.1)} \qquad T_{g_j}(r,s)/T_f(r,s) \longrightarrow 0 \quad \text{for} \quad r \longrightarrow \infty$$

for $j = 1, \dots, q$. Nevanlinna proved the conjecture if $q = 3$. He defined

$$\text{(B.2)} \qquad h = \frac{f - g_1}{f - g_3} \cdot \frac{g_2 - g_3}{g_2 - g_1} .$$

Then $\delta_f(g_j) = \delta_h(a_j)$ where $a_1 = 0$, $a_2 = 1$, and $a_3 = \infty$. Thus the defect relation for fixed targets yields the defect relation for moving targets.

Dufresnoy [25] proved a defect relation for polynomials of degree $\leqslant d$, with a defect bound $d + 2$. If the meromorphic functions $g_1, \dots, g_q$ span a vector space of dimension p over $\mathbb{C}$, then C.T. Chuang [15] obtained a defect relation with bound $p(1 - \delta_f(\infty)) + 1$. Hence he proved the conjecture of Nevanlinna if f is entire. The general case is still open. If f has finite lower order, Lo Yang [127] shows that there are at most countable many deficient target functions satisfying (B.1).

In several complex variables, B. Shiffman [83], [84] proved the Nevanlinna conjecture under the assumptions (B.1) and

$$\text{(B.3)} \qquad \operatorname{rank}(f, g_1, \dots, g_q) = \operatorname{rank}(g_1, \dots, g_q) + 1$$

which reduces in one variable to the case of constant g_j. In Section 9, we will obtain Shiffman's defect relation as a special case of our general theory. However, we have to make a slight modification in the assumptions.

S. Mori [63] extends the original Nevanlinna method (B.2) to meromorphic maps $f : \mathbb{C}^m \longrightarrow \mathbb{P}(V)$ and to meromorphic target maps $g_j : \mathbb{C}^m \longrightarrow \mathbb{P}(V^*)$ where V is a complex vector space of dimension $n + 1 > 1$ and where $g_1, \ldots, g_{n+2}$ are in general position satsifying (B.1). With a non-degeneracy condition he obtains

$$\text{(B.4)} \qquad \sum_{j=1}^{n+2} \delta_f(g_j) \leqslant n + 1.$$

In Section 11, we will extend Mori's method to meromorphic maps $f : M \longrightarrow \mathbb{P}(V)$ and $g_j : M \longrightarrow \mathbb{P}(V^*)$ for $j = 1, \ldots, n + 2$, where (M,τ) is a parabolic covering manifold of $(\mathbb{C}^m, \tau_0)$.

In this monograph, we will establish a value distribution theory for meromorphic maps $f : M \longrightarrow \mathbb{P}(V)$ and target maps $g_j : M \longrightarrow \mathbb{P}(V^*)$ for $j = 1, \ldots, q$. We will encounter new and surprising concepts, methods and results. Perhaps the journey is more valuable than its destination.

The small Griffiths conjecture was the prime motive for these investigations. Clearly, they would not solve the conjecture, but would shed new light on the problematic. This is the connection: Let H be the hyperplane section bundle on $\mathbb{P}(V)$. Take $1 < p \in \mathbb{N}$. A section in H^p can be identified with a homogeneous polynomial $\alpha : V \longrightarrow \mathbb{C}$ of degree p. We assume that the zero divisor of $\alpha \not\equiv 0$ is simple and smooth, that is, that $d\alpha(\mathfrak{x}) \neq 0$ for all $\mathfrak{x} \in V_*$ with $\alpha(\mathfrak{x}) = 0$. Since $d\alpha(\mathfrak{x},\mathfrak{x}) = p\alpha(\mathfrak{x})$ for $\mathfrak{x} \in V$, we have $d\alpha(\mathfrak{x}) \neq 0$ for all $\mathfrak{x} \in V_*$, which means $d\alpha(\mathfrak{x}) \in V_*^*$ if $\mathfrak{x} \in V_*$. Let $\mathfrak{v} : U \longrightarrow V$ be a reduced representation of f. Then $\mathfrak{w} \neq d\alpha \circ \mathfrak{v} : U \longrightarrow V^*$ is a representation of a meromorphic map $g : M \longrightarrow \mathbb{P}(V^*)$. The intersection divisor for the incidence $f(z) \in \ddot{E}[g(z)]$ is given on U by the divisor of $\langle \mathfrak{v}, \mathfrak{w} \rangle = d\alpha(\mathfrak{v}, \mathfrak{v}) = p\alpha \circ \mathfrak{v}$, which is the intersection divisor of f with the hypersurface defined by α. Since T_g grows as $(p - 1)T_f$ we cannot hope to resolve the Griffiths Conjecture, but we may gain new insights.

B2. Hermitian geometry. Basic notations and definitions were given in Section A4a). Here we introduce additional concepts which will be helpful in our investigations.

Let $V_1, \ldots, V_q$ and W be hermitian vector spaces. A p-fold operation

(B.5) $$\Theta : V_1 \times \ldots \times V_p \longrightarrow W$$

is said to be projective, if there are maps $\rho_j : \mathbb{C}_* \longrightarrow \mathbb{C}_*$ such that

(B.6) $$\mathfrak{r}_1 \Theta \ldots \Theta \mathfrak{r}_{j-1} \Theta \lambda\mathfrak{r}_j \Theta \mathfrak{r}_{j+1} \Theta \ldots \Theta \mathfrak{r}_p = \rho_j(\lambda)(\mathfrak{r}_1 \Theta \ldots \Theta \mathfrak{r}_p)$$

for $\lambda \in \mathbb{C}_*$ and $\mathfrak{r}_j \in V_j$ with $j = 1, \ldots, p$. If $x_j = \mathbb{P}(\mathfrak{r}_j) \in \mathbb{P}(V)$ and if $\mathfrak{r}_1 \Theta \ldots \Theta \mathfrak{r}_p \neq 0$, then

(B.7) $$x_1 \Theta \ldots \Theta x_p = \mathbb{P}(\mathfrak{r}_1 \Theta \ldots \Theta \mathfrak{r}_p)$$

is well-defined. The projective operation is said to be <u>unitary</u> of degree $(q_1, \ldots, q_p) \in \mathbb{Z}^p$ if $|\rho_j(\lambda)| = |\lambda|^{q_j}$ for all $\lambda \in \mathbb{C}_*$. If so, the <u>distance</u>

(B.8) $$\square\, x_1 \dot{\Theta} \ldots \dot{\Theta} x_p \,\square = \frac{\|\mathfrak{r}_1 \Theta \cdots \Theta \mathfrak{r}_p\|}{\|\mathfrak{r}_1\|^{q_1} \cdots \|\mathfrak{r}_p\|^{q_p}}$$

is well-defined, but not a function of $x_1 \Theta \ldots \Theta x_p$ which is indicated by the dot over the operation symbol. The projective operation is said to be <u>unitary</u> if it is unitary of degree $(1, \ldots, 1)$. The projective operation is said to be <u>homogeneous of degree</u> $(q_1, \ldots, q_p) \in \mathbb{Z}^p$ if $\rho_j(\lambda) = \lambda^{q_j}$ for all $\lambda \in \mathbb{C}_*$ and all $j \in \mathbb{N}[1,p]$ and if $\mathfrak{r}_1 \Theta \ldots \Theta \mathfrak{r}_p$ is a holomorphic function of $\mathfrak{r}_1, \ldots, \mathfrak{r}_p$ on $V_1 \times \ldots \times V_p$. Such an operation is unitary of degree $(q_1, \ldots, q_p)$. A projective operation Θ is said to be <u>homogeneous</u>, if it is homogeneous of degree $(1, \ldots, 1)$.

A homogeneous, projective operation is said to be p-<u>linear</u> if it is additive in each variable $\mathfrak{x}_j$. A p-linear operation extends to a linear map

$$\Theta : V_1 \otimes \ldots \otimes V_p \longrightarrow W \tag{B.9}$$

An example of a p-linear operation is the <u>**exterior product**</u>, which defines □ $x_1 \dot{\wedge} \ldots \dot{\wedge} x_p$ □. Examples of bilinear operations are the <u>**interior product**</u> which defines □ $x \dot{\llcorner} y$ □, and the <u>**inner product**</u>, which defines □ $x;y$ □ = □ $\langle x;y \rangle$ □ with $0 \leqslant$ □ $x;y$ □ $\leqslant 1$. An example of an unitary operation is the <u>**hermitian product**</u> which defines □ $x \dot{|} y$ □ = □ $(x \dot{|} y)$ □ with $0 \leqslant$ □ $x \dot{|} y$ □ $\leqslant 1$.

Important for our investigation is the <u>**contraction product**</u>

$$\boxplus : (\bigwedge_{p+1} V) \times (\bigwedge_{q+1} V^*) \longrightarrow (\bigwedge_p V) \times (\bigwedge_q V^*) \tag{B.10}$$

which is defined as follows: Put $\mu = \mathrm{Min}(p+1, q+1)$. Take $\mathfrak{x} = \mathfrak{x}_0 \wedge \ldots \wedge \mathfrak{x}_p$ in $\tilde{G}_p(V)$ and $\mathfrak{y} = \mathfrak{y}_0 \wedge \ldots \wedge \mathfrak{y}_q$ in $\tilde{G}_q(V^*)$. Define

$$\hat{\mathfrak{x}}_j = (-1)^j \mathfrak{x}_0 \wedge \ldots \wedge \mathfrak{x}_{j-1} \wedge \mathfrak{x}_{j+1} \wedge \ldots \wedge \mathfrak{x}_p \tag{B.11}$$

$$\hat{\mathfrak{y}}_k = (-1)^k \mathfrak{y}_0 \wedge \ldots \wedge \mathfrak{y}_{k-1} \wedge \mathfrak{y}_{k+1} \wedge \ldots \wedge \mathfrak{y}_q \tag{B.12}$$

$$\mathfrak{x} \boxplus \mathfrak{y} = \frac{1}{\mu} \sum_{j=0}^{p} \sum_{k=0}^{q} \langle \mathfrak{x}_j, \mathfrak{y}_k \rangle \hat{\mathfrak{x}} \otimes \hat{\mathfrak{y}}_k . \tag{B.13}$$

Then $\boxplus$ extends uniquely to a well-defined bilinear map. A linear map

$$\boxplus : (\bigwedge_{p+1} V) \otimes (\bigwedge_{q+1} V^*) \longrightarrow (\bigwedge_p V) \otimes (\bigwedge_q V^*) \tag{B.14}$$

results, which can be iterated $\boxplus^\rho = \boxplus \circ \boxplus^{\rho-1}$ for $\rho = 1, \dots, \mu$, where $\boxplus^0$ is the identity and $\boxplus^\mu = L$. Thus the iterations of $\boxplus$ reach from the tensor product to the interior product. If $v = \mathbb{P}(\mathfrak{v}) \in G_p(V)$ and $w = \mathbb{P}(\mathfrak{w}) \in G_q(V^*)$, then $\mathfrak{v} \boxplus \mathfrak{w} = 0$ if and only if $E(v) \subseteq E[w]$. Also $\mathfrak{v} \boxplus^\rho \mathfrak{w} = 0$ if and only if $\dim E(v) \cap E[w] > p + 1 - \rho$. Moreover

$$(B.15) \qquad \sqrt{\binom{\mu}{\rho}} \; \square \; v \; \dot\boxplus^\rho \; w \; \square \leqslant 1 \qquad \sqrt{\mu} \; \square \; v \; \dot\boxplus \; w \; \square \leqslant 1.$$

B3. The First Main Theorem. Hermitian line bundles and their Chern forms are explained in A4c). Meromorphic maps and their representations are defined in A4d). Divisors are discussed in A4e). Parabolic manifolds are introduced in A5. The value distribution functions on parabolic manifolds are expounded in A6. Here we outline the First Main Theorem for a homogeneous operation of degree $(q_1, \dots, q_p)$ and specialize to specific operations.

Let $V_1, \dots, V_k$ and W be hermitian vector spaces. Consider a homogeneous projective operation $\Theta : V_1 \times \dots V_k \longrightarrow W$ of degree $(q_1, \dots, q_p) \in \mathbb{Z}^k$. Let (M, τ) be a parabolic manifold of dimension m. For $j = 1, \dots, k$, let $f_j : M \longrightarrow \mathbb{P}(V_j)$ be a meromorphic map with hyperplane section bundle L_{f_j} and representation section f_{f_j}. (See A4d)). The operation extends to a fiber preserving, holomorphic map

$$(B.16) \qquad \Theta : (V_{1M} \otimes L_{f_1}) \oplus \dots \oplus (V_{kM} \otimes L_{f_k}) \longrightarrow W_M \otimes L_{f_1}^{q_1} \otimes \dots \otimes L_{f_k}^{q_k}$$

which defines a global, holomorphic section

$$(B.17) \qquad F = F_{f_1} \Theta \dots \Theta F_{f_k} \in \Gamma(M, W_M \otimes L_{f_1}^{q_1} \otimes \dots \otimes L_{f_k}^{q_k}).$$

If $F \not\equiv 0$, then $(f_1, \dots, f_k)$ is said to be **free** for Θ, in which case the **operation divisor**

(B.18) $$\mu_{f_1 \dot{\Theta} \ldots \dot{\Theta} f_k} = \mu_F$$

exists. Its <u>counting function</u> and <u>valence function</u> are abbreviated to

(B.19) $$n_{f_1\dot{\Theta}\ldots\dot{\Theta}f_k} = n_{\mu_F} \qquad N_{f_1\dot{\Theta}\ldots\dot{\Theta}f_f} = N_{\mu_F}$$

If $(f_1, \ldots, f_k)$ is free for Θ, a meromorphic map $f_1 \Theta \ldots \Theta f_k : M \longrightarrow \mathbb{P}(W)$ is defined by

(B.20) $$(f_1 \Theta \ldots \Theta f_k)(x) = f_1(x) \Theta \ldots \Theta f_k(x) \qquad \text{for} \quad x \in M - Z(F)$$

If $\mathfrak{v}_j : U \longrightarrow V_j$ is a reduced representation of f_j for $j = 1, \ldots, k$, then $\mathfrak{v} = \mathfrak{v}_1 \Theta \ldots \Theta \mathfrak{v}_k : U \longrightarrow W$ is a representation of $f_1 \Theta \ldots \Theta f_k$ with

(B.21) $$\mu_{f_1\dot{\Theta}\ldots\dot{\Theta}f_k} | U = \mu_{\mathfrak{v}} .$$

The <u>compensation function</u> is defined for $r \in \hat{\ell}_\tau$ by

(B.22) $$m_{f_1\dot{\Theta}\ldots\dot{\Theta}f_k}(r) = \int_{M\langle r\rangle} \log \frac{1}{\Box f_1 \dot{\Theta} \ldots \dot{\Theta} f_k \Box} \sigma .$$

In general the sign is undetermined, but this can easily be rectified. Since the distance function is continuous on the compact space $\mathbb{P}(V_1) \times \ldots \times \mathbb{P}(V_k)$ a constant $c \geqslant 1$ exists such that

$\Box x_1 \dot{\Theta} \ldots \dot{\Theta} x_k \Box \leqslant c$ for all $x_j \in \mathbb{P}(V_j)$ and $j = 1, \ldots, k$. The modified compensation function

(B.23) $$m^c_{f_1\dot{\Theta}\ldots\dot{\Theta}f_k}(r) = \int_{M\langle r\rangle} \log \frac{c}{\Box f_1 \dot{\Theta} \ldots \dot{\Theta} f_k \Box} \sigma \geqslant 0$$

is non-degenerate with

(B.24) $$m^{c}_{f_1\dot{\Theta}\ldots\dot{\Theta}f_k}(r) = m_{f_1\dot{\Theta}\ldots\dot{\Theta}f_k}(r) + C_1 \log c \geq 0.$$

In most applications, our maps f_j map into subspaces on which the distance function is bounded by 1 such that we can use (B.22).

<u>The General First Main Theorem</u> (Theorem 3.4) holds

(B.25)

$$\sum_{j=1}^{k} q_j T_{f_j}(r,s) = T_{f_1\Theta\ldots\Theta f_k}(r,s) + N_{f_1\dot{\Theta}\ldots\dot{\Theta}f_k}(r,s) + m_{f_1\dot{\Theta}\ldots\dot{\Theta}f_k}(r) - m_{f_1\dot{\Theta}\ldots\dot{\Theta}f_k}(s)$$

where $0 < s < r \in \hat{\mathcal{E}}_\tau$ and $s \in \hat{\mathcal{E}}_\tau$. The identity extends the definition of the compensation function continuously to all $r > 0$ such that (B.25) holds for all $0 < s < r$. Since the constant cancels in the difference, the modified compensation function can be used. The First Main Theorem contains only "essentially" non–negative terms, and it is a balancing statement. If $W = \mathbb{C}$, then $f_1 \Theta \ldots \Theta f_k : M \longrightarrow \mathbb{P}_0$ is constant and the term $T_{f_1\Theta\ldots\Theta f_k}$ vanishes. Now we can discuss special cases:

1. **<u>SPECIAL CASE</u>**. Let V be a hermitian vector space of dimension $n + 1$. Take $n_j \in \mathbb{N}$ with $p = n_1 + \ldots + n_k \leq n + 1$. Define $V_j = \bigwedge_{n_j} V$ and $q_j = 1$. Define $W = \bigwedge_p V$. The exterior product $\wedge : V_1 \times \ldots \times V_k \longrightarrow W$ is a homogeneous projective operation. We obtain the <u>First Main Theorem</u> for the <u>exterior product</u>

(B.26)

$$\sum_{j=1}^{k} T_{f_j}(r,s) = T_{f_1\wedge\ldots\wedge f_k}(r,s) + N_{f_1\dot{\wedge}\ldots\dot{\wedge}f_k}(r,s) + m_{f_1\dot{\wedge}\ldots\dot{\wedge}f_k}(r) - m_{f_1\dot{\wedge}\ldots\dot{\wedge}f_k}(s)$$

If $p = n+1$, then $T_{f_1 \wedge \ldots \wedge f_k}(r,s) \equiv 0$. If $f_j : M \longrightarrow G_{n_{j-1}}(V)$ for $j = 1, \ldots, k$, then $m_{f_1 \dot{\wedge} \ldots \dot{\wedge} f_k} \geq 0$. If $n_1 = \ldots = n_k = 1$, then $f_1, \ldots, f_k$ are said to be in **general position** if and only if $(f_1, \ldots, f_k)$ is free for $\wedge$. This case will be discussed later.

If $k = 2$, if $a \in G_q(V)$ is constant and if $f_1 \equiv a$, the case was already proved by Ahlfors [1] for $M = \mathbb{C}$. He called it the First Law of Equidistribution. He used an integral averaging method.

2. **SPECIAL CASE.** Take $k = 2$, $p \in \mathbb{Z}[0,n]$, $q \in \mathbb{Z}[0,n]$. Let V be a hermitian vector space of dimension $n+1$. Define $V_1 = \bigwedge_{p+1} V$ and $V_2 = \bigwedge_{q+1} V^*$. Let $f : M \longrightarrow \mathbb{P}(\bigwedge_{p+1} V)$ and $g : M \longrightarrow \mathbb{P}(\bigwedge_{q+1} V^*)$ be meromorphic maps. Define $\mu = \mathrm{Min}(p+1, q+1)$. Take $\rho \in \mathbb{N}[1,\mu]$. As operation take the iterated contraction $\boxplus^\rho$. Then $q_1 = q_2 = 1$ and $W = (\bigwedge_{p+1-\rho} V) \otimes (\bigwedge_{q+1-\rho} V^*)$. We obtain the **First Main Theorem for contraction**

(B.27)
$$\boxed{\begin{gathered} T_f(r,s) + T_g(r,s) - T_{f \boxplus^\rho g}(r,s) \\ N_{f \dot{\boxplus}^\rho g}(r,s) + m_{f \dot{\boxplus}^\rho g}(r) - m_{f \dot{\boxplus}^\rho g}(s) \end{gathered}}$$

If $f : M \longrightarrow G_p(V)$ and $g : M \longrightarrow G_q(V^*)$, then $m_{f \dot{\boxplus}^\rho g} \geq 0$.

Write $\boxplus = \boxplus^1$.

3. **SPECIAL CASE.** Take the situation of the 2. Special Case, but assume that $q \leq p$ and $\rho = \mu = q+1$. Our operation becomes the interior product $\llcorner = \boxplus^\mu$ with $q_1 = q_2 = 1$ and $W = \bigwedge_{p-q} V$. Now (B.27) becomes the **First Main Theorem for the interior product.**

(B.28)

$$T_f(r,s) + T_g(r,s) = N_{f\dot{L}g}(r,s) + m_{f\dot{L}g}(r) - m_{f\dot{L}g}(s) + T_{fLg}(r,s)$$

If $f : M \longrightarrow G_p(V)$ and $g : M \longrightarrow G_q(V^*)$, then $m_{f\dot{L}g} \geqslant 0$. If g is constant, $T_g \equiv 0$ and we obtain Theorem 4.9 of Wu [126] page 112, which is proved by taking integral averages. Se also (A.155).

4. SPECIAL CASE. Take $k = 2$. Let V be a hermitian vector space of dimension $n + 1$. Take $V_1 = V$ and $V_2 = V^*$. Let $f : M \longrightarrow \mathbb{P}(V)$ and $g : M \longrightarrow \mathbb{P}(V^*)$ be meromorphic maps. As operation take the inner product. Then $q_1 = q_2 = 1$ and $W = \mathbb{C}$. We obtain the First Main Theorem for the inner product

(B.29)

$$T_f(r,s) + T_g(r,s) = N_{f;g}(r,s) + m_{f;g}(r) - m_{f;g}(s)$$

If g is constant, then $T_g \equiv 0$ and we obtain the classical FMTH (A.117).

Our proof of Mori's, Theorem [63], Section 11, uses a special operation which will be helpful, but as an operation is not interesting.

B4. Notations. Associated maps were defined in A.8. The Plücker Difference Formula was stated in A.9. Here we introduce some additional notations concerning the interaction of two meromorphic maps.

Let (M,τ) be a parabolic manifold of dimension m. Define

(B.30) $$M^+ = \{x \in M \mid \upsilon(x) > 0\} .$$

Let B be a holomorphic form of bidegree $(m - 1, 0)$ on M. Let V be a hermitian vector space of dimension $n + 1$. The Fubini Study form on $G_q(V)$ is denoted by Ω_q and on $G_p(V^*)$ is denoted by Ω_p^*. Let

$f : M \longrightarrow \mathbb{P}(V)$ and $g : M \longrightarrow \mathbb{P}(V^*)$ be meromorphic maps with generality index ℓ_f for f and ℓ_g for g in respect to B. Let $F_q = F_{qf}$ be the qth representation section of f and $G_p = F_{pg}$ the pth representation section of g. For $q \in \mathbb{Z}[0,\ell_f]$ and $p \in \mathbb{Z}[0,\ell_g]$ the associated maps $f_q : M \longrightarrow G_q(V)$ and $g_p : M \longrightarrow G_p(V^*)$ exist. Define

(B.31) $$\mathbb{H}_q = m i_{m-1} f_q^*(\Omega_q) \wedge B \wedge \bar{B} \geq 0$$

(B.32) $$\mathbb{K}_p = m i_{m-1} g_p^*(\Omega_p^*) \wedge B \wedge \bar{B} \geq 0\ .$$

Then $\mathbb{H}_{\ell_f} \equiv 0 \equiv \mathbb{K}_{\ell_g}$. If $0 \leq q < \ell_f$ and $0 \leq p < \ell_g$, then $\mathbb{H}_q > 0$ and $\mathbb{K}_p > 0$ outside a thin analytic on M. On $M^+ - I_{f_g}$ define $h_q \geq 0$ by $\mathbb{H}_q = h_q^2 \upsilon^m$. On $M^+ - I_{g_p}$ define $k_p \geq 0$ by $\mathbb{K}_p = k_p^2 \upsilon^m$. On $M - (I_{f_q} \cup I_{g_p})$ define $\mathbb{K}_{pq} = \sqrt{\mathbb{K}_p \mathbb{H}_q}$. Then $\mathbb{K}_{pq} = k_p h_q \upsilon^m$ on $M^+ - (I_{f_q} \cup I_{q_p})$. If $\mathfrak{z} : U \longrightarrow U'$ is a chart and if χ is defined by (A.126), we have

(B.33) $$\mathbb{H}_q | U = H_q^2 \chi^m \qquad \mathbb{K}_p | U = K_p^2 \chi^m \qquad \mathbb{K}_{pq} | U = K_p H_q \chi^m$$

where $H_q \geq 0$ and $K_p \geq 0$.

Define $\mu = \mathrm{Min}(p+1,q+1)$. Take $\rho \in \mathbb{Z}[0,\mu]$. Let Ω_{qp} be the Fubini Study Form on $\mathbb{P}((\bigwedge_{q+1} V) \otimes (\bigwedge_{p+1} V^*))$. Define

(B.34) $$\Phi_{pq}^{(\rho)} = \binom{\mu}{\rho}^2 \square f_g \mathbin{\dot{\boxplus}}^{\rho} g_p \square^2 \qquad \Phi_{pq} = \Phi_{pq}^{(1)}\ .$$

Then $0 \leqslant \Phi_{pq}^{(\rho)} \leqslant \binom{\mu}{\rho}$ and $0 \leqslant \Phi_{pq} \leqslant \mu$. If (f_g, g_p) is free for $\boxplus^\rho$, then $\Phi_{pq}^{(\rho)} > 0$ outside a thin analytic subset of M and $f_q \boxplus^\rho g_p$ exists. Define

(B.35) $$\mathbb{H}_{pq}^{(\rho)} = m i_{m-1}(f_q \boxplus^\rho g_p)^*(\Omega_{q-\rho,p-\rho}) \wedge B \wedge \bar{B} \geqslant 0 .$$

Abbreviate $\mathbb{H}^1_{pq} = \mathbb{H}_{pq}$.

B5. __Frenet frames and stress.__ By the method of Frenet frames, we will obtain curvature formulas and estimates which will lead to the Ahlfors estimates and the defect relation. We have reached a crucial point in the proof. If a is not constant, $dd^c\Phi_p(a)$ and $d\Phi_p(a) \wedge d^c\Phi_p(a)$ are difficult to calculate and to interpret and the identities (A.159) and (A.160) become invalid. The Frenet frames and the operator $\boxplus$ will save us. A large amount of Frenet identities is proved in Section 5. As far as f alone is involved, they are known, Stoll [100] and [108]. In this introduction, we shall avoid this horde of formulas as much as possible.

From the start, we will assume that f is general for B, that is, $\ell_f = n$. Denote $\ell_g = s$. Define $\mathfrak{Z} = Z(F_n) \cup Z(G_s)$. Our calculations occur on $M - \mathfrak{Z}$ or subsets thereof.

Let $\{U_\lambda, \mathfrak{a}_\lambda, \mathfrak{v}_\lambda, \mathfrak{w}_\lambda\}_{\lambda \in \Lambda}$ be a representation atlas of f and g. Here $\{U_\lambda\}_{\lambda\in\Lambda}$ is an open covering of M. For $\lambda = (\lambda_0, \ldots, \lambda_p)$ in Λ^{p+1} define U_λ by (A.129). Define $\Lambda[p]$ by (A.130). For $\lambda \in \Lambda$, the map $\mathfrak{a}_\lambda : U_\lambda \longrightarrow U'_\lambda$ is a chart, see (A.125). Define χ_λ and $\mathfrak{z}_\lambda$ by (A.126) and (A.127). We have reduced representations $\mathfrak{v}_\lambda : U_\lambda \longrightarrow V$ of f and $\mathfrak{w}_\lambda : U_\lambda \longrightarrow V^*$ of g. If $(\lambda,\mu) \in \Lambda[1]$, then $\Delta_{\lambda\mu}$ is defined by (A.131). Also zero free holomorphic functions $v_{\lambda\mu}$ and $w_{\lambda\mu}$ are determined on $U_{\lambda\mu}$ by

(B.36) $$\mathfrak{v}_\lambda = v_{\lambda\mu}\mathfrak{v}_\mu \qquad \mathfrak{w}_\lambda = w_{\lambda\mu}\mathfrak{w}_\mu .$$

For $p \in \mathbb{Z}[0,s]$ and $q \in \mathbb{Z}[0,n]$ we have

$$\text{(B.37)} \qquad \mathfrak{v}_{\lambda\underline{q}} = (v_{\lambda\mu})^{q+1}(\Delta_{\mu\lambda})^{\frac{q(q+1)}{2}} \mathfrak{v}_{\mu\underline{q}}$$

$$\text{(B.38)} \qquad \mathfrak{w}_{\lambda\underline{p}} = (w_{\lambda\mu})^{p+1}(\Delta_{\mu\lambda})^{\frac{p(p+1)}{2}} \mathfrak{w}_{\mu\underline{p}} \,.$$

Vector functions $\mathfrak{e}_q^\lambda : (U_\lambda - \mathfrak{Z}) \longrightarrow V$ and $\mathfrak{g}_p^\lambda : (U_\lambda - \mathfrak{Z}) \longrightarrow V^*$ of class C^∞ are uniquely determined by

(B.39) $\quad \|\mathfrak{e}_q^\lambda\| = 1 \qquad$ for $q \in \mathbb{Z}[0,n]$

(B.40) $\quad (\mathfrak{e}_q^\lambda | \mathfrak{e}_j^\lambda) = 0 \qquad$ for $0 \leqslant q < j \leqslant n$

(B.41) $\quad \mathfrak{v}_{\lambda\underline{q}} = \|\mathfrak{v}_{\lambda\underline{q}}\| \mathfrak{e}_0^\lambda \wedge \dots \wedge \mathfrak{e}_q^\lambda \qquad$ for $q \in \mathbb{Z}[0,n]$

(B.42) $\quad \|\mathfrak{g}_p^\lambda\| = 1 \qquad$ for $p \in \mathbb{Z}[0,s]$

(B.43) $\quad (\mathfrak{g}_p^\lambda | \mathfrak{g}_k^\lambda) = 0 \qquad$ for $0 \leqslant p < k \leqslant s$

(B.44) $\quad \mathfrak{w}_{\lambda\underline{p}} = \|\mathfrak{w}_{\lambda\underline{p}}\|^2 \mathfrak{g}_0^\lambda \wedge \dots \wedge \mathfrak{e}_p^\lambda \qquad$ for $p \in \mathbb{Z}[0,s]$.

Then $\mathfrak{e}_0^\lambda, \dots, \mathfrak{e}_n^\lambda$ is called a **Frenet frame** of f on $U_\lambda - \mathfrak{Z}$. At every point $x \in U_\lambda - \mathfrak{Z}$, the vectors $\mathfrak{e}_0^\lambda(x), \dots, \mathfrak{e}_n^\lambda(x)$ constitute an orthonormal base of V. Let $\mathfrak{e}_0^{*\lambda}(x), \dots, \mathfrak{e}_n^{*\lambda}(x)$ be the dual base, which is also orthonormal. Then $\mathfrak{e}_0^{*\lambda}, \dots, \mathfrak{e}_n^{*\lambda}$ are vector functions of class C^∞ on $U_\lambda - \mathfrak{Z}$ called the **dual Frenet frame**. For $p \in \mathbb{Z}[0,s]$ and $q \in \mathbb{Z}[0,n]$ and $\lambda \in \Lambda$ define the **stress coefficient**

$$\Lambda^{\lambda}_{pq} = (\mathfrak{y}^{\lambda}_{p} | \mathfrak{e}^{*\lambda}_{q}) \tag{B.45}$$

where $|\Lambda^{\lambda}_{pq}| \leqslant 1$. We have

$$\mathfrak{y}^{\lambda}_{p} = \sum_{q=0}^{n} \Lambda^{\lambda}_{pq} \mathfrak{e}^{*\lambda}_{q} \quad \text{for} \quad p \in \mathbb{Z}[o,s] \,. \tag{B.46}$$

The stress coefficients describe the relative movement of one Frenet frame against another. This ought to be a fundamental problem in geometry, but I know no literature concerning these coefficients. Λ^{λ}_{pq} can be calculated in terms of $\mathfrak{v}_{\lambda,\underline{q}}$, $\mathfrak{v}_{\lambda,\underline{q-1}}$, $\mathfrak{w}_{\lambda\underline{p}}$ and $\mathfrak{w}_{\lambda,\underline{p-1}}$ (Lemma 5.8), which provides a definition of the stress coefficients without Frenet frames, but the method lacks geometric meaning. The stress coefficient matrix is unitary

$$\sum_{j=0}^{n} \Lambda^{\lambda}_{pj} \bar{\Lambda}^{\lambda}_{qj} = \begin{cases} 1 & \text{if} \quad p = q \\ 0 & \text{if} \quad p \neq q \,. \end{cases} \tag{B.47}$$

Recall (B.34). For $p \in \mathbb{Z}[0,s]$ and $q \in \mathbb{Z}[0,n]$ we obtain

$$\Phi_{pq} | U_{\lambda} = \sum_{j=0}^{p} \sum_{k=0}^{q} |\Lambda^{\lambda}_{jk}|^{2} \,. \tag{B.48}$$

Surprisingly, this unusual operation on a matrix, locally defined, gives the restriction of a global invariant. Take $(\lambda,\mu) \in \Lambda[1]$. On $U_{\lambda\mu}$ we have

$$\mathfrak{e}^{\lambda}_{q} = \frac{v_{\lambda\mu}}{|v_{\lambda\mu}|} \left[\frac{\Delta_{\mu\lambda}}{|\Delta_{\mu\lambda}|} \right]^{q} \mathfrak{e}^{\mu}_{q} \tag{B.49}$$

$$\mathfrak{e}^{*\lambda}_{q} = \frac{v_{\mu\lambda}}{|v_{\mu\lambda}|} \left[\frac{\Delta_{\lambda\mu}}{|\Delta_{\lambda\mu}|} \right]^{q} \mathfrak{e}^{*\mu}_{q} \tag{B.50}$$

(B.51) $$\varphi_p^\lambda = \frac{w_{\mu\lambda}}{|w_{\mu\lambda}|}\left[\frac{\Delta_{\mu\lambda}}{|\Delta_{\mu\lambda}|}\right]^p \varphi_p^\mu$$

(B.52) $$\Lambda_{pq}^\lambda = \frac{v_{\lambda\mu}}{|v_{\lambda\mu}|}\frac{w_{\lambda\mu}}{|w_{\lambda\mu}|}\left[\frac{\Delta_{\mu\lambda}}{|\Delta_{\mu\lambda}|}\right]^{p+q} \Lambda_{pq}^\mu .$$

If $a \in \mathbb{Z}[0,s]$, $p \in \mathbb{Z}[0,s]$, $b \in \mathbb{Z}[0,n]$ and $q \in \mathbb{Z}[0,n]$ with $p + q = a + b$, then a function $S_{pq}^{ab} = S_{ab}^{pq}$ of class C^∞, called a **primary stress invariant**, is uniquely defined on $M - \mathfrak{J}$ by

(B.53) $$S_{pq}^{ab} | U_\lambda = \tfrac{1}{2}(\Lambda_{pq}^\lambda \bar{\Lambda}_{ab}^\lambda + \Lambda_{ab}^\lambda \bar{\Lambda}_{pq}^\lambda)$$

where $-1 \leqslant S_{pq}^{ab} \leqslant +1$.

B6. The Stress Formulas. For $p \in \mathbb{Z}[0,s]$ and $q \in \mathbb{Z}[0,n]$ we have the **Curvature Stress Formula**

(B.54)
$$\boxed{\begin{aligned} & m i_{m-1}\, dd^c \Phi_{pq} \wedge B \wedge \bar{B} \\ &= (\Phi_{p+1,q} - 2\Phi_{pq} + \Phi_{p-1,q})\mathbb{K}_p \\ &\quad + (\Phi_{p,q+1} - 2\Phi_{pq} + \Phi_{p,q-1})\mathbb{H}_q \\ &\quad + 2 S_{p+1\ q}^{p\ q+1}\, \mathbb{K}_{pq} \end{aligned}}$$

The Curvature Stress formula surprises by its elegance, balance and simplicity. The proof is difficult though it has been simplified several times. In order to compute the gradient, higher order stress invariants have to be introduced.

For $j = 1, \ldots, k$, take $a_j \in \mathbb{R}[0,s]$, $b_j \in \mathbb{Z}[0,n]$, $c_j \in \mathbb{Z}[0,s]$, $d_j \in \mathbb{Z}[0,n]$ with

(B.55)
$$\sum_{j=1}^{k} (a_j + b_j) = \sum_{j=1}^{k} (c_j + d_j) .$$

A function $S^{c_1 d_1 \ldots c_k d_k}_{a_1 b_1 \ldots a_k b_k}$ of class C^{∞}, called a <u>**stress invariant of level**</u> k, is uniquely defined on $M - \mathfrak{Z}$ such that

(B.56)
$$S^{c_1 d_1 \ldots c_k d_k}_{a_1 b_1 \ldots a_k b_k} = \frac{1}{2}\left[\prod_{j=1}^{k} \Lambda^{\lambda}_{a_j b_j} \bar{\Lambda}^{\lambda}_{c_j d_j} + \prod_{j=1}^{k} \Lambda^{\lambda}_{c_j d_j} \bar{\Lambda}_{a_j b_j}\right]$$

(B.57)
$$-1 \leqslant S^{c_1 d_1 \ldots c_k d_k}_{a_1 b_1 \ldots a_k b_k} = S^{a_1 b_1 \ldots a_k b_k}_{c_1 d_1 \ldots c_k d_k} \leqslant +1$$

(B.58)
$$0 \leqslant S^{a_1 b_1 \ldots a_k b_k}_{a_1 b_1 \ldots a_k b_k} = \prod_{j=1}^{k} S^{a_j b_j}_{a_j b_j} \leqslant 1 .$$

We introduce a repeatable summation convention

(B.59)
$$\sum_{j=0}^{r} S^{\ldots j \ldots}_{\ldots j \ldots} = S_{\begin{smallmatrix} \ldots r \ldots \\ | \\ \ldots r \ldots \end{smallmatrix}} .$$

For instance we have $\mu^2 \, \square \, f_q \, \dot{\boxplus} \, g_p \, \square^2 = S_{\begin{smallmatrix} p & q \\ | & | \\ p & q \end{smallmatrix}}$. For $p \in \mathbb{Z}[0,s]$ and $q \in \mathbb{Z}[0,n]$ we have the <u>**Stress Gradient Formula**</u>

(B.60)
$$\boxed{\begin{aligned} & m i_{m-1}\, d\Phi_{pq} \wedge d^{c}\Phi_{pq} \wedge B \wedge \bar{B} \\ & = S_{\begin{smallmatrix} p & q & p+1 & q \\ & | & & | \\ p+1 & q & p & q \end{smallmatrix}} \mathbb{K}_p + S_{\begin{smallmatrix} p & q & p & q+1 \\ | & & | & \\ p & q+1 & p & q \end{smallmatrix}} \mathbb{H}_q \\ & + 2 S_{\begin{smallmatrix} p & q & p & q+1 \\ & | & | & \\ p+1 & q & p & q \end{smallmatrix}} \mathbb{K}_{pq} \end{aligned}}$$

While (B.54) and (B.60) are difficult to proof, $dd^c \log \Phi_{pq}$ is easily calculated

(B.61) $$dd^c \log \Phi_{pq} = (f_q \boxplus g_p)^*(\Omega_{q-1,p-1}) - g_p^*(\Omega_p^*) - f_q^*(\Omega_q)$$

(B.62) $$m i_{m-1} dd^c \log \Phi_{pq} \, B \wedge \bar{B} = \mathbb{H}_{pq} - \mathbb{K}_p - \mathbb{H}_q$$

B7. __The Ahlfors Estimates.__ For $p \in \mathbb{Z}[0,s]$ and $q \in \mathbb{Z}[0,n]$ define

(B.63) $$\Xi_{pq} = \Phi_{p+1,q}\mathbb{K}_p + \Phi_{p,q+1}\mathbb{H}_q + 2S_{p+1}^{p}\,{}_{,}{}_{q}^{q+1}\,\mathbb{K}_{pq} \geqslant 0 .$$

Assume that (f_q, g_p) is free for $\boxplus$. Then $\Phi_{pq} > 0$ outside a thin analytic set. Define $\mu = \operatorname{Min}(p+1, q+1)$. Take $\beta \in \mathbb{R}(0,1]$. Then we have the estimate

(B.64) $$\beta^2 \left[\frac{1}{\mu^2}\,\Phi_{pq}\right]^{\beta-1} \Xi_{pq} \leqslant 4(n+1)^2 (m i_{m-1} dd^c \log(1 + (\Phi_{pq})^{\beta}) \wedge B \wedge \bar{B} + \mathbb{K}_p + \mathbb{H}_q)$$

If τ majorizes B with majorant Y (see A.10), we obtain the __Ahlfors Estimate__

(B.65)

$$\boxed{\begin{aligned} &\beta^2 \int_s^r \int_{M[t]} \left[\frac{1}{\mu^2}\,\Phi_{pq}\right]^{\beta-1} \Xi_{pq} \,\frac{dt}{t^{2m-1}} \\ &\leqslant 4(n+1)^3 Y(r)(T_{f_q}(r,s) + T_{g_p}(r,s) + C_1) \end{aligned}}$$

which in particular implies the integrability of $(\Phi_{pq})^{\beta-1}\Xi_{pq}$ over $M[t]$ for all $t > 0$, which is remarkable. As seen in (A.162), the classical Ahlfors estimate involves the term $\Phi_{p,q+1}\mathbb{H}_q$ only (for $p = 0$). In view of (B.63) it seems to be reasonable to drop $\Phi_{p+1,q}\mathbb{K}_p \geqslant 0$ in (B.65) and take the term $2S^{p\ \ q+1}_{p+1,\ q}\mathbb{K}_{pq}$ to the otherside as a remainder term measuring the movement of g in relation to f. This fine idea fails since each term

(B.66) $$\Phi^{\beta-1}_{pq}\Phi_{p+1,q}\mathbb{K}_p,\quad \Phi^{\beta-1}_{pq}\Phi_{p,q+1}\mathbb{H}_q,\quad 2\Phi^{\beta-1}_{pq}S^{p\ \ q+1}_{p+1\ \ q}\mathbb{K}_{pq}$$

may not be integrable over $M[t]$, as an example in Section 10 shows. So it is not easy to proceed to the Second Main Theorem without running into a disaster. After long considerations, the following procedure was chosen, which avoids divergent integrals.

Define $\xi_{pq} \geqslant 0$ on M^+ by $\Xi_{pq} = \xi_{pq}\upsilon^m$. We obtain

(B.67) $$\xi_{pq} = \Phi_{p+1,q}k_p^2 + \Phi_{p,q+1}h_q^2 + 2S^{p\ \ ,q+1}_{p+1,\ \ q}k_ph_q\ .$$

On M^+ define

(B.68) $$0 \leqslant \rho_{pq} = \frac{\Phi_{p,q+1}h_q^2}{\xi_{pq}}\ .$$

Then the integrals

(B.69) $$P_{pq}(r) = P_{pq}(r,g) = \frac{1}{2}\int_{M\langle r\rangle}(\log \rho_{pq})\sigma$$

(B.70) $$R_{pq}(r) = R_{pq}(r,g) = \frac{1}{2}\int_{M\langle r\rangle}\log\left[1+\frac{k_p}{h_q}\right]\sigma \geqslant 0$$

exist for almost all $r > 0$. The terms P_{pq} and R_{pq} are undesirable, but measure the divergence and convergence of the jets of f and g. In one particular case there is good news:

(B.71)

$$\boxed{\text{If } p = \ell_g, \text{ then } k_p = 0,\ \rho_{pq} = 1 \text{ and } P_{pq} \equiv 0 \equiv R_{pq}}$$

After the difficulties of general position are cleared up, we proceed to the Second Main Theorem as in the case of constant targets, which was outlined in Part A.

B8. **General position.** The classical expositions of general position are not easy to understand. With the help of the exterior product, norms and the operation $\boxplus^{\rho}$ we hope to present a simpler and clearer picture. Also the constant $c_p(A)$ in the product to sum estimates depends on A in a non–constructive manner. Here we construct a constant whose dependency on the target set A is explicitly known, which permits us to handle the moving target case.

a) **The definition of general position.** Let V be a hermitian vector space of dimension $n + 1 > 1$. For $\emptyset \neq A \subseteq \mathbb{P}(V)$ let $\operatorname{span}(A, \mathbb{P}(V))$ be the intersection of all projective planes in $\mathbb{P}(V)$ containing A. Then $\operatorname{span}(A, \mathbb{P}(V))$ is a projective plane. The following four conditions are equivalent.

α) If L is a proper plane in $\mathbb{P}(V)$, then $\#(A \cap L) \leq 1 + \dim L$.

β) If $\emptyset \neq B \subseteq A$ with $\#B \leq n + 1$, then $\#B = 1 + \dim \operatorname{span}(B, \mathbb{P}(V))$.

γ) Every subset $B \neq \emptyset$ of A with $\#B \leq n + 1$ is linearly independent.

δ) Take $\emptyset \neq B \subseteq \mathbb{P}(V)$ with $\#B \leq n + 1$, then each $a \in B$ defines a hyperplane $\ddot{E}[a]$ in $\mathbb{P}(V^*)$. Then

(B.72)
$$\dim \bigcap_{a \in B} \ddot{E}[a] = n - \#B .$$

A subset $A \neq \emptyset$ of $\mathbb{P}(V)$ is said to be in **general position** if one and hence all of these four conditions are satisfied.

b) **The gauge**. Take $A \subseteq \mathbb{P}(V)$ with $0 < \#A = k < \infty$. If $k \leqslant n + 1$, take an enumeration $A = \{a_1, \ldots, a_k\}$. The **gauge**

$$\Gamma(a) = \square\, a_1 \mathbin{\dot\wedge} \ldots \mathbin{\dot\wedge} a_k\, \square \tag{B.73}$$

does not depend on the enumeration of A. Then $0 \leqslant \Gamma(A) \leqslant 1$. If $k > n + 1$, the **gauge** is defined by

$$\Gamma(A) = \inf\{\Gamma(B) \mid B \subseteq A, \#B = n + 1\} . \tag{B.74}$$

Since A is finite, $\Gamma(A) > 0$ if and only if A is in **general position**.

c) **Incidence space**. Let p, q, n be integers with $0 \leqslant s \leqslant p \leqslant n$ and $0 \leqslant s \leqslant q \leqslant n$. Define

$$D(p,q,s) = \binom{n+1}{p+1} - \sum_{j=0}^{p-s} \binom{q+1}{s+j+1}\binom{n-q}{p-s-j} . \tag{B.75}$$

Abbreviate $D(p,q,0) = D(p,q)$. Then $0 \leqslant D(p,q,s) \leqslant \binom{n+1}{p+1}$. Take $b \in G_q(v)$. Then $D(p,q,s)$ is the dimension of the linear subspace

(B.76)

$$S[b,p,s] = \left\{ \mathfrak{x} \in \bigwedge_{p+1} V^* \mid \langle \mathfrak{y} \wedge \mathfrak{z} ; \mathfrak{x} \rangle = 0 \quad \forall \left\{ \begin{matrix} \mathfrak{y} \in \tilde{G}_s(E(b)) \\ \mathfrak{z} \in \tilde{G}_{p-s-1}(V) \end{matrix} \right\} \right\}$$

of $\bigwedge_{p+1} V^*$ (Lemma 7.6). Define $\ddot{S}[b,p,s] = \mathbb{P}(S[b,p,s])$. Then (Lemma 7.7)

$$\ddot{S}[b,p,s] \wedge G_p(V^*) = \{ a \in G_p(V^*) \mid \square\, b \mathbin{\dot\boxplus}^{s+1} a\, \square = 0\} . \tag{B.77}$$

Let $\mathfrak{G} \neq \emptyset$ be a finite subset of $G_p(V^*)$. Then $\mathfrak{G} \subseteq \mathbb{P}(\bigwedge_{p+1} V^*)$. Therefore it is defined when $\mathfrak{G}$ is in general position. Define $k = \#\mathfrak{G}$. If $a \in G_p(V^*)$, then $E(a) \subseteq \mathbb{P}(V^*)$ and $G_s(E(a)) \subseteq G_s(V^*)$. Take $b \in G_q(V)$. Since $0 \leqslant s \leqslant \mathrm{Min}(p,q)$, we can define

(B.78) $$\mathfrak{A}(b,s) = \{a \in \mathfrak{A} \mid \square\, b \dot{\llcorner} x \,\square = 0 \quad \forall \quad x \in G_s(E(a))\}$$

(B.79) $$\mathfrak{A}(b,s,r) = \{a \in \mathfrak{A} \mid \square\, b \;\dot{\boxplus}^{s+1}\; a \,\square \leqslant r\} \;.$$

Then $\mathfrak{A}(b,s,0) = \mathfrak{A}(b,s)$ (Lemma 7.8). Form these rather difficult results, we easily derive Theorem 7.8: "If $\mathfrak{A}$ is in general position, then $\#\mathfrak{A}(b,s) \leqslant D(p,q,s)$." The theorem is known, Wu [126], but the proof here is easier to understand. Also we obtain a new result, Theorem 7.11: "If $\mathfrak{A}$ is in general position, then $\#\mathfrak{A}(b,s,r) \leqslant D(p,q,s)$ for all $r \in \mathbb{R}$ with $0 < r \leqslant 2^{-3n-3}\Gamma(\mathfrak{A})$." The explicit domain for r is crucial in the proof of the Second Main Theorem. Classically, only the existence of some interval $0 < r \leqslant r_0$ is determined by Heine–Borel. If $\mathfrak{A}$ moves, the dependence of r_0 on $\mathfrak{A}$ has to be known.

d) <u>Product to sum estimate</u>. Take p,q and s as in c). For $x \in \mathbb{R}$, define $x^+ = \mathrm{Max}(x,0)$. Let $\mathfrak{A} \neq \emptyset$ be a finite subset of $G_p(V^*)$. Define $k = \#\mathfrak{A}$. Assume that $\mathfrak{A}$ is in general position. Abbreviate

(B.80) $$v = \frac{2(k - D(p,q,s))^+}{D(p,q,s)}$$

(B.81) $$c_{pq}(\mathfrak{A}) = (2^{-3n-3}\Gamma(\mathfrak{A}))^v \;.$$

Let $\beta : \mathfrak{A} \longrightarrow \mathbb{R}(0,1)$ and $P : \mathfrak{A} \longrightarrow \mathbb{R}[0,1]$ be functions on $\mathfrak{A}$. Take $b \in G_q(V)$. Theorem 7.12 asserts

(B.82) $$D(p,q,s)c_{pq}(\mathfrak{A}) \left[\prod_{a \in \mathfrak{A}} \frac{P(a)}{\square\, b \;\dot{\boxplus}^{s+1}\; a \,\square^{2-2\beta(a)}} \right]^{\frac{1}{D(p,q,s)}}$$

$$\leqslant \sum_{a \in \mathfrak{A}} \frac{P(a)}{\square\, b \;\dot{\boxplus}^{s+1}\; a \,\square^{2-2\beta(a)}} + (D(p,q,s) - k)^+$$

which generalizes the theorem that the arithemetic mean majorizes the geometric mean.

e) General position for moving targets. Let (M,τ) be a parabolic manifold of dimension m. Let $\mathfrak{g} \neq \emptyset$ be a finite set of meromorphic maps $g : M \longrightarrow \mathbb{P}(V^*)$. Define $k = \#\mathfrak{g}$. The indeterminacy of $\mathfrak{g}$ is defined by

(B.83)
$$I_{\mathfrak{g}} = \bigcup_{g \in \mathfrak{g}} I_g$$

For $z \in M - I_{\mathfrak{g}}$ define $\mathfrak{g}(z) = \{g(z) \mid g \in \mathfrak{g}\}$. Then

(B.84)
$$\mathfrak{J}_{\mathfrak{g}} = I_{\mathfrak{g}} \cup \{z \in M - I_{\mathfrak{g}} \mid \#\mathfrak{g}(z) < k\}$$

is a thin analytic subset of M. The gauge of $\mathfrak{g}$ is the function $\Gamma(\mathfrak{g})$ defined by $\Gamma(\mathfrak{g})(z) = \Gamma(\mathfrak{g}(z))$ for $z \in M - \mathfrak{J}_{\mathfrak{g}}$. Then $0 \leq \Gamma(\mathfrak{g}) \leq 1$. The degeneracy

(B.85)
$$\Delta(\mathfrak{g}) = \mathfrak{J}_{\mathfrak{g}} \cup \{z \in M - \mathfrak{J}_{\mathfrak{g}} \mid \Gamma(\mathfrak{g})(z) = 0\}$$

of $\mathfrak{g}$ is analytic in M. The set $\mathfrak{g}$ is said to be in general position if and only if $\mathfrak{g}(z_0)$ is in general position for some point $z_0 \in M - \mathfrak{J}_{\mathfrak{g}}$, which is the case if and only if $\Delta(\mathfrak{g}) \neq M$.

If $\mathfrak{g}$ is in general position, the gauge measure

(B.86)
$$\Gamma_{\mathfrak{g}}(r) = \int_{M\langle r\rangle} \log \frac{1}{\Gamma(\mathfrak{g})}\, \sigma \geq 0$$

is defined for all $r \in \hat{\ell}_\tau$. It measures the decline of general position for $r \longrightarrow \infty$. We shall estimate $\Gamma_{\mathfrak{g}}(r)$.

f) The First Main Theorem for general position. For $h \in \mathbb{N}[1,k]$ define $\mathfrak{P}_h(\mathfrak{g})$ as the set of all subsets $\mathfrak{h}$ of $\mathfrak{g}$ with $\#\mathfrak{h} = h$. Assume that $\mathfrak{g}$ is in general position, then $\mathfrak{h}$ is in general position. Enumerate $\mathfrak{h} = \{g^1, \ldots, g^h\}$. Then

(B.87) $$\Gamma(\mathfrak{h}) = \square\, \mathfrak{h}\, \square = \square\, g^1 \ddot{\wedge} \ldots \ddot{\wedge} g^h \,\square .$$

Define $m_{\mathfrak{h}} = m_{g^1 \ddot{\wedge} \ldots \ddot{\wedge} g^h}$ and $N_{\mathfrak{h}} = N_{g^1 \ddot{\wedge} \ldots \ddot{\wedge} g^h}$.

Abbreviate $\wedge\mathfrak{h} = g^1 \wedge \ldots \wedge g^h$. The First Main Theorem (B.26) for the exterior product implies

(B.88) $$\sum_{g \in \mathfrak{g}} T_g(r,s) = N_{\mathfrak{h}}(r,s) + m_{\mathfrak{h}}(r) - m_{\mathfrak{h}}(s) + T_{\wedge\mathfrak{h}}(r,s)$$

where $0 < s \leqslant r \in \hat{\ell}_\tau$ and $s \in \hat{\ell}_\tau$. The formula continuously extends $m_{\mathfrak{h}}(r)$ to all $r > 0$. such that (B.88) holds for all $0 < s < r$. If $h = \#\mathfrak{h} = n + 1$, then $T_{\wedge\mathfrak{h}}(r,s) = 0$ and we obtain

(B.89) $$\sum_{g \in \mathfrak{h}} T_g(r,s) = N_{\mathfrak{h}}(r,s) + m_{\mathfrak{h}}(r) - m_{\mathfrak{h}}(s) .$$

If $k < n + 1$, we have (B.88) with $\mathfrak{h} = \mathfrak{g}$. If $k = n + 1$, we have (B.89) with $\mathfrak{h} = \mathfrak{g}$. In both cases we obtain the estimate

(B.90) $$\Gamma_{\mathfrak{g}}(r) = m_{\mathfrak{g}}(r) \leqslant \sum_{g \in \mathfrak{g}} T_g(r,s) + m_{\mathfrak{g}}(s)$$

If $k > n + 1$, define

(B.91) $$N_{\mathfrak{g}}(r,s) = \sum_{\mathfrak{h} \in \mathfrak{P}_{n+1}(\mathfrak{g})} N_{\mathfrak{h}}(r,s)$$

(B.92) $$m_{\mathfrak{g}}(r) = \sum_{\mathfrak{h} \in \mathfrak{P}_{n+1}(\mathfrak{g})} m_{\mathfrak{h}}(r) .$$

Addition implies the <u>First Main Theorem for General Position</u>: If $k > n + 1$ and if $0 < s < r$, then we have

(B.93) $$\binom{k-1}{n} \sum_{g \in \mathfrak{g}} T_g(r,s) = N_{\mathfrak{g}}(r,s) + m_{\mathfrak{g}}(r) - m_{\mathfrak{g}}(s)$$

and the estimate

$$(B.94)\qquad 0 \leqslant \Gamma_{\mathfrak{g}}(r) \leqslant m_{\mathfrak{g}}(r) \leqslant \binom{k-1}{n} \sum_{g\in\mathfrak{g}} T_g(r,s) + m_{\mathfrak{g}}(s)\,.$$

Surprisingly, the First Main Theorem solves the problem of general position. Up to here, only sets $\mathfrak{g}$ were considered. A family $\mathfrak{g} = \{g_\lambda\}_{\lambda\in\Lambda}$ of meromorphic maps $g_\lambda : M \longrightarrow \mathbb{P}(V^*)$ is said to be <u>finite</u> if Λ is finite, and is said to be in <u>general position</u> if the map defined by $\lambda \longrightarrow g_\lambda$ is injective and if $\tilde{\mathfrak{g}} = \{g_\lambda \mid \lambda \in \Lambda\}$ is in general position in which case $\mathfrak{g} = \tilde{\mathfrak{g}}$ can be identified.

B8. <u>The Second Main Theorem</u>. We make these <u>general assumptions</u>

[B1] Let M be a connected, complex manifold of dimension m.
[B2] Let τ be a parabolic exhaustion of M.
[B3] Let V be a hermitian vector space of dimension $n + 1 > 1$.
[B4] Let $f : M \longrightarrow \mathbb{P}(V)$ be a meromorphic map.
[B5] Let $\mathfrak{g}$ be a finite set of meromorphic maps $g : M \longrightarrow \mathbb{P}(V^*)$.
[B6] Let B be a holomorphic form of bidegree $(m - 1,0)$ on M.
[B7] Assume that τ majorizes B with majorant Y.
[B8] Assume that f is general for B.

Then $\ell_f = n$. Define $\ell_{\mathfrak{g}} = \mathrm{Min}\{\ell_g \mid g \in \mathfrak{g}\}$ and $k = \#\mathfrak{g} > 0$. Take $p \in \mathbb{Z}[0,\ell_{\mathfrak{g}}]$. Define $\mathfrak{g}_p = \{g_p\}_{g\in\mathfrak{g}}$. Assume that $\mathfrak{g}_p$ is in general position. Assume that (f,g_p) is free for L for all $g \in \mathfrak{g}$. Take $0 < s < r$. Abbreviate

$$(B.95)\qquad \Theta_p(r,s) = \sum_{q=1}^{n-1} \binom{n-q-1}{p-1} [N_{F_q}(r,s) + T_{f_q}(r,s)]$$

$$(B.96)\qquad P_p(r,\mathfrak{g}) = \sum_{q=0}^{n-1} \sum_{g\in\mathfrak{g}} P_{pq}(r,g)$$

$$(B.97)\qquad R_p(r,\mathfrak{g}) = \sum_{q=0}^{n-1} (k - D(p,q))^+ \sum_{g\in\mathfrak{g}} R_{pq}(r,g)$$

$$\text{(B.98)} \qquad S_p(r,\mathfrak{G}) = 3\begin{bmatrix} n+1 \\ p+2 \end{bmatrix} C_n k(\log T_f(r,s) + \log^+ \mathrm{Ric}_\tau(r,s))$$

$$+ 3\begin{bmatrix} n+1 \\ p+2 \end{bmatrix} C_n \sum_{g\in\mathfrak{G}} \log^+ T_g(r,s)$$

$$\text{(B.99)} \qquad \Upsilon(p,k) = \sum_{q=0}^{n-1} (k - D(p,q))^+ .$$

Take $\varepsilon > 0$ and $S > 0$. Recall that $\lessdot$ means that the inquality holds for all value $r > 0$ outside some set of finite measure. Then we have the <u>Second</u> <u>Main</u> <u>Theorem</u> **(Theorem 8.31).**

(B.100)

$$\boxed{\begin{aligned} &\Theta_p(r,s) + \begin{bmatrix} 1 \\ p+1 \end{bmatrix} N_{F_n}(r,s) + \sum_{g\in\mathfrak{G}} m_{g_p} \dot{L}f(r,s) \\ &\lessdot \left(\begin{bmatrix} n \\ p+1 \end{bmatrix} + \begin{bmatrix} n-1 \\ p \end{bmatrix}\right) T_f(r,s) + \begin{bmatrix} n+1 \\ p+2 \end{bmatrix} \mathrm{Ric}_\tau(r,s) \\ &+ 3\begin{bmatrix} n+1 \\ p+1 \end{bmatrix} kC_n(1+\varepsilon)\log Y(r) + \Upsilon(p,k)\Gamma_{\mathfrak{G}_p}(r) \\ &+ P_p(r,\mathfrak{G}) + R_p(r,\mathfrak{G}) + S_p(r,\mathfrak{G}) + \varepsilon \log r \end{aligned}}$$

If $p = 0$, then $\Theta_p(r,s) = 0$ and $\begin{bmatrix} 1 \\ p+1 \end{bmatrix} = 1$. The remaining term $N_{F_n}(r,s)$ is the classical ramification term which can be used to truncate the intersection divisors at the level n. If $p > 0$, then $\begin{bmatrix} 1 \\ p+1 \end{bmatrix} = 0$ and $\Theta_p(r,s) > 0$ is a new ramification term whose significance is not yet known. In any case, the terms can be dropped. The sum of the compensation functions, the characteristic, the Ricci function, the majorant are well-known classical terms. The term $\Lambda_{\mathfrak{G}_p}$ measures general position and can be estimated by (B.90) respectively (B.93) and (A.153). The terms P_p and R_p are new remainder terms which account for the movement of the targets in respect to f. They are difficult to estimate, but vanish if $p = \ell_g$ for all $g \in \mathfrak{G}$. The terms $S_p(r,\mathfrak{G})$ is a classical remainder term. These terms exist under natural assumptions.

B9. <u>The Defect relation</u>. Assume that [B.1] – [B.8] hold. Take $p \in \mathbb{Z}[0,\ell_{\mathfrak{G}}]$. Assume that $\mathfrak{G}_p$ is in general position and that (f,g_p) is free for L for all $g \in \mathfrak{G}$. Then $T_f(r,s) \longrightarrow \infty$ for $r \longrightarrow \infty$. Define the <u>Nevanlinna defect</u> of f for g_p by

(B.101) $$\delta_f(g_p) = \liminf_{r\to\infty} \frac{m_{\dot{g}Lf}(r)}{T_f(r,s) + T_{g_p}(r,s)}.$$

The First Main Theorem implies

(B.102) $$0 \leqslant \delta_f(g_p) = 1 - \limsup_{r\to\infty} \frac{N_{\dot{g}_pLf}(r,s) + T_{g_pLf}(r,s)}{T_f(r,s) + T_{g_p}(r,s)} \leqslant 1$$

We also assume

(B.103) $$\frac{\mathrm{Ric}_\tau(r,s)}{T_f(r,s)} \longrightarrow 0 \qquad \text{for } r \longrightarrow \infty$$

(B.104) $$\frac{\log Y(r)}{T_f(r,s)} \longrightarrow 0 \qquad \text{for } r \longrightarrow \infty$$

(B.105) $$\frac{T_g(r,s)}{T_f(r,s)} \longrightarrow 0 \qquad \text{for } r \longrightarrow \infty \text{ for all } g \in \mathfrak{G}$$

(B.106) $$\frac{P_p(r,\mathfrak{G})}{T_f(r,s)} \longrightarrow 0 \qquad \text{for } r \longrightarrow \infty$$

(B.107) $$\frac{R_p(r,\mathfrak{G})}{T_f(r,s)} \longrightarrow 0 \qquad \text{for } r \longrightarrow \infty .$$

Here (B.106) and (B.107) are trivially satisfied if $p = \ell_g = \ell_{\mathfrak{G}}$ for all $g \in \mathfrak{G}$. Assumptions (B.103) and (B.104) are classical conditions on the growth of f, while (B.105) is the customary assumption.

Then we obtain the Defect relation (Theorem 8.12).

(B.108)
$$\boxed{\sum_{g\in\mathfrak{G}} \delta_f(g_p) \leqslant \binom{n}{p+1} + \binom{n}{p-1}}$$

B.10 The case $p = 0 = \ell_{\mathfrak{G}}$. We assume that [B.1] – [B.8] are satisfied. We assume that $\mathfrak{G}$ is in general position and that f, g are free for each $g \in \mathfrak{G}$. We assume that $p = 0 = \ell_{\mathfrak{G}} = \ell_g$ for all $g \in \mathfrak{G}$. Let $\mu_{f,g}^{(n)}$ the divisor $\mu_{f,g}$ truncated at level n and let $N_{f,g}^{(n)}$ be the valence function of $\mu_{f,g}^{(n)}$. Recall (B.91). Then we have (Theorem 8.7):

(B.109)
$$\sum_{g\in\mathfrak{G}} N_{f,g}(r,s) \leqslant N_{F_n}(r,s) + N_{\mathfrak{G}}(r,s) + \sum_{g\in\mathfrak{G}} N_{f,g}^{(n)}(r,s) .$$

Thus we obtain another version of the Second Main Theorem

(B.110)
$$\boxed{\begin{aligned} &(k-n-1)T_f(r,s) \\ &\leqslant \sum_{g\in\mathfrak{G}} N_{f,g}^{(n)}(r,s) + \frac{n(n+1)}{2}\, \mathrm{Ric}_\tau(r,s) \\ &\quad + nk\binom{k-1}{n} \sum_{g\in\mathfrak{G}} T_g(r,s) + 2n(n+1)kC_1 \log Y(r) \\ &\quad + 2n(n+1)kC_1(\log T_f(r,s) + \log^+ \mathrm{Ric}_\tau(r,s)) + \varepsilon \log r \end{aligned}}$$

The Nevanlinna defects are given by

(B.111) $$0 \leqslant \delta_f(g) = \liminf_{r\to\infty} \frac{m_{f,g}(r)}{T_f(r,s) + T_g(r,s)}$$

$$= 1 - \limsup_{r\to\infty} \frac{N_{f,g}(r,s)}{T_f(r,s) + T_g(r,s)} \leqslant 1$$

(B.112) $$\delta_f(g) \leqslant \theta_f(g) = 1 - \limsup_{r\to\infty} \frac{N_{f,g}^{(n)}(r,s)}{T_f(r,s) + T_g(r,s)} \leqslant 1 .$$

If in addition we assume that (B.103), (B.104) and (B.105) hold, we obtain the <u>Defect relation</u> (Theorem 8.13)

(B.113) $$\boxed{\sum_{g\in\mathfrak{G}} \delta_f(g) \leqslant \sum_{g\in\mathfrak{G}} \theta_f(g) \leqslant n + 1}$$

The Second Main Theorem (B.110) and the Defect Relation (B.113) can be viewed as an extension of B. Shiffman's defect relation [83], [84] from meromorphic functions $f : \mathbb{C}^m \longrightarrow \mathbb{P}_1$ to meromorphic maps $f : M \longrightarrow \mathbb{P}(V)$ on parabolic manifolds M.

B11. <u>Analytic dependence</u>.

a) <u>Functions</u>. Let M be a connected, complex manifold of dimension $m > 1$. Let $\mathfrak{K} = \mathfrak{K}_M$ be the field of meromorphic functions on M. If B is a meromorphic form on M,

(B.114) $$\mathfrak{K}_M[B] = \mathfrak{K}[B] = \{\varphi \in \mathfrak{K} \mid d\varphi \wedge B = 0\}$$

is a subfield of $\mathfrak{K}$. The meromorphic functions $\varphi_1, \ldots, \varphi_k$ on M are said to be <u>analytically dependent</u> if $d\varphi_1 \wedge \ldots \wedge d\varphi_k \equiv 0$ and <u>analytically independent</u> if $d\varphi_1 \wedge \ldots \wedge d\varphi_k \not\equiv 0$. If they are analytically independent, then $k \leqslant m$. A single function φ is analytically dependent if and only if φ is constant. Take $\emptyset \neq \Phi \subseteq \mathfrak{K}$. Then $\varphi \in \mathfrak{K}$ is said to be <u>analytically dependent</u> on Φ if either φ is constant or if there are analytically independent functions $\varphi_1, \ldots, \varphi_k$ in Φ such that

$\varphi, \varphi_1, \ldots, \varphi_k$ are analytically dependent. The set $\mathfrak{K}(\Phi) = \mathfrak{K}_M(\Phi)$ of all $\varphi \in \mathfrak{K}$ analytically dependent on Φ is a subfield of $\mathfrak{K}$ with $\Phi \subseteq \mathfrak{K}(\Phi) = \mathfrak{K}(\mathfrak{K}(\Phi))$. If $\emptyset \neq \Phi \subseteq \Psi \subseteq \mathfrak{K}$, then $\mathfrak{K}(\Phi) \subseteq \mathfrak{K}(\Psi)$. If $\varphi \in \mathfrak{K}(\Phi)$, a finite subset Ψ of φ exists such that $\varphi \in \mathfrak{K}(\Psi)$. If $\psi \in \mathfrak{K}$ and $\Phi \subseteq \mathfrak{K}$ and if $\varphi \in \mathfrak{K}(\Phi \cup \{\psi\})$ but $\varphi \notin \mathfrak{K}(\Phi)$, then $\psi \in \mathfrak{K}(\Phi \cup \{\varphi\})$. Hence the assignment $\Phi \longrightarrow \mathfrak{K}(\Phi)$ is a dependency relation in the sense of Van Der Waerden [114] and Zariski–Samuel [128]. Therefore the concept of a <u>base</u> of $\mathfrak{K}(\Phi) \neq \mathbb{C}$ is defined. Here analytically independent functions $\varphi_1, \ldots, \varphi_k$ constitute a base of $\mathfrak{K}(\Phi)$ if and only if $\mathfrak{K}(\Phi) = \mathfrak{K}(\varphi_1, \ldots, \varphi_k)$. The number k is independent of the choice of the base and is called the rank of Φ. Then $\mathfrak{K}(\Phi)$ has the same rank as Φ. Analytically independent functions $\varphi_1, \ldots, \varphi_k$ constitute a base of $\mathfrak{K}(\Phi)$ if and only if

(B.115)
$$\mathfrak{K}(\Phi) = \mathfrak{K}[d\varphi_1 \wedge \ldots \wedge d\varphi_k] .$$

A meromorphic form B is said to <u>define</u> $\mathfrak{K}(\Phi)$ if and only if $\mathfrak{K}(\Phi) = \mathfrak{K}[B]$. If $\varphi_1, \ldots, \varphi_k$ is a base of $\mathfrak{K}(\Phi)$ and if there are holomorphic functions $u_j \not\equiv 0 \not\equiv v_j$ with $v_j \varphi_j = u_j$ on M for $j = 1, \ldots, k$, the holomorphic form

(B.116)
$$B = v_1^2 \ldots v_k^2 d\varphi_1 \wedge \ldots \wedge d\varphi_k = \bigwedge_{j=1}^{k} (v_j du_j - u_j dv_j)$$

defines $\mathfrak{K}(\Phi)$. If $\Phi \subseteq \mathbb{C}$, then $\mathfrak{K}(\Phi) = \mathbb{C}$ and the rank of Φ and $\mathfrak{K}(\Phi)$ is defined to be zero.

b) <u>Meromorphic vector functions</u>. Let V be a complex space of dimension $n + 1 > 1$. Let S be an analytic subset of M with $A = M - S \neq \emptyset$. A holomorphic vector function $\mathfrak{v} : A \longrightarrow V$ is said to be <u>meromorphic</u> on M if for each point $x \in S$ there is an open, connected neighborhood U of x in M and a holomorphic vector function $\mathfrak{w} : U \longrightarrow V$ and a holomorphic function $h \not\equiv 0$ on U such that $h(x)\mathfrak{v}(x) = \mathfrak{w}(x)$ for all $x \in U \cap A$. We can take A maximal, in which case S is called the <u>pole set</u> of $\mathfrak{v}$.

c) **Admissible bases.** Let $f : M \longrightarrow \mathbb{P}(V)$ be a meromorphic map. Let $\mathfrak{a}_0, \dots, \mathfrak{a}_n$ be a base of V. Let $\mathfrak{b}_0, \dots, \mathfrak{b}_n$ be the dual base. Define $b_j = \mathbb{P}(\mathfrak{b}_j)$. The base $\mathfrak{a}_0, \dots, \mathfrak{a}_n$ is said to be **admissible** for f if $f(M) \not\subseteq \ddot{E}[b_0]$. If so, then $\langle \mathfrak{v}, \mathfrak{b}_0 \rangle \not\equiv 0$ for any representation $\mathfrak{v} : U \longrightarrow V$ of f. Let $\mathfrak{a}_0, \dots, \mathfrak{a}_n$ be an admissible base of f. Then there are uniquely meromorphic functions $f_1, \dots, f_n$, called the **coordinate functions** of f in respect to the admissible base $\mathfrak{a}_0, \dots, \mathfrak{a}_n$, such that

(B.117) $$f_j | U = \frac{\langle \mathfrak{v}, \mathfrak{b}_j \rangle}{\langle \mathfrak{v}, \mathfrak{b}_0 \rangle} \quad \text{for } j = 1, \dots, n$$

for any representation $\mathfrak{v} : U \longrightarrow V$. Then

(B.118) $$\mathfrak{w} = \mathfrak{a}_0 + f_1 \mathfrak{a}_1 + \dots + f_n \mathfrak{a}_n$$

is a meromorphic vector function on M. Let S be its pole set. Then, $\mathfrak{w} : (M - S) \longrightarrow V_*$ is a reduced representation of f.

d) **Maps.** Let $\Phi \neq \emptyset$ be a set of meromorphic functions on M. The meromorphic map $f : M \longrightarrow \mathbb{P}(V)$ is said to be **analytically dependent** on Φ, if and only if there is an admissible base $\mathfrak{a}_0, \dots, \mathfrak{a}_n$ of V for f such that the coordinate functions $f_1, \dots, f_n$ in respect to this base belong to $\mathfrak{K}(\Phi)$. If so, this is the case for any other choice of an admissible base. The meromorphic map f is said to be **analytically independent** of Φ if and only if f is not analytically dependent on Φ. In $\mathfrak{a}_0, \dots, \mathfrak{a}_n$ is an admissible base of V for f with coordinate functions $f_1, \dots, f_n$, then f is analytically independent of Φ if and only if $f_j \notin \mathfrak{K}(\Phi)$ for at least one index $j \in \mathbb{N}[1,n]$. If B is a meromorphic form defining $\mathfrak{K}(\Phi)$ and if $\mathfrak{v} : U \longrightarrow V$ is a representation of f, then f is analytically dependent on Φ if and only if $(\mathfrak{v} \wedge d\mathfrak{v}) \wedge B \equiv 0$ on U.

The meromorphic map $f : M \longrightarrow \mathbb{P}(V)$ is said to be **free** of Φ if and only if (f,g) is free for every meromorphic map $g : M \longrightarrow \mathbb{P}(V^*)$ which depends analytically on Φ. If f is free of Φ, then f is analytically independent of Φ. If $\mathfrak{a}_0, \ldots, \mathfrak{a}_n$ is an admissible base of V for f with coordinate functions $f_1, \ldots, f_n$, then f is free of Φ if and only if $f_1 g_1 + \ldots + f_n g_n \notin \mathfrak{K}(\Phi)$ for all $g_j \in \mathfrak{K}(\Phi)$ and $j = 1, \ldots, n$ with $(g_1, \ldots, g_n) \not\equiv 0$.

Let $f : M \longrightarrow \mathbb{P}(V)$ be a meromorphic map. Let $\Phi \neq \emptyset$ be a set of meromorphic functions of rank $m - 1$. Let B be a meromorphic form of bidegree $(m - 1, 0)$ on M which defines $\mathfrak{K}(\Phi)$. Let P be the pole set of B. Then B is holomorphic on $M - P$. Hence, we can use B, at least on $M - P$, to define our differential operator. Then f is analytically dependent on Φ if and only if $\ell_f(B) = 0$ (Proposition 9.1) and f is free of Φ if and only if $\ell_f(B) = n$, that is, if and only if f is general for B (Theorem 9.4). Thus if we can find a reasonable holomorphic form B on M defining $\mathfrak{K}(\Phi)$ we obtain a Second Main Theorem and a Defect Relation over the function field $\mathfrak{K}(\Phi)$. A Poincaré Theorem with growth conditions proved by Skoda [85] and (B.116) will provide a suitable form B in the covering case. Of course the general case can also be formulated, see Theorems 9.8 and 9.9. Unfortunately the theorem of Miles [55] has not been proved for several variables.

B.12 __A Second Main Theorem over function fields.__ We assume:

[D1] Let M be a connected, complex manifold of dimension $m > 1$.

[D2] Let $\Phi \neq \emptyset$ be a set of meromorphic functions on M of rank $m - 1$.

[D3] Let $\pi : M \longrightarrow \mathbb{C}^m$ be a proper, surjective, holomorphic map with generic sheet number ς and branching divisor ρ.

[D4] There are meromorphic functions φ_j on $\mathbb{C}^m$ such that $\psi_j = \varphi_j \circ \pi$ belongs to $\mathfrak{K}(\Phi)$ for $j = 1, \ldots, m - 1$ and such that $\psi_1, \ldots, \psi_{m-1}$ is a base of $\mathfrak{K}(\Phi)$.

[D5] $\mathfrak{G} \neq \emptyset$ is a finite set of meromorphic maps $g : M \longrightarrow \mathbb{P}(V^*)$.

[D6] Each $g \in \mathfrak{G}$ is analytically dependent on Φ.

[D7] $\mathfrak{G}$ is in general position with $\#\mathfrak{G} = k > n + 1$.

[D8] $f : M \longrightarrow \mathbb{P}(V)$ is a meromorphic map free of Φ.

These assumptions can be formulated and understood without any knowledge of value distribution theory. The function $\tau = \tau_0 \circ \pi = \|\pi\|^2$ is a parabolic exhaustion of M and value distribution applies. Assuming [D1] – [D8] we have the **Second Main Theorem** (Theorem 9.15)

(B.119)

$$(k - n - 1)T_f(r,s)$$

$$\leqq \sum_{g \in \mathfrak{G}} N_{f,g}^{(n)}(r,s) + nk\binom{k-1}{n} \sum_{g \in \mathfrak{G}} T_g(r,s) + \frac{n(n+1)}{2} N_\rho(r,s)$$

$$+ 2n(n+1)k\varsigma k(\log T_f(r,s) + \log^+ N_\rho(r,s)) + S(r)$$

where the remainder term $S(r)$ can be estimated as follows.

(A) For each $\varepsilon > 0$ and $s > 0$ there is a constant $c_0(\varepsilon,s) > 0$ such that

(B.120) $$S(r) \leqq c_0(\varepsilon,s)(1 + r)^{2m-1} \sum_{j=1}^{m-1} T_{\psi_j}(r + \varepsilon,s) \qquad \text{for all } r > s .$$

(B) For each $\varepsilon > 0$ and $s > 0$ there is a constant $c_1(\varepsilon,s) > 0$ such that

(B.121)

$$S(r) \leqq c_1(\varepsilon,s)(1 + (\log(1 + r^2))^2) \sum_{j=1}^{m-1} T_{\psi_j}(r + \varepsilon r,s) \qquad \text{for all } r > s .$$

(C) If $\operatorname{Ord} \psi_j < \lambda < \infty$ for $j = 1, \ldots, m - 1$, then there is a constant $c_2(\lambda) > 0$ such that

(B.122) $$S(r) \leqq c_2(\lambda)r^\lambda \qquad \text{for all } r > 1 .$$

(D) If the functions $\varphi_1, \dots, \varphi_{m-1}$ are rational, there is a constant $c_3 > 0$ such that

(B.123) $$S(r) \leqslant c_3 \log r \quad \text{for all} \quad r \geqslant 2 .$$

The Ricci (or branching) Defect of f is defined by

(B.124) $$R_f = \limsup_{r\to\infty} \frac{N_\rho(r,s)}{T_f(r,s)} \geqslant 0 .$$

If f separates the fibers of π, then $R_f < \infty$ by a theorem of Noguchi [70]. Now we assume in addition:

[D9] $R_f < \infty$.

[D10] If $g \in \mathfrak{G}$, then $T_g(r,s)/T_f(r,s) \longrightarrow 0$ for $r \longrightarrow \infty$.

[D11] At least one of the following cases is satisfied.

(A) A number $\varepsilon > 0$ exists such that

$$\frac{(1+r)^{4m-1}T_{\psi_j}(r+\varepsilon,s)}{T_f(r,s)} \longrightarrow 0 \quad \text{for} \quad r \longrightarrow \infty \quad \text{and} \quad j \in \mathbb{N}[1,m-1] .$$

(B) A number $\varepsilon > 0$ exists such that

$$\frac{(1+(\log(1+r^2))^2)T_{\psi_j}(r+\varepsilon r,s)}{T_f(r,s)} \longrightarrow 0 \quad \text{for} \quad r \longrightarrow \infty .$$

(C) $\operatorname{Ord} \psi_j < \underline{\operatorname{Ord}}\, f \leqslant \infty$ for $j = 1, \dots, m-1$ where $\underline{\operatorname{Ord}}\, f$ is the lower order of f.

(D) The functions $\varphi_1, \dots, \varphi_{m-1}$ are rational and f has transcendental growth, that is $T_f(r,s)/\log r \longrightarrow \infty$ for $r \longrightarrow \infty$.

Assumtion [D11] implies the transcendental growth of f. Therefore, if the (m−1)–dimensional component of $\pi(\mathrm{supp}\ \rho)$ is affine algebraic in $\mathbb{C}^m$, then [D11] implies $R_f = 0$. If there is a function $g \in \mathfrak{G}$ which separates the fibers of π, then [D10] and Noguchi's Theorem [70] imply $R_f = 0$.

Assume that [D1] – [D11] are satisfied. Then we have the **Defect Relation** (Theorem 9.15)

(B.125)
$$\boxed{\sum_{g \in \mathfrak{G}} \delta_f(g_p) \leq \sum_{g \in \mathfrak{G}} \theta_f(g) \leq n + 1 + \frac{n(n+1)}{2} R_f}$$

If $n = 1$ and $M = \mathbb{C}^m$ and π the identity, the Second Main Theorem (B.119) and the Defect Relation (B.125) were obtained by B. Shiffman [83], [84] under a bit weaker assumptions.

B13. The Theorem of Mori. A defect relation for $n + 2$ moving target functions $g : \mathbb{C}^m \longrightarrow \mathbb{P}(V)$ in general position was proved by Mori [63]. In Section 11, we improve Mori's result and extend them to meromorphic maps on parabolic manifolds. Following Nevanlinna's method, Mori transforms the given situation for moving targets into one for fixed targets without changing the value of the defects. We assume

[E1] Let (M,τ) be a parabolic manifold of dimension m.
[E2] Let V be a hermitian vector space of dimension $n + 1 > 1$.
[E3] Let $f : M \longrightarrow \mathbb{P}(V)$ be a non–constant, meromorphic map.
[E4] Let $\mathfrak{G} \neq \emptyset$ be a set of meromorphic maps $g : M \longrightarrow \mathbb{P}(V^*)$.
[E5] Assume that $\mathfrak{G}$ is in general position with $\#\mathfrak{G} = n + 2$.
[E6] Let (f,g) be free for each $g \in \mathfrak{G}$. Enumerate $\mathfrak{G} = \{g_0, \ldots, g_{n+1}\}$.
[E7] Let $\alpha_0, \ldots, \alpha_n$ be an orthonormal base of V^*. Define
$\alpha_{n+1} = -\alpha_0 - \ldots - \alpha_n$. Define $a_j = \mathbb{P}(\alpha_j)$ for $j = 0, \ldots, n + 1$.

Let $\mathfrak{n}_0, \ldots, \mathfrak{n}_n$ be the dual base of $\alpha_0, \ldots, \alpha_n$. Put $\mathfrak{n} = \mathfrak{n}_0 \wedge \ldots \wedge \mathfrak{n}_n$. If $\mathfrak{z}_0, \ldots, \mathfrak{z}_{n+1}$ are vectors in V^* define

(B.126) $$\hat{\mathfrak{a}}_j = (-1)^j \mathfrak{a}_0 \wedge \dots \wedge \mathfrak{a}_{j-1} \wedge \mathfrak{a}_{j+1} \wedge \dots \wedge \mathfrak{a}$$

for $j = 0,1, \dots ,n + 1$. A homogeneous projective operation

(B.127) $$\nabla : V \times V^* \times \dots \times V^* \longrightarrow V$$

is defined by

(B.128) $$\mathfrak{x} \nabla \mathfrak{x}_0 \nabla \dots \nabla \mathfrak{a}_{n+1} = \frac{1}{\sqrt{n+1}} \sum_{j=0}^{n} \langle \mathfrak{x}, \mathfrak{a}_j \rangle \langle \hat{\mathfrak{a}}_j, \mathfrak{n} \rangle \mathfrak{n}_j$$

Now, [E.4] and [E.5] imply that the Mori map

(B.129) $$h = f \nabla g_0 \nabla \dots \nabla g_{n+1} : M \longrightarrow \mathbb{P}(V)$$

exists and is meromorphic, (of course Mori [63] does not define this map by an operation). Abbreviate

(B.130) $$\eta = \mu_{f \dot{\nabla} g_0 \dot{\nabla} \dots \dot{\nabla} g_{n+1}} \geqslant 0$$

(B.131) $$m(r) = m_{f \dot{\nabla} g_0 \dot{\nabla} \dots \dot{\nabla} g_{n+1}} \geqslant 0 \, .$$

If $0 < s < r$, the General First Main Theorem (B.25) implies

(B.132) $$T_f(r,s) + \sum_{j=0}^{n+1} T_{g_j}(r,s) = N_\eta(r,s) + m(r) - m(s) + T_h(r,s)$$

which yields the estimate

(B.133) $$\boxed{T_h(r,s) \leqslant T_f(r,s) + \sum_{j=0}^{n+1} T_{g_j}(r,s) + m(s)}$$

Define $\mathfrak{G}_j = (g_0, \dots ,g_{j-1},g_{j+1}, \dots ,g_{n+1})$ for $j = 0,1, \dots ,n + 1$. Then (B.89) reads (Lemma 11.8))

(B.134)
$$\sum_{\substack{k=0\\k\neq j}}^{n+1} T_{g_k}(r,s) = N_{\mathfrak{g}_j}(r,s) + m_{\mathfrak{g}_j}(r) - m_{\mathfrak{g}_j}(s) .$$

We estimate directly (Lemmata 11.9 and 11.11)

(B.135)
$$N_{\eta}(r,s) \leqslant \sum_{j=0}^{n+1} N_{\mathfrak{g}_j}(r,s)$$

(B.136)
$$m(r) \leqslant \sum_{j=0}^{n+1} m_{\mathfrak{g}_j}(r) + \log(n + 1) .$$

Thus (B.132), (B.134), (B.135) and (B.136) combine to (Corollary 11.12).

(B.137)

$$\boxed{T_f(r,s) \leqslant T_h(r,s) + n \sum_{j=0}^{n+1} T_{g_j}(r,s) + \sum_{j=0}^{n+1} m_{\mathfrak{g}_j}(s) + \log(n + 1)}$$

For $j \in \mathbb{Z}[0,n + 1]$ we obtain similar estimates

(B.138)
$$N_{h,a_j}(r,s) \leqslant N_{f,g_j}(r,s) + \sum_{\substack{k=0\\k\neq j}}^{n+1} T_{g_p}(r,s) + m_{\mathfrak{g}_j}(s)$$

(B.139)
$$N_{f,g_j}(r,s) \leqslant N_{h,a_j}(r,s) + (n + 1) \sum_{k=0}^{n+1} T_{g_k}(r,s) + \sum_{k=0}^{n+1} m_{\mathfrak{g}_k}(s) .$$

Now we assume in addition

[E8] If $j \in \mathbb{Z}[0,n + 1]$, then $T_{g_j}(r,s)/T_f(r,s) \longrightarrow 0$ for $r \longrightarrow \infty$.

Then we have

(B.140)
$$T_h(r,s)/T_f(r,s) \longrightarrow 1 \quad \text{for } r \longrightarrow \infty$$

(B.141)
$$\boxed{\delta_f(g_j) = \delta_h(a_j)} \quad \text{for } j = 0,1, \ldots ,n + 1 .$$

The diagonal $D = \{(b, \ldots ,b) \in \mathbb{C}^{n+2} \mid b \in \mathbb{C}\}$ is a complex line in $\mathbb{C}^{n+2}$. Let $E = \mathbb{C}^{n+2}/D$ be the quotient vector space and let $\rho : \mathbb{C}^{n+2} \longrightarrow E$ be the residual map. For each $c \in \mathbb{P}(E)$ one and only one meromorphic map $g_c : M \longrightarrow \mathbb{P}(V^*)$ is defined by

"Take $\mathfrak{b} \in \mathbb{C}^{n+2}$ with $c = \mathbb{P}(\rho(\mathfrak{b}))$. Then $\mathfrak{b} = (b_0, \ldots ,b_{n+1})$. Let $\mathfrak{w}_j : U \longrightarrow V^*$ be a reduced representation of g_j for $j = 0,1, \ldots ,n + 1$. Then

$$\mathfrak{y}_{\mathfrak{b}} = \sum_{j=0}^{n+1} b_j \langle \widehat{\mathfrak{w}}_j, \mathfrak{n} \rangle \mathfrak{w}_j : U \longrightarrow V^*$$

is a representation of g_c on U."

The map $c \longrightarrow g_c$ is injective.

The meromorphic map $f : M \longrightarrow \mathbb{P}(V)$ is said to be <u>linearly non-degenerate over</u> $\mathfrak{G}$ if and only if (f,g_c) is free for every $c \in \mathbb{P}(E)$. (Mori's, non-degeneracy condition seems to be more restrictive.) The meromorphic map $f : M \longrightarrow \mathbb{P}(V)$ is linearly non-degenerate over $\mathfrak{G}$ if and only if the associated Mori map is not linearly degenerate (Theorem 11.15).

In addition to [E1] – [E8] we assume:

[E9] A proper surjective holomorphic map $\pi : M \longrightarrow \mathbb{C}^m$ exists such that $\tau = \|\pi\|^2$.

[E10] Let ρ be the branching divisor of π. Define R_f by (B.124).

[E11] Let f be linearly non-degenerate over $\mathfrak{G}$.

[E12] At least one map $g \in \mathfrak{G}$ is not constant.

Then [E8] and [E12] imply the transcendental growth of f. Then (A.176), (B.140) and (B.141) imply the <u>Defect Relation</u> of <u>Nevanlinna–Mori</u> (Theorem 11.16)

(B.142)
$$\boxed{\sum_{j=0}^{n+1} \delta_f(g_j) \leqq n + 1 + \frac{n(n+1)}{2} R_f}\ .$$

§1. Hermitian Geometry

Let V be a complex vector space of dimension $n + 1$. Define $V_* = V - \{0\}$. The **dual vector space** V^* consists of all linear functions $\mathfrak{w} : V \longrightarrow \mathbb{C}$. For $\mathfrak{v} \in V$ and $\mathfrak{w} \in V^*$ the **inner product** is defined by $<\mathfrak{v}, \mathfrak{w}> = \mathfrak{w}(\mathfrak{v})$. Identify $V^{**} = V$ by $<\mathfrak{v}, \mathfrak{w}> = <\mathfrak{w}, \mathfrak{v}>$ and $(\bigwedge_p V)^* = \bigwedge_p V^*$ by

$$(1.1) \qquad <\mathfrak{v}_1 \wedge \ldots \wedge \mathfrak{v}_p, \mathfrak{w}_1 \wedge \ldots \wedge \mathfrak{w}_p> = \det(<\mathfrak{v}_j, \mathfrak{w}_k>)$$

For $\mathfrak{v} \in \bigwedge_p V$ and $\mathfrak{w} \in \bigwedge_q V^*$ with $1 \leqslant q \leqslant p \leqslant n + 1$, the **interior product** $\mathfrak{v} \llcorner \mathfrak{w} \in \bigwedge_{p-q} V$ is defined by $<\mathfrak{v} \llcorner \mathfrak{w}, \mathfrak{z}> = <\mathfrak{v}, \mathfrak{w} \wedge \mathfrak{z}>$ for all $\mathfrak{z} \in \bigwedge_{p-q} V^*$. If $p = q$, then $<\mathfrak{v}, \mathfrak{w}> = \mathfrak{v} \llcorner \mathfrak{w} \in \mathbb{C} = \bigwedge_0 V$.

Each $\mathfrak{v} \in V_*$ spans a complex line $\mathbb{P}(\mathfrak{v}) = \mathbb{C}\mathfrak{v}$. For $A \subseteq V$ define

$$(1.2) \qquad \mathbb{P}(A) = \{\mathbb{P}(\mathfrak{v}) \mid 0 \neq \mathfrak{v} \in A\}$$

Then $\mathbb{P}(V)$ is the **complex projective space** associated to V and $\mathbb{P} : V_* \longrightarrow \mathbb{P}(V)$ is the natural projection. If $n = 0$, then $\mathbb{P}(\mathbb{C}) = \{\infty\}$.

For $p \in \mathbb{Z}[0,n]$, the **Grassmann cone** of order p is defined by

$$(1.3) \qquad \widetilde{G}_p(V) = \{\mathfrak{v}_0 \wedge \ldots \wedge \mathfrak{v}_p \mid \mathfrak{v}_j \in V\}$$

and $G_p(V) = \mathbb{P}(\tilde{G}_p(V))$ is the **Grassmann manifold** of order p embedded into $\mathbb{P}(\bigwedge_{p+1} V)$. Observe that $G_n(V) = \{\infty\}$ and $G_0(V) = \mathbb{P}(V)$. Define $\tilde{G}_{-1}(V) = \mathbb{C}$ and $G_{-1}(V) = \{\infty\}$.

Every $x = \mathbb{P}(\mathfrak{x}) \in G_p(V)$ with $\mathfrak{x} = \mathfrak{x}_0 \wedge \ldots \wedge \mathfrak{x}_p$ defines a (p+1)-dimensional linear subspace $E(x)$ of V with base $\mathfrak{x}_0, \ldots, \mathfrak{x}_p$ by

(1.4) $$E(x) = E(\mathfrak{x}) = \mathbb{C}\mathfrak{x}_0 + \ldots + \mathbb{C}\mathfrak{x}_p = \{\mathfrak{x} \in V \mid \mathfrak{w} \wedge \mathfrak{x} = 0\}$$

Then $\ddot{E}(x) = \ddot{E}(\mathfrak{x}) = \mathbb{P}(E(x))$ is a p-dimensional **projective plane** in $\mathbb{P}(V)$. Put $q = n - p - 1$. For $a = \mathbb{P}(\alpha) \in G_q(V^*)$ with $\alpha = \alpha_0 \wedge \ldots \wedge \alpha_q$ define a (p+1)-dimensional linear subspace $E[a]$ of V by

(1.5) $$E[a] = E[\alpha] = \ker \alpha_0 \cap \ldots \cap \ker \alpha_q$$
$$= \{\mathfrak{w} \in V \mid \alpha \llcorner \mathfrak{w} = 0\}.$$

Then $\ddot{E}[a] = \ddot{E}[\alpha] = \mathbb{P}(E[a])$ is a p-dimensional **projective plane** in $\mathbb{P}(V)$. A biholomorphic map $\delta : G_p(V) \longrightarrow G_q(V^*)$, called the **duality** is uniquely defined by $E(x) = E[\delta(x)]$. By symmetry $\delta : G_q(V^*) \longrightarrow G_p(V)$ is defined with $\delta \circ \delta = \text{Identity}$.

These definitions extend to Grassmann manifolds. Let p and q be non-negative integers with $p + q < n$. For $x = \mathbb{P}(\mathfrak{x}) \in G_p(V)$ define

(1.6) $$E_q(x) = E_q(\mathfrak{x}) = \{\mathfrak{w} \in \tilde{G}_q(V) \mid \mathfrak{w} \wedge \mathfrak{x} = 0\}.$$

Then $\ddot{E}_q(x) = \ddot{E}_q(\mathfrak{x}) = \mathbb{P}(E_q(x))$ is a thin analytic subset of $G_q(V)$. Obviously $\ddot{E}_0(x) = \ddot{E}(x)$. For $v = \mathbb{P}(\mathfrak{w}) \in G_q(V) - \ddot{E}_q(x)$ the **exterior product**

(1.7) $$v \wedge x = \mathbb{P}(\mathfrak{v} \wedge \mathfrak{x}) \in G_{p+q+1}(V)$$

is defined. Now, assume $0 \leqslant q \leqslant p \leqslant n$. For $w = \mathbb{P}(\mathfrak{w}) \in G_q(V^*)$ define

(1.8) $$E_p[w] = E_p[\mathfrak{w}] = \{\mathfrak{v} \in \widetilde{G}_p(V) \mid \mathfrak{v} \llcorner \mathfrak{w} = 0\}.$$

Then $\ddot{E}_p[w] = \ddot{E}_p[\mathfrak{w}] = \mathbb{P}(E_p[w])$ is a thin analytic subset of $G_p(V)$. Obviously, $\ddot{E}_0[w] = \ddot{E}[w]$. For $v = \mathbb{P}(\mathfrak{v}) \in G_p(V) - \ddot{E}_p[w]$ the <u>interior</u> <u>product</u> is defined by

(1.9) $$v \llcorner w = \mathbb{P}(\mathfrak{v} \llcorner \mathfrak{w}) \in G_{p-q-1}(V)$$

A positive definite hermitian form $(\ |\ )$ on V is called a <u>hermitian</u> <u>product</u> or a <u>hermitian</u> <u>metric</u> on V and V together with $(\ |\ )$ is said to be a <u>hermitian</u> <u>vector</u> <u>space</u>. A <u>norm</u> on V is defined by $\|\mathfrak{x}\| = \sqrt{(\mathfrak{x}|\mathfrak{x})}$. For each $\mathfrak{x} \in V$, one and only one <u>dual</u> <u>vector</u> $\mathfrak{x}^* \in V^*$ is defined by $(\mathfrak{z}|\mathfrak{x}) = \langle\mathfrak{z},\mathfrak{x}^*\rangle$ for all $\mathfrak{z} \in V$. The map $\mathfrak{x} \longrightarrow \mathfrak{x}^*$ is an anti-linear isomorphism of V onto V^*. Here V^* becomes a hermitian vector space by setting $(\mathfrak{x}^*|\mathfrak{y}^*) = (\mathfrak{y}|\mathfrak{x})$ for all $\mathfrak{y}$ and $\mathfrak{x}$ in V. Then $\mathfrak{x}^{**} = \mathfrak{x}$ and $V^{**} = V$ as hermitian vector spaces. Also $(\mathfrak{w}|\mathfrak{y}^*) = \langle\mathfrak{y},\mathfrak{w}\rangle$ if $\mathfrak{y} \in V$ and $\mathfrak{w} \in V^*$. If $x = \mathbb{P}(\mathfrak{x}) \in \mathbb{P}(V)$, then $x^* = \mathbb{P}(\mathfrak{x}^*) \in \mathbb{P}(V^*)$ is defined.

On $\bigwedge_p V$ a hermitian product is uniquely defined by the requirement

(1.10) $$(\mathfrak{x}_1 \wedge \ldots \wedge \mathfrak{x}_p | \mathfrak{y}_1 \wedge \ldots \wedge \mathfrak{y}_p) = \det((\mathfrak{x}_j|\mathfrak{y}_k))$$

for all $\mathfrak{x}_j$ and $\mathfrak{y}_j$ in V.

Let V and W be hermitian vector spaces. Then $V \oplus W$ and $V \otimes W$ become hermitian vector spaces such that for $\mathfrak{v} \in V$, $\mathfrak{x} \in V$, $\mathfrak{w} \in W$ and $\mathfrak{y} \in W$ we have

$$(1.11) \qquad ((\mathfrak{x},\mathfrak{w})|(\mathfrak{z},\mathfrak{y})) = (\mathfrak{x}|\mathfrak{z}) + (\mathfrak{w}|\mathfrak{y})$$

$$(1.12) \qquad (\mathfrak{x} \otimes \mathfrak{w}|\mathfrak{z} \otimes \mathfrak{y}) = (\mathfrak{x}|\mathfrak{z})(\mathfrak{w}|\mathfrak{y})$$

Then

$$(1.13) \qquad \|(\mathfrak{x},\mathfrak{w})\|^2 = \|\mathfrak{x}\|^2 + \|\mathfrak{w}\|^2$$

$$(1.14) \qquad \|\mathfrak{x} \otimes \mathfrak{w}\| = \|\mathfrak{x}\|\,\|\mathfrak{w}\|$$

$(V \oplus W)^* = V^* \oplus W^*$ and $(V \otimes W)^* = V^* \otimes W^*$ are identified as hermitian vector spaces by .

$$(1.15) \qquad <(\mathfrak{x},\mathfrak{w}),(\mathfrak{z},\mathfrak{y})> = <\mathfrak{x},\mathfrak{z}> + <\mathfrak{w},\mathfrak{y}>$$

$$(1.16) \qquad <\mathfrak{x} \otimes \mathfrak{w}, \mathfrak{z} \otimes \mathfrak{y}> = <\mathfrak{x},\mathfrak{z}><\mathfrak{w},\mathfrak{y}>$$

where $\mathfrak{x} \in V$, $\mathfrak{w} \in V$, $\mathfrak{z} \in V^*$, $\mathfrak{y} \in V^*$.

We shall define an operation $\boxplus$ called <u>contraction</u>. Take $p \in \mathbb{Z}[0,n]$ and $q \in \mathbb{Z}[0,n]$. Define

$$(1.17) \qquad \mu = \mu(p,q) = \mathrm{Min}(p+1,q+1)$$

If $\mathfrak{x} = \mathfrak{x}_0 \wedge \ldots \wedge \mathfrak{x}_p \in \tilde{G}_p(V)$ and $j \in \mathbb{Z}[0,p]$, if $\mathfrak{w} = \mathfrak{w}_0 \wedge \ldots \wedge \mathfrak{w}_q \in \tilde{G}_q(V^*)$ and $k \in \mathbb{Z}[0,q]$ abbreviate

$$(1.18) \qquad \hat{\mathfrak{x}}_j = (-1)^j \mathfrak{x}_0 \wedge \ldots \wedge \mathfrak{x}_{j-1} \wedge \mathfrak{x}_{j+1} \wedge \ldots \wedge \mathfrak{x}_p$$

$$(1.19) \qquad \hat{\mathfrak{w}}_k = (-1)^k \mathfrak{w}_0 \wedge \ldots \wedge \mathfrak{w}_{k-1} \wedge \mathfrak{w}_{k+1} \wedge \ldots \wedge \mathfrak{w}_q$$

If $p = 0$, this shall mean $\hat{\mathfrak{x}}_j = 1$; if $q = 0$, this shall mean $\hat{\mathfrak{w}}_k = 1$. Define

$$(1.20) \qquad \mathfrak{v} \boxplus \mathfrak{w} = \frac{1}{\mu} \sum_{j=0}^{p} \sum_{k=0}^{q} \langle \mathfrak{v}_j, \mathfrak{w}_k \rangle \, \widehat{\mathfrak{v}}_j \otimes \widehat{\mathfrak{w}}_k .$$

Then

$$(1.21) \qquad \mathfrak{v} \boxplus \mathfrak{w} = \frac{1}{\mu} \sum_{k=0}^{q} (\mathfrak{v} \llcorner \mathfrak{w}_k) \otimes \widehat{\mathfrak{w}}_k$$

$$= \frac{1}{\mu} \sum_{j=0}^{p} \widehat{\mathfrak{v}}_j \otimes (\mathfrak{w} \llcorner \mathfrak{v}_j)$$

which shows that $\mathfrak{v} \boxplus \mathfrak{w}$ is well defined and extends to a linear map

$$(1.22) \qquad \boxplus : \bigwedge_{p+1} V \otimes \bigwedge_{q+1} V^* \longrightarrow \bigwedge_p V \otimes \bigwedge_q V^*$$

called <u>contraction</u>. If $r \in \mathbb{Z}[1,\mu]$, the operator can be iterated $\boxplus^r = \boxplus \circ \boxplus^{r-1}$. If $r = 0$, we let $\boxplus^0$ be the identity. Then $\boxplus^r = \boxplus^s \circ \boxplus^{r-s}$ for $0 \leqslant s \leqslant r$.

<u>LEMMA</u> 1.1. <u>If</u> $p \geqslant q$, <u>then</u> $\boxplus^{q+1} = \llcorner : \bigwedge_{p+1} V \otimes \bigwedge_{q+1} V^* \longrightarrow \bigwedge_{p-q} V$ <u>is the interior product</u>. <u>If</u> $p \leqslant q$, <u>then</u>

$\boxplus^{p+1} = \llcorner : \bigwedge_{p+1} V \otimes \bigwedge_{q+1} V^* \longrightarrow \bigwedge_{q-p} V^*$ <u>is the interior product, where we use the convention</u> $\llcorner(\mathfrak{v} \otimes \mathfrak{w}) = \mathfrak{w} \llcorner \mathfrak{v}$.

<u>PROOF</u>. If $p \geqslant q$, then $\mu = q + 1$. If $q = 0$, then $\mathfrak{v} \boxplus \mathfrak{w} = \mathfrak{v} \llcorner \mathfrak{w}$ by (1.21). Assume that the Lemma is proved for $q - 1$ with $0 \leqslant q \leqslant p$. Take $\mathfrak{w} = \mathfrak{w}_0 \wedge \ldots \wedge \mathfrak{w}_q \in \widetilde{G}_q(V^*)$ and $\mathfrak{v} \in \widetilde{G}_p(V)$. Then

$$\mathfrak{v} \boxplus^{q+1} \mathfrak{w} = \boxplus^q (\mathfrak{v} \boxplus \mathfrak{w})$$

$$= \frac{1}{q+1} \sum_{k=0}^{q} \boxplus^q ((\mathfrak{v} \llcorner \mathfrak{w}_k) \otimes \widehat{\mathfrak{w}}_k)$$

$$= \frac{1}{q+1} \sum_{k=0}^{q} (\mathfrak{v} \llcorner \mathfrak{w}_k) \llcorner \widehat{\mathfrak{w}}_k$$

$$= \frac{1}{q+1} \sum_{k=0}^{q} \mathfrak{v} \llcorner (\mathfrak{w}_k \wedge \widehat{\mathfrak{w}}_k) = \mathfrak{v} \llcorner \mathfrak{w} .$$

By linearity, we see that $\boxplus^{q+1} = \llcorner$. If $p \leqslant q$, we proceed likewise, q.e.d.

Consequently if $\mathfrak{v} \llcorner \mathfrak{w} \neq 0$, respectively, $\mathfrak{w} \llcorner \mathfrak{v} \neq 0$, then $\mathfrak{v} \boxplus \mathfrak{w} \neq 0$. If $v = \mathbb{P}(\mathfrak{v}) \in \mathbb{P}(\underset{p+1}{\wedge} V)$ and $w \in \mathbb{P}(\mathfrak{w}) \in \mathbb{P}(\underset{q+1}{\wedge} V^*)$ with $\mathfrak{v} \boxplus \mathfrak{w} \neq 0$, define

$$v \boxplus w = \mathbb{P}(\mathfrak{v} \boxplus \mathfrak{w}) \in \mathbb{P}(\underset{p}{\wedge} V \otimes \underset{q}{\wedge} V^*) \tag{1.23}$$

If $\mathfrak{v} \boxplus^r \mathfrak{w} \neq 0$, we define

$$v \boxplus^r w = \mathbb{P}(\mathfrak{v} \boxplus^r \mathfrak{w}) \in \mathbb{P}(\underset{p+1-r}{\wedge} V \otimes \underset{q+1-r}{\wedge} V^*) .$$

<u>**LEMMA**</u> **1.2.** If $\mathfrak{v} \in \widetilde{G}_p(V)$ and $\mathfrak{w} \in \underset{q+1}{\wedge} V^*$, and if $\mathfrak{e}_0, \ldots, \mathfrak{e}_p$ is an orthonormal system such that $\mathfrak{v} = \|\mathfrak{v}\| \mathfrak{e}_0 \wedge \ldots \wedge \mathfrak{e}_p$, then

$$\mu^2 \|\mathfrak{v} \boxplus \mathfrak{w}\|^2 = \|\mathfrak{v}\|^2 \sum_{j=0}^{p} \|\mathfrak{w} \llcorner \mathfrak{e}_j\|^2 \tag{1.24}$$

<u>**PROOF**</u>. Abbreviate $\mathfrak{e} = \mathfrak{e}_0 \wedge \ldots \wedge \mathfrak{e}_p$. Then

$$\mu^2 \|\mathfrak{v} \boxplus \mathfrak{w}\|^2 = \|\mathfrak{v}\|^2 (\mathfrak{e} \boxplus \mathfrak{w} \mid \mathfrak{e} \boxplus \mathfrak{w})$$

$$= \|\mathfrak{v}\|^2 \sum_{j=0}^{p} \sum_{k=0}^{p} (\widehat{\mathfrak{e}}_j \otimes (\mathfrak{w} \llcorner \mathfrak{e}_j) \mid \widehat{\mathfrak{e}}_k \otimes (\mathfrak{w} \llcorner \mathfrak{e}_k))$$

$$= \|\mathfrak{v}\|^2 \sum_{j=0}^{p} \sum_{k=0}^{p} (\widehat{\mathfrak{e}}_j \mid \widehat{\mathfrak{e}}_k)(\mathfrak{w} \llcorner \mathfrak{e}_j \mid \mathfrak{w} \llcorner \mathfrak{e}_k)$$

$$= \|\mathfrak{v}\|^2 \sum_{j=0}^{p} \|\mathfrak{w} \llcorner \mathfrak{e}_j\|^2, \quad \text{q.e.d.}$$

LEMMA 1.3. *If* $v = \mathbb{P}(\mathfrak{w}) \in G_p(V)$ *and* $w = \mathbb{P}(\mathfrak{v}) \in G_p(V^*)$, *then* $\mathfrak{w} \boxplus \mathfrak{v} = 0$ *if and only if* $E(v) \subseteq E[w]$.

PROOF. Let $\mathfrak{t}_0, \ldots, \mathfrak{t}_p$ be the orthonormal base of $E(v)$ such that $\mathfrak{w} = \|\mathfrak{w}\| \mathfrak{t}_0 \wedge \ldots \wedge \mathfrak{t}_p$. Then $\mathfrak{w} \boxplus \mathfrak{v} = 0$ iff $\mathfrak{v} \llcorner \mathfrak{t}_j = 0$ for $j = 0, \ldots, p$ which is the case iff $\mathfrak{t}_j \in E[w]$ for $j = 0, \ldots, p$ which is the case if and only if $E(v) \subseteq E[w]$, q.e.d.

We shall provide an explicit formula for $\boxplus^r$ and extend the last two Lemmata to this operator $\boxplus^r$, which is much more difficult.

Take $q \in \mathbb{Z}[0,n]$. Let $\mathfrak{J}[q,n]$ be the set of all increasing, injective maps $\lambda : \mathbb{Z}[0,q] \longrightarrow \mathbb{Z}[0,n]$. Let ι be the inclusion map. If $q = n$, then $\mathfrak{J}[n,n] = \{\iota\}$. If $q < n$, there exists one and only one $\lambda^\perp \in \mathfrak{J}[n-q-1,n]$ for each $\lambda \in \mathfrak{J}[q,n]$ such that $\operatorname{Im} \lambda \cap \operatorname{Im} \lambda^\perp = \emptyset$. Then $\lambda^{\perp\perp} = \lambda$. The map $\perp$ is bijective. A permutation $(\lambda, \lambda^\perp)$ of $\mathbb{Z}[0,n]$ is defined by

$$(\lambda,\lambda^\perp)(x) = \begin{cases} \lambda(x) & \text{if } x \in \mathbb{Z}[0,q] \\ \lambda^\perp(x-q-1) & \text{if } x \in \mathbb{Z}[q+1,n]\,. \end{cases}$$

The signature of the permutation is denoted by $\operatorname{sign}(\lambda,\lambda^\perp)$. Observe that

$$\#\mathfrak{J}[q,n] = \begin{bmatrix} n+1 \\ q+1 \end{bmatrix}. \tag{1.25}$$

If $q = n$, let $\iota^\perp$ be the "empty" map. If $\mathfrak{t}_0, \ldots, \mathfrak{t}_n$ are vectors and if $\lambda \in \mathfrak{J}[q,n]$, define

$$\mathfrak{t}_\lambda = \mathfrak{t}_{\lambda(0)} \wedge \cdots \wedge \mathfrak{t}_{\lambda(q)}\,. \tag{1.26}$$

LEMMA 1.4. Take $p \in \mathbb{Z}[0,n]$ and $q \in \mathbb{Z}[0,n]$. Define $\mu = \operatorname{Min}(p+1,q+1)$. Take $\mathfrak{t} = \mathfrak{t}_0 \wedge \ldots \wedge \mathfrak{t}_p \in \widetilde{G}_p(V)$ and $\mathfrak{v} \in \bigwedge_{q+1} V^*$. Take $r \in \mathbb{Z}[0,\mu]$. Then

$$(1.27)\qquad \begin{bmatrix}\mu\\ r\end{bmatrix}\mathfrak{x}\boxplus^{r}\mathfrak{w} = \sum_{\lambda\in\mathfrak{S}[r-1,p]} \operatorname{sign}(\lambda,\lambda^{\perp})\mathfrak{x}_{\lambda^{\perp}} \otimes (\mathfrak{w} \,L\, \mathfrak{x}_{\lambda}) .$$

PROOF. If $r = 1$, this is true by (1.21). Assume that the Lemma is proved for $r < p + 1$. It shall be proved for $r + 1$. We have

$$\begin{bmatrix}\mu\\ r+1\end{bmatrix}\mathfrak{x}\boxplus^{r+1}\mathfrak{w} = \frac{\mu - r}{r+1}\begin{bmatrix}\mu\\ r\end{bmatrix}\mathfrak{x}\boxplus^{r+1}\mathfrak{w}$$

$$= \frac{\mu - r}{r+1}\sum_{\lambda\in\mathfrak{S}[r-1,p]} \operatorname{sign}(\lambda,\lambda^{\perp})\mathfrak{x}_{\lambda^{\perp}} \boxplus (\mathfrak{w} \,L\, \mathfrak{x}_{\lambda})$$

$$= \frac{1}{r+1}\sum_{\lambda\in\mathfrak{S}[r-1,p]}\sum_{j=0}^{p-r} \operatorname{sign}(\lambda,\lambda^{\perp})(\widehat{\mathfrak{x}_{\lambda^{\perp}}})_j \boxplus (\mathfrak{w} \,L\, \mathfrak{x}_{\lambda}) L\, \mathfrak{x}_{\lambda^{\perp}(j)} .$$

A map $\alpha : \mathfrak{S}[r-1,p] \times \mathbb{Z}[0,p-r] \longrightarrow \mathfrak{S}[r,p]$ is defined by $\alpha(\lambda,j) = \tau$ if $\lambda \in \mathfrak{S}[r-1,p]$ and $j \in \mathbb{Z}[0,p-r]$ where τ is defined in this manner: One and only one number $\rho \in \mathbb{Z}[0,r-1]$ exists such that

$$\lambda(0) < \lambda(1) < \ldots < \lambda(\rho) < \lambda^{\perp}(j) < \lambda(\rho+1) < \ldots < \lambda(r-1)$$

Define

$$\tau(x) = \begin{cases} \lambda(x) & \text{if } 0 \leqslant x \leqslant \rho + 1 \\ \lambda^{\perp}(j) & \text{if } x = \rho + 1 \\ \lambda(x-1) & \text{if } \rho + 1 < x \leqslant r . \end{cases}$$

Then

$$\tau^{\perp}(x) = \begin{cases} \lambda^{\perp}(x) & \text{if } 0 \leqslant x \leqslant j - 1 \\ \lambda^{\perp}(x+1) & \text{if } j \leqslant x \leqslant p - r - 1 . \end{cases}$$

Therefore

$$(\mathfrak{w} \llcorner \mathfrak{a}_\lambda) \llcorner \mathfrak{a}_{\lambda^\perp(j)} = \mathfrak{w} \llcorner \mathfrak{a}_\lambda \wedge \mathfrak{a}_{\lambda^\perp(j)} = (-1)^{r-\rho-1} \mathfrak{w} \llcorner \mathfrak{a}_\tau$$

$$(\widehat{\mathfrak{a}_{\lambda^\perp}})_j = (-1)^j \mathfrak{a}_{\tau^\perp}$$

$$\operatorname{sign}(\lambda,\lambda^\perp) = (-1)^{r-\rho-1+j} \operatorname{sign}(\tau,\tau^\perp).$$

Also $\#\alpha^{-1}(\tau) = r + 1$. Therefore we obtain

$$\begin{bmatrix} \mu \\ r+1 \end{bmatrix} \mathfrak{a} \boxplus^{r+1} \mathfrak{w} = \sum_{\tau \in \mathfrak{T}[r,p]} \operatorname{sign}(\tau,\tau^\perp) \mathfrak{a}_{\lambda^\perp} \otimes \mathfrak{w} \llcorner \mathfrak{a}_\tau. \qquad \text{q.e.d.}$$

LEMMA 1.5. Take $p \in \mathbb{Z}[0,n]$ and $q \in \mathbb{Z}[0,n]$. Define $\mu = \operatorname{Min}(p+1,q+1)$. Take $\mathfrak{a} = \mathfrak{a}_0 \wedge \ldots \wedge \mathfrak{a}_p \in \tilde{G}_p(V)$ and $\mathfrak{w} \in \bigwedge_{q+1} V^*$. Assume that $\mathfrak{a}_0, \ldots, \mathfrak{a}_p$ is an orthonormal system. Take $r \in \mathbb{Z}[0,\mu]$. Then

$$(1.28) \qquad \begin{bmatrix} \mu \\ r \end{bmatrix}^2 \|\mathfrak{a} \boxplus^r \mathfrak{w}\|^2 = \sum_{\lambda \in \mathfrak{T}[r-1,p]} \|\mathfrak{w} \llcorner \mathfrak{a}_\lambda\|^2 .$$

PROOF. Abbreviate $\sigma(\lambda) = \operatorname{sign}(\lambda,\lambda^\perp)$ for $\lambda \in \mathfrak{T}[r-1,p]$. From (1.27) we have

$$\begin{bmatrix} \mu \\ r \end{bmatrix}^2 \|\mathfrak{a} \boxplus^r \mathfrak{w}\|^2 = \sum_\lambda \sum_\tau \sigma(\lambda)\sigma(\tau)(\mathfrak{a}_{\lambda^\perp} \otimes \mathfrak{w} \llcorner \mathfrak{a}_\lambda \mid \mathfrak{a}_{\tau^\perp} \otimes \mathfrak{w} \llcorner \mathfrak{a}_\tau)$$

$$= \sum_\lambda \sum_\tau \sigma(\lambda)(\sigma(\tau)(\mathfrak{a}_{\lambda^\perp} \mid \mathfrak{a}_{\tau^\perp})(\mathfrak{w} \llcorner \mathfrak{a}_\lambda \mid \mathfrak{w} \llcorner \mathfrak{a}_\tau)$$

$$= \sum_\lambda \|\mathfrak{w} \llcorner \mathfrak{a}_\lambda\|^2$$

where λ and τ run through $\mathfrak{T}[r-1,p]$. q.e.d.

Naturally, the operation $\boxplus$ is also defined when we regard V^* as the initially given vector space and $V^{**} = V$ as its dual space. Hence we obtain the diagram

$$\begin{array}{ccc} \bigwedge_{p+1} V \otimes \bigwedge_{q+1} V^* & \xrightarrow{\sigma} & \bigwedge_{q+1} V^* \otimes \bigwedge_{p+1} V \\ \downarrow \boxplus & & \downarrow \boxplus \\ \bigwedge_p V \otimes \bigwedge_q V^* & \xrightarrow{\sigma} & \bigwedge_q V^* \otimes \bigwedge_p V \end{array} \tag{1.29}$$

Here σ is an isometry given by $\sigma(\mathfrak{v} \otimes \mathfrak{w}) = \mathfrak{w} \otimes \mathfrak{v}$. The diagram commutes by (1.21). Hence the last two Lemmata yield:

LEMMA 1.6. Take $p \in \mathbb{Z}[0,n]$ and $q \in \mathbb{Z}[0,n]$. Define $\mu = \mathrm{Min}(p+1,q+1)$. Take $\mathfrak{v} \in \bigwedge_{p+1} V$ and $\varphi = \varphi_0 \wedge \ldots \wedge \varphi_q \in \tilde{G}_q(V^*)$. Take $r \in \mathbb{Z}[0,\mu]$. Then

$$\binom{\mu}{r} \mathfrak{v} \boxplus^r \varphi = \sum_{\lambda \in \mathfrak{T}[r-1,q]} \mathrm{sign}(\lambda,\lambda^\perp)(\mathfrak{v} \llcorner \varphi_\lambda) \otimes \varphi_{\lambda^\perp} . \tag{1.30}$$

LEMMA 1.7. Take $p \in \mathbb{Z}[0,n]$ and $q \in \mathbb{Z}[0,n]$. Define $\mu = \mathrm{Min}(p+1,q+1)$. Take $\mathfrak{v} \in \bigwedge_{p+1} V$ and $\varphi = \varphi_0 \wedge \ldots \wedge \varphi_q \in \tilde{G}_q(V^*)$. Assume that $\varphi_0, \ldots, \varphi_q$ is an orthonormal system. Take $r \in \mathbb{Z}[0,\mu]$. Then

$$\binom{\mu}{r}^2 \|\mathfrak{v} \boxplus^r \varphi\|^2 = \sum_{\lambda \in \mathfrak{T}[r-1,q]} \|\mathfrak{v} \llcorner \varphi_\lambda\|^2 \tag{1.31}$$

LEMMA 1.8. Take $\mathfrak{v} \in \tilde{G}_p(V)$ and $\mathfrak{w} \in \tilde{G}_q(V^*)$. Define $\mu = \mathrm{Min}(p+1,q+1)$. Take $r \in \mathbb{Z}[0,\mu]$. Then

$$\binom{\mu}{r}^{1/2} \|\mathfrak{v} \boxplus^r \mathfrak{w}\| \leq \|\mathfrak{v}\| \, \|\mathfrak{w}\| . \tag{1.32}$$

PROOF. W.l.o.g. we can assume that $p \leq q$. Then $\mu = p + 1$. Take an orthonormal base $\mathfrak{t}_0 \wedge \ldots \wedge \mathfrak{t}_p$. Lemma 1.5 implies

$$\binom{\mu}{r}^2 \|\mathfrak{v} \boxplus^r \mathfrak{w}\|^2 = \binom{\mu}{r}^2 \|\mathfrak{v}\|^2 \|\mathfrak{e} \boxplus^r \mathfrak{w}\|^2$$

$$= \|\mathfrak{v}\|^2 \sum_{\lambda \in \mathfrak{T}[r-1,p]} \|\mathfrak{w} \llcorner \mathfrak{e}_\lambda\|^2$$

$$\leqq \binom{p+1}{r} \|\mathfrak{v}\|^2 \|\mathfrak{w}\|^2 = \binom{\mu}{r} \|\mathfrak{v}\|^2 \|\mathfrak{w}\|^2$$

q.e.d.

LEMMA 1.9. *If* $v = \mathbb{P}(\mathfrak{v}) \in G_p(V)$ *and* $w = \mathbb{P}(\mathfrak{w}) \in G_q(V^*)$, *if* $\mu = \mathrm{Min}(p+1, q+1)$ *and if* $r \in \mathbb{Z}[0,\mu]$, *then* $\mathfrak{v} \boxplus^r \mathfrak{w} = 0$ *if and only if*

$$\dim E(v) \cap E[w] > p + 1 - r .$$

PROOF. Define $s + 1 = \dim E(v) \cap E[w]$. Abbreviate

$$A = E(v) \cap E[w] .$$

Then we obtain the following exact diagram of vector spaces, where we have written the dimension under each vector space and where we abbreviate $t = q + s + 1$.

$$\begin{array}{ccccccccc}
 & & 0 & & 0 & & 0 & & \\
 & & \downarrow & & \downarrow & & \downarrow & & \\
0 & \longrightarrow & \underset{s+1}{A} & \longrightarrow & \underset{p+1}{E(v)} & \longrightarrow & \underset{p-s}{B} & \longrightarrow & 0 \\
 & & \downarrow & & \downarrow & & \downarrow & & \\
0 & \longrightarrow & \underset{n-q}{E[w]} & \longrightarrow & \underset{n+1}{V} & \longrightarrow & \underset{q+1}{C} & \longrightarrow & 0 \\
 & & \downarrow & & \downarrow & & \downarrow & & \\
0 & \longrightarrow & \underset{n-t}{D} & \longrightarrow & \underset{n-p}{E} & \longrightarrow & \underset{t-p}{F} & \longrightarrow & 0 \\
 & & \downarrow & & \downarrow & & \downarrow & & \\
 & & 0 & & 0 & & 0 & &
\end{array}$$

We can consider each of the exact sequences as an orthogonal splitting. Therefore we have

$$E(\mathfrak{v}) = A \oplus B \qquad V = E[\mathfrak{w}] \oplus C = E(\mathfrak{v}) \oplus E$$

$$E[\mathfrak{w}] = A \oplus D \qquad C = B \oplus F \qquad E = D \oplus F$$

$$V = A \oplus B \oplus D \oplus F \ .$$

Hence we can take an orthonormal base $\mathfrak{e}_0, \dots, \mathfrak{e}_n$ of V such that

$$\mathfrak{e}_0, \dots, \mathfrak{e}_s \text{ spans } A \qquad \mathfrak{e}_{p+1}, \dots, \mathfrak{e}_{n-t+p} \text{ spans } D$$

$$\mathfrak{e}_{s+1}, \dots, \mathfrak{e}_p \text{ spans } B \qquad \mathfrak{e}_{n-t+p+1}, \dots, \mathfrak{e}_n \text{ spans } F \ .$$

Let $\mathfrak{e}_0^*, \dots, \mathfrak{e}_n^*$ be the dual base of $\mathfrak{e}_0, \dots, \mathfrak{e}_n$. The linear subspace $E(\mathfrak{w})$ contains the vectors $\mathfrak{e}_{s+1}^*, \dots, \mathfrak{e}_p^*, \mathfrak{e}_{n-t+p+1}^*, \dots, \mathfrak{e}_n^*$ which are linearly independent and $p - s + t - p = q + 1$ in number. Hence, they consititute a base of $E(\mathfrak{w})$. Also $\mathfrak{e}_0, \dots, \mathfrak{e}_p$ is a base of $E(\mathfrak{v})$. Numbers $a \neq 0$ and $b \neq 0$ exist such that

$$\mathfrak{v} = a\mathfrak{e}_0 \wedge \dots \wedge \mathfrak{e}_p \tag{1.33}$$

$$\mathfrak{w} = b\mathfrak{e}_{s+1}^* \wedge \dots \wedge \mathfrak{e}_p^* \wedge \mathfrak{e}_{n-t+p+1}^* \wedge \dots \wedge \mathfrak{e}_n^* \tag{1.34}$$

where $n - t \geq 0$. Assume that $s + 1 > p + 1 - r$. Thus $r > p - s$. Take any $\lambda \in \mathfrak{T}[r - 1, p]$. Then

$$\mathfrak{e}_\lambda = \mathfrak{e}_{\lambda(0)} \wedge \dots \wedge \mathfrak{e}_{\lambda(r-1)}$$

$$0 \leq \lambda(0) < \lambda(1) < \dots < \lambda(r - 1) \leq p \ .$$

Since $r > p - s$, we see that $\mathfrak{w} \llcorner \mathfrak{e}_\lambda = 0$. Now (1.27) and (1.33) imply $\mathfrak{v} \boxplus^r \mathfrak{w} = 0$.

Assume that $\mathfrak{v} \boxplus^r \mathfrak{w} = 0$. Then (1.28) and (1.33) imply $\mathfrak{w} \llcorner \mathfrak{x}_\lambda = 0$ for all $\lambda \in \mathfrak{S}[r - 1, p]$. Suppose that $s + 1 \leqslant p + 1 - r$. Then $r \leqslant p - s$. Define $\lambda : \mathbb{Z}[0, r - 1] \longrightarrow \mathbb{Z}[0, p]$ by $\lambda(x) = s + x + 1$. Then λ is injective and increases. We have

$$\mathfrak{w} \llcorner \mathfrak{x}_\lambda = b\mathfrak{x}^*_{s+r+1} \wedge \cdots \wedge \mathfrak{x}^*_p \wedge \mathfrak{x}^*_{n-t+p+1} \wedge \cdots \wedge \mathfrak{x}^*_n \neq 0$$

which is a contradiction. Hence $s + 1 > p + 1 - r$. q.e.d.

If $\mathfrak{x} \in \tilde{G}_p(V)$ and $\mathfrak{y} \in \tilde{G}_q(V)$, then $\mathfrak{x} \wedge \mathfrak{y} \in \tilde{G}_{p+q+1}(V)$ with

$$\|\mathfrak{x} \wedge \mathfrak{y}\| \leqslant \|\mathfrak{x}\| \, \|\mathfrak{y}\| \; . \tag{1.35}$$

If $\mathfrak{x} \in \tilde{G}_p(V)$ and $\mathfrak{y} \in \tilde{G}_q(V^*)$ with $0 \leqslant q \leqslant p \leqslant n$, then $\mathfrak{x} \llcorner \mathfrak{y} \in G_{p-q-1}(V)$ and

$$\|\mathfrak{x} \llcorner \mathfrak{y}\| \leqslant \|\mathfrak{x}\| \, \|\mathfrak{y}\| \tag{1.36}$$

which follows from Lemma 1.1 and Lemma 1.8.

If $\mathfrak{y} \in \bigwedge_{q+1} V$, then $\mathfrak{y}^* \in \bigwedge_{q+1} V^*$ was defined with $\|\mathfrak{y}^*\| = \|\mathfrak{y}\|$. For $\mathfrak{x} \in \bigwedge_{p+1} V$ and $\mathfrak{y} \in \bigwedge_{q+1} V$ with $0 \leqslant q \leqslant p \leqslant n$, define the <u>metric contraction</u>.

$$\mathfrak{x} \llcorner^* \mathfrak{y} = \mathfrak{x} \llcorner \mathfrak{y}^* \in \bigwedge_{p-q} V. \tag{1.37}$$

If $p = q$, then $\mathfrak{x} \llcorner^* \mathfrak{y} = (\mathfrak{x} \mid \mathfrak{y}) \in \mathbb{C}$. If $\mathfrak{x} \in \tilde{G}_p(V)$ and $\mathfrak{y} \in \tilde{G}_q(V)$ with $0 \leqslant q \leqslant p \leqslant n$, then $\mathfrak{y}^* \in \tilde{G}_q(V^*)$ and $\mathfrak{x} \llcorner^* \mathfrak{y} \in \tilde{G}_{p-q-1}(V)$ with

$$\|\mathfrak{x} \llcorner^* \mathfrak{y}\| \leqslant \|\mathfrak{x}\| \, \|\mathfrak{y}\| \; . \tag{1.38}$$

Let $V_1, \dots, V_p$ and W be vector spaces. A p-fold operation

$$\Theta : V_1 \times \dots \times V_p \longrightarrow W \tag{1.39}$$

is said to be **projective**, if for each $j \in \mathbb{N}[1,p]$ there is a map $\rho_j : \mathbb{C}_* \longrightarrow \mathbb{C}_*$ such that

$$\mathfrak{x}_1 \Theta \dots \Theta \mathfrak{x}_{j-1} \Theta \lambda \mathfrak{x}_j \Theta \mathfrak{x}_{j+1} \Theta \dots \Theta \mathfrak{x}_p = \rho_j(\lambda) \mathfrak{x}_1 \Theta \dots \Theta \mathfrak{x}_p$$

for all $\lambda \in \mathbb{C}_*$, all $\mathfrak{x}_j \in V_j$ and $j = 1, \dots, p$. If $x_j = \mathbb{P}(\mathfrak{x}_j) \in \mathbb{P}(V_j)$ for $j = 1, \dots, p$, then the condition $\mathfrak{x}_1 \Theta \dots \Theta \mathfrak{x}_p \neq 0$ is independent of the choices of $\mathfrak{x}_1, \dots, \mathfrak{x}_p$ and if the condition is satisfied,

$$x_1 \Theta \dots \Theta x_p = \mathbb{P}(\mathfrak{x}_1 \Theta \dots \Theta \mathfrak{x}_p) \in \mathbb{P}(W) \tag{1.40}$$

is well defined independent of the choices of $\mathfrak{x}_1, \dots, \mathfrak{x}_p$. We say that $x_1 \Theta \dots \Theta x_p$ exists.

The operation Θ is called **unitary of degree** $(q_1, \dots, q_p) \in \mathbb{Z}^p$ if $|\rho_j(\lambda)| = |\lambda|^{q_j}$ for all $\lambda \in \mathbb{C}_*$ and $j = 1, \dots, p$. Assume that $V_1, \dots, V_p$ and W are hermitian vector spaces. Take $x_j = \mathbb{P}(\mathfrak{x}_j)$ in V_j for $j = 1, \dots, p$. Then

$$\Box\, x_1 \dot{\Theta} \dots \dot{\Theta} x_p \,\Box = \frac{\|\mathfrak{x}_1 \Theta \dots \Theta \mathfrak{x}_p\|}{\|\mathfrak{x}_1\|^{q_1} \cdots \|\mathfrak{x}_p\|^{q_p}} \tag{1.41}$$

is well defined. Observe that $\Box \dots \Box$ is not a function of $x_1 \Theta \dots \Theta x_p$ which we indicate by the dot over the operation sign. Here $\Box x_1 \dot{\Theta} \dots \dot{\Theta} x_p \Box \neq 0$ if and only if $x_1 \Theta \dots \Theta x_p$ exists. The operation Θ is called **unitary** if it is unitary of degree $(1, 1, \dots, 1)$.

The operation Θ is called <u>**homogeneous of degree**</u> $(q_1, \ldots, q_p) \in \mathbb{Z}^p$ if $\rho_j(\lambda) = \lambda^{q_j}$ for all $\lambda \in \mathbb{C}_*$ and all $j = 1, \ldots, p$ and if $\mathfrak{x}_1 \Theta \ldots \Theta \mathfrak{x}_p$ is a holomorphic function of $(\mathfrak{x}_1, \ldots, \mathfrak{x}_p)$ on $V_1 \times \ldots \times V_p$. Such an operation is also unitary of degree $(q_1, \ldots, q_p)$.

A number of examples shall be given. If $x = \mathbb{P}(\mathfrak{x}) \in \mathbb{P}(\bigwedge_{p+1} V)$ and $y = \mathbb{P}(\mathfrak{y}) \in \mathbb{P}(\bigwedge_{q+1} V)$, then

(1.42) $$\square\, x \dot{\wedge} y\, \square = \frac{\|\mathfrak{x} \wedge \mathfrak{y}\|}{\|\mathfrak{x}\|\;\|\mathfrak{y}\|}$$

If $x \in G_p(V)$ and $y \in G_q(V)$, then

(1.43) $$0 \leqslant \square\, x \dot{\wedge} y\, \square \leqslant 1$$

If $x = \mathbb{P}(\mathfrak{x}) \in \mathbb{P}(\bigwedge_{p+1} V)$ and $y = \mathbb{P}(\mathfrak{y}) \in \mathbb{P}(\bigwedge_{q+1} V^*)$ with $0 \leqslant q \leqslant p \leqslant n$, then

(1.44) $$\square\, x \dot{\llcorner} y\, \square = \frac{\|\mathfrak{x} \llcorner \mathfrak{y}\|}{\|\mathfrak{x}\|\;\|\mathfrak{y}\|}$$

If $x \in G_p(V)$ and $y \in G_q(V^*)$, then

(1.45) $$0 \leqslant \square\, x \dot{\llcorner} y\, \square \leqslant 1$$

If $x = \mathbb{P}(\mathfrak{x}) \in \mathbb{P}(\bigwedge_{p+1} V)$ and $y = \mathbb{P}(\bigwedge_{p+1} V^*)$, then

(1.46) $$0 \leqslant \square\, x ; y\, \square = \frac{|\langle \mathfrak{x}, \mathfrak{y} \rangle|}{\|\mathfrak{x}\|\;\|\mathfrak{y}\|} = \square\, x \dot{\llcorner} y\, \square \leqslant 1.$$

if $x \in \mathbb{P}(\mathfrak{x}) \in \mathbb{P}(\bigwedge_{p+1} V)$ and $y = \mathbb{P}(\mathfrak{y}) \in \mathbb{P}(\bigwedge_{q+1} V)$ with $0 \leqslant q \leqslant p \leqslant n$, then

(1.47) $$\square\, x \dot{L}^* y \,\square = \frac{\|\mathfrak{x} L^* \mathfrak{y}\|}{\|\mathfrak{x}\|\;\|\mathfrak{y}\|}$$

(1.48) $$0 \leqslant \square\, x \dot{L}^* y \,\square \leqslant 1 \qquad \text{if } x \in G_p(V) \text{ and } y \in G_q(V).$$

If $x = \mathbb{P}(\mathfrak{x}) \in \mathbb{P}(\bigwedge_{p+1} V)$ and $y = \mathbb{P}(\mathfrak{y}) \in \mathbb{P}(\bigwedge_{p+1} V)$, then

(1.49) $$0 \leqslant \square\, x \,\dot{|}\, y \,\square = \frac{|(\mathfrak{x}|\mathfrak{y})|}{\|\mathfrak{x}\|\;\|\mathfrak{y}\|} = \square x \dot{L}^* y \,\square \leqslant 1 .$$

If $x = \mathbb{P}(\mathfrak{x}) \in \mathbb{P}(\bigwedge_{p+1} V)$ and $y = \mathbb{P}(\mathfrak{y}) \in \mathbb{P}(\bigwedge_{q+1} V^*)$ and $\mu = \mathrm{Min}(p+1, q+1)$, then

(1.50) $$\square\, x \,\dot{\boxplus}\, y \,\square = \frac{\|\mathfrak{x} \boxplus \mathfrak{y}\|}{\|\mathfrak{x}\|\;\|\mathfrak{y}\|}$$

(1.51) $$0 \leqslant \square\, x \,\dot{\boxplus}\, y \,\square \leqslant \frac{1}{\sqrt{\mu}} \qquad \text{if } x \in G_p(V) \text{ and } y \in G_q(V^*) .$$

If $x = \mathbb{P}(\mathfrak{x}) \in \mathbb{P}(\bigwedge_{p+1} V)$ and $y = \mathbb{P}(\mathfrak{y}) \in \mathbb{P}(\bigwedge_{q+1} V^*)$, if $0 \leqslant \rho \leqslant \mathrm{Min}(p+1, q+1)$, then

(1.52) $$\square\, x \,\dot{\boxplus}^\rho\, y \,\square = \frac{\|\mathfrak{x} \boxplus^\rho \mathfrak{y}\|}{\|\mathfrak{x}\|\;\|\mathfrak{y}\|}$$

(1.53) $$0 \leqslant \square\, x \,\dot{\boxplus}^\rho\, y \,\square \leqslant \binom{\mu}{\rho}^{-1/2} \qquad \text{if } x \in G_p(V) \text{ and } y \in G_q(V^*) .$$

<u>LEMMA</u> 1.10. <u>If</u> $x \in \mathbb{P}(V)$ <u>and</u> $y \in \mathbb{P}(V)$, <u>then</u>

$$\square\, x \,\dot{|}\, y \,\square^2 + \square\, x \,\dot{\wedge}\, y \,\square^2 = 1 .$$

<u>PROOF</u>. Take $0 \neq \mathfrak{x} \in V$ and $0 \neq \mathfrak{y} \in V$ such that $x = \mathbb{P}(\mathfrak{x})$ and $y \in \mathbb{P}(\mathfrak{y})$. Then

$$\square\, x \dot{|} y\, \square^2 + \square\, x \dot{\wedge} y\, \square^2$$

$$= \frac{1}{\|\mathfrak{z}\|^2\ \|\mathfrak{w}\|^2}\left[\,|(\mathfrak{z}|\mathfrak{w})|^2 + \|\mathfrak{z}\|^2\ \|\mathfrak{w}\|^2 - |(\mathfrak{z}|\mathfrak{w})|^2\right] = 1 .$$

q.e.d.

Observe that $G_n(V) = \{\infty\} = G_n(V^*)$.

LEMMA 1.11. Take $v \in G_p(V)$ and $w \in G_q(V^*)$. Then

$$(1.54) \qquad \square\, v \dot{\boxplus} \infty\, \square^2 = \frac{1}{p+1} \qquad \square\, \infty \dot{\boxplus} w\, \square^2 = \frac{1}{q+1} .$$

PROOF. Let $\mathfrak{t}_0, \dots, \mathfrak{t}_n$ be an orthonormal base of V. Define $\mathfrak{t} = \mathfrak{t}_0 \wedge \dots \wedge \mathfrak{t}_n$. Then $\infty = \mathbb{P}(\mathfrak{t}) \in G_n(V)$. Take $\mathfrak{w}_j \in V^*$ such that $\mathfrak{w} = \mathfrak{w}_0 \wedge \dots \wedge \mathfrak{w}_q \neq 0$ and $w = \mathbb{P}(\mathfrak{w})$ and such that $\mathfrak{w}_0, \dots, \mathfrak{w}_q$ is an orthonormal system. Then

$$\square\, \infty \dot{\boxplus} w\, \square^2 = \|\mathfrak{t} \boxplus \mathfrak{w}\|^2 = \frac{1}{(q+1)^2}\, \Big\| \sum_{j=0}^{q} \mathfrak{t} \llcorner \mathfrak{w}_j \otimes \widehat{\mathfrak{w}}_j \Big\|^2$$

$$= \frac{1}{(q+1)^2} \sum_{j=0}^{q} \sum_{k=0}^{q} (\mathfrak{t} \llcorner \mathfrak{w}_j \otimes \widehat{\mathfrak{w}}_j | \mathfrak{t} \llcorner \mathfrak{w}_k \otimes \widehat{\mathfrak{w}}_k)$$

$$= \frac{1}{(q+1)^2} \sum_{j=0}^{q} \sum_{k=0}^{q} (\mathfrak{t} \llcorner \mathfrak{w}_j | \mathfrak{t} \llcorner \mathfrak{w}_k)(\widehat{\mathfrak{w}}_j | \widehat{\mathfrak{w}}_k)$$

$$= \frac{1}{(q+1)^2} \sum_{j=0}^{q} \|\mathfrak{t} \llcorner \mathfrak{w}_j\|^2$$

$$= \frac{1}{(q+1)^2} \sum_{j=0}^{q} \Big\| \sum_{k=0}^{n} \langle \mathfrak{t}_k, \mathfrak{w}_j \rangle \widehat{\mathfrak{t}}_k \Big\|^2$$

$$= \frac{1}{(q+1)^2} \sum_{j=0}^{q} \sum_{k=0}^{n} |\langle \mathfrak{n}_k, \mathfrak{w}_j \rangle|^2$$

$$= \frac{1}{(q+1)^2} \sum_{j=0}^{q} \|\mathfrak{w}_j\|^2$$

$$= \frac{1}{q+1} .$$

By symmetry, we obtain $\square\, v \,\dot{\boxplus}\, \infty\, \square^2 = \frac{1}{p+1}$. q.e.d.

LEMMA 1.12. Take $v \in \mathbb{P}(\bigwedge_{p+1} V)$, $w \in \mathbb{P}(\bigwedge_{q+1} V^*)$, $x \in \mathbb{P}(\bigwedge_{k+1} V^*)$ where $0 \leqslant q < p \leqslant n$ and $0 \leqslant k \leqslant p - q - 1$. Assume that $v \llcorner w$ and $w \wedge x$ exist. Then

$$\square(v \llcorner w) \dot{\llcorner} x \,\square\, \square\, v \dot{\llcorner} w \,\square = \square\, v \dot{\llcorner}(w \wedge x) \,\square\, \square\, w \,\dot{\wedge}\, x \,\square . \tag{1.59}$$

PROOF. Take $0 \neq \mathfrak{v} \in \bigwedge_{p+1} V$, $0 \neq \mathfrak{w} \in \bigwedge_{q+1} V^*$ and $\mathfrak{x} \in \bigwedge_{k+1} V^*$ such that $v = \mathbb{P}(\mathfrak{v})$, $w = \mathbb{P}(\mathfrak{w})$ and $x = \mathbb{P}(\mathfrak{x})$. Then we have

$$\square\, v \llcorner w \dot{\llcorner} x \,\square\, \square\, v \dot{\llcorner} w \,\square = \frac{\|(\mathfrak{v} \llcorner \mathfrak{w}) \llcorner \mathfrak{x}\| \; \|\mathfrak{v} \llcorner \mathfrak{w}\|}{\|\mathfrak{v} \llcorner \mathfrak{w}\| \; \|\mathfrak{x}\| \; \|\mathfrak{v}\| \; \|\mathfrak{w}\|}$$

$$= \frac{\|\mathfrak{v} \llcorner (\mathfrak{w} \wedge \mathfrak{x})\| \; \|\mathfrak{w} \wedge \mathfrak{x}\|}{\|\mathfrak{v}\| \; \|\mathfrak{w} \wedge \mathfrak{x}\| \; \|\mathfrak{w}\| \; \|\mathfrak{x}\|}$$

$$= \square\, v \dot{\llcorner}(w \wedge x) \,\square\, \square\, w \,\dot{\wedge}\, x \,\square . \qquad \text{q.e.d.}$$

On any complex manifold the exterior **derivative** $d = \partial + \bar{\partial}$ twists to $d^c = (i/2\pi)(\bar{\partial} - \partial)$ and dd^c is the **Levi operator**.

On $\bigwedge_{p+1} V$ define τ_p by $\tau_p(\mathfrak{x}) = \|\mathfrak{x}\|^2$. Then there is one and only one position form Ω_p of bidegree (1,1) on $\mathbb{P}(\bigwedge_{p+1} V)$ such that

(1.56) $$\mathbb{P}^*(\Omega_p) = dd^c \log \tau_p \quad \text{on} \quad (\textstyle\bigwedge_{p+1} V)_*$$

The form Ω_p is called the **Fubini Study Form**. The **Fubini Study Form** on $\mathbb{P}(\bigwedge_{p+1} V^*)$ is also denoted by Ω_p.

Take $\mathfrak{a} \in (\bigwedge_{n+1} V)^*_*$. Take $q \in \mathbb{Z}[0,n]$ and $\mathfrak{x} \in \bigwedge_{q+1} V$. A linear map $D_{\mathfrak{a}}\mathfrak{x} : \bigwedge_{n-q} V \longrightarrow \mathbb{C}$ is defined by

(1.57) $$\langle D_{\mathfrak{a}}\mathfrak{x}, \mathfrak{y}\rangle = \langle \mathfrak{x} \wedge \mathfrak{y}, \mathfrak{a}\rangle \quad \text{for all } \mathfrak{y} \in \textstyle\bigwedge_{n-q} V$$

Therefore $D_{\mathfrak{a}}\mathfrak{x} \in \bigwedge_{n-q} V^*$. The map $D_{\mathfrak{a}} : \bigwedge_{q+1} V \longrightarrow \bigwedge_{n-q} V^*$ is a linear isomorphism. If V is a hermitian vector space and if $\|\mathfrak{a}\| = 1$, then $D_{\mathfrak{a}}$ is an isometry. If $\mathfrak{n} \in (\bigwedge_{n+1} V)_*$ is dual to $\mathfrak{a}$, that is, if $\langle \mathfrak{n}, \mathfrak{a}\rangle = 1$, then

(1.58) $$D_{\mathfrak{n}} \circ D_{\mathfrak{a}}\mathfrak{x} = (-1)^{(q+1)(n-q)}\mathfrak{x} \quad \text{for } \mathfrak{x} \in \textstyle\bigwedge_{q+1} V$$

Let $\mathfrak{n}_0, \ldots, \mathfrak{n}_n$ be a base of V with $\mathfrak{n}_0 \wedge \ldots \wedge \mathfrak{n}_n = \mathfrak{n}$. Let $\mathfrak{a}_0, \ldots, \mathfrak{a}_n$ be the dual base. Then $\mathfrak{a}_0 \wedge \ldots \wedge \mathfrak{a}_n = \mathfrak{a}$. For $\mu \in \mathfrak{J}[q,n]$ and $\mathfrak{x} \in \bigwedge_{q+1} V$ define $x_\mu = \langle \mathfrak{x}, \mathfrak{a}_\mu\rangle$. Then

(1.59) $$\mathfrak{x} = \sum_{\mu \in \mathfrak{J}[q,n]} x_\mu \mathfrak{n}_\mu$$

(1.60) $$D_{\mathfrak{a}}\mathfrak{x} = \sum_{\mu \in \mathfrak{J}[q,n]} \operatorname{sign}(\mu, \mu^\perp) x_\mu \mathfrak{a}_{\mu^\perp}$$

If $\mathfrak{x} \in \bigwedge_{q+1} V$ and $\mathfrak{z} \in \bigwedge_{p+1} V$ with $0 \leq p + q + 1 < n$, then

(1.61) $$D_{\mathfrak{a}}(\mathfrak{x} \wedge \mathfrak{z}) = (D_{\mathfrak{a}}\mathfrak{x}) \llcorner \mathfrak{z}.$$

If $\mathfrak{x} \in \bigwedge_{q+1} V$ and $\mathfrak{z} \in \bigwedge_{p+1} V^*$ with $0 \leqslant q < p \leqslant n$, then

(1.62) $$D_{\mathfrak{n}}(\mathfrak{z} \llcorner \mathfrak{x}) = (-1)^{(n-p)(q+1)}(D_{\mathfrak{n}}\mathfrak{z}) \wedge \mathfrak{x} .$$

The map $D_{\mathfrak{a}}$ maps Grassmann cones into Grassmann cones:

(1.63) $$D_{\mathfrak{a}} : \tilde{G}_q(V) \longrightarrow \tilde{G}_{n-q-1}(V^*) .$$

If $\mathfrak{b} \in (\bigwedge_{n+1} V)^*_*$, then $\lambda \in \mathbb{C}_*$ exists such that $\lambda\mathfrak{b} = \mathfrak{a}$. Then $D_{\mathfrak{a}} = \lambda D_{\mathfrak{b}}$. Therefore a biholomorphic map

(1.64) $$\delta : \mathbb{P}(\bigwedge_{q+1} V) \longrightarrow \mathbb{P}(\bigwedge_{n-q} V^*)$$

is uniquely defined such that $\delta \circ \mathbb{P} = \mathbb{P} \circ D_{\mathfrak{a}}$. The map δ does not depend on $\mathfrak{a}$. The map $\delta \circ \delta$ is the identity; also

(1.65) $$\delta : G_q(V) \longrightarrow G_{n-q-1}(V^*)$$

such that $E(x) = E[\delta(x)]$ for $x \in G_q(V)$. Hence δ coincides with the previously defined duality map δ. We have the following identities.

If $x \in \mathbb{P}(\bigwedge_{q+1} V)$ and $z \in \mathbb{P}(\bigwedge_{p+1} V)$ with $0 \leqslant p + q + 1 \leqslant n$, then

(1.66) $$\Box\, \delta x \dot{\llcorner} z \,\Box = \Box\, x \dot{\wedge} z \,\Box = \Box\, \delta x \dot{\wedge} \delta z \,\Box = \Box\, \delta z \dot{\llcorner} x \,\Box .$$

Moreover if $0 \leqslant p + q + 1 < n$, then $(\delta x)\llcorner z$ exists if and only if $x \wedge z$ exists and

(1.67) $$\delta(x \wedge z) = \delta x \llcorner z .$$

However if $n \leqslant p + q + 1$, then

(1.68) $$\square\, x \dot{L} \delta z\, \square = \square\, \delta x \dot{\wedge} \delta z\, \square = \square\, z \dot{L} \delta x\, \square\,.$$

If $x \in \mathbb{P}(\underset{q+1}{\wedge} V)$ and $z \in \mathbb{P}(\underset{p+1}{\wedge} V^*)$ with $0 \leqslant q \leqslant p \leqslant n$, then

(1.69) $$\square\, \delta x \dot{L} \delta z\, \square = \square\, z \dot{L} x\, \square = \square\, \delta z \dot{\wedge} x\, \square\,.$$

Moreover if $0 \leqslant q < p \leqslant n$, then $(\delta z) \wedge x$ exists if and only if zLx exists and

(1.70) $$\delta(zLx) = (\delta z) \wedge x\,.$$

If $q \in \mathbb{Z}[0,n)$, then

(1.71) $$\delta^*(\Omega_{n-q-1}) = \Omega_q\,.$$

The projective space $\overline{\overline{V}} = \mathbb{P}(\mathbb{C} \oplus V)$ is called the **projective closure** of V. We identify $V = \mathbb{P}(\{1\} \times V)$ by $\mathfrak{x} = \mathbb{P}((1,\mathfrak{x}))$ for all $\mathfrak{x} \in V$ and $\mathbb{P}(V) = \mathbb{P}(\{0\} \times V)$ by $\mathbb{P}(\mathfrak{x}) = \mathbb{P}((0,\mathfrak{x}))$ for all $\mathfrak{x} \in V_*$. Then $\overline{\overline{V}} = V \cup \mathbb{P}(V)$ is the disjoint union, in which $\mathbb{P}(V)$ is called the **infinite plane**.

Define $\delta_{\mu\nu} = 0$ if $\mu \neq \nu$ and $\delta_{\mu\mu} = 1$. Then $\mathfrak{n}_\mu = (\delta_{\mu 0}, \ldots, \delta_{\mu n}) \in \mathbb{C}^{n+1}$. We choose the hermitian metric on $\mathbb{C}^{n+1}$ such that $\mathfrak{n}_0, \ldots, \mathfrak{n}_n$ is an orthonormal base and we identify $(\mathbb{C}^{n+1})^* = \mathbb{C}^{n+1}$ such that $\mathfrak{n}_0, \ldots, \mathfrak{n}_n$ is a self dual base. Now, $\mathbb{C}^{n+1}$ obtains a conjugation by setting $\bar{\mathfrak{x}} = \mathfrak{x}^*$. The associated projective space is

(1.72) $$\mathbb{P}_n = \mathbb{P}(\mathbb{C}^{n+1}) = \overline{\overline{\mathbb{C}}}^n = \mathbb{C}^n \cup \mathbb{P}_{n-1}$$

Here $\mathbb{P}_0 = \{\infty\}$ and $\mathbb{P}_1 = \mathbb{C} \cup \{\infty\}$, which we realize as a sphere of diameter 1 in $\mathbb{R}^3$. For $x \in \mathbb{C}$ and $a \in \mathbb{C}$ we have $\mathbb{P}(1,x) = x$ and $\mathbb{P}(1,a) = a$. Let $j(a) = \mathbb{P}(-a,1)$ be the inversion. Here $\mathbb{P}(0,1) = \infty$ and $0 = j(\infty) = \mathbb{P}(1,0)$. The cordal distance is given by

(1.73) $$\Box\, x \curlywedge a\, \Box = \frac{|a - x|}{\sqrt{1 + |a|^2}\,\sqrt{1 + |x|^2}} = \Box\, x;j(a)\, \Box$$

(1.74) $$\Box\, x \curlywedge \infty\, \Box = \frac{1}{\sqrt{1 + |x|^2}} = \Box\, x;j(\infty)\, \Box$$

(1.75) $$\Box\, x\, \dot{L}\, a\, \Box = \frac{|1 + \bar{a}x|}{\sqrt{1 + |a|^2}\,\sqrt{1 + |x|^2}}$$

If W is any holomorphic vector bundle over a complex space M, and if s is a holomorphic section, the zero set

(1.76) $$Z(s) = \{x \in M \mid s(x) = 0\}$$

is analytic. Let V be a hermitian vector space of dimension $n + 1 > 1$. Then trivial bundle $\mathbb{P}(V) \times V$ has the tautological bundle

(1.77) $$\mathcal{O}(-1) = \{(x, \xi) \in \mathbb{P}(V) \times V \mid \xi \in E(x)\}$$

as a subbundle, where $E(x) = \mathcal{O}(-1)_x$ is the fiber over $x \in \mathbb{P}(V)$. The dual bundle $\mathcal{O}(1) = \mathcal{O}(-1)^*$ is called the **hyperplane section bundle**. It is a quotient bundle $\varepsilon : \mathbb{P}(V) \times V^* \longrightarrow \mathcal{O}(1)$ where $\varepsilon_x(\alpha) = \alpha \mid E(x)$ for $\alpha \in V^*$ and $x \in \mathbb{P}(V)$. The quotient map ε is dual to the inclusion map $\mathcal{O}(-1) \longrightarrow \mathbb{P}(V) \times V$. If $\alpha \in V^*$, a global holomorphic section $\hat{\alpha}$ of $\mathcal{O}(1)$ over $\mathbb{P}(V)$ is defined by

(1.78) $$\hat{\alpha}(x) = \varepsilon(x, \alpha) = \varepsilon_x(\alpha) = \alpha \mid E(x) \qquad \forall\, x \in \mathbb{P}(V).$$

Then $Z(\hat{\alpha}) = E[w]$ where $w = \mathbb{P}(\alpha)$ if $\alpha \neq 0$. Sometimes we write $\hat{\alpha} = \alpha$.

Define $\mathcal{O}(0) = \mathbb{P}(V) \times \mathbb{C}$. For $p > 0$ define $\mathcal{O}(p) = \bigotimes_p \mathcal{O}(1)$ and $\mathcal{O}(-p) = \bigotimes_p \mathcal{O}(-1)$. Then $\mathcal{O}(p)^* = \mathcal{O}(-p)$ for all $p \in \mathbb{Z}$. The hermitian product on V defines a hermitian metric along the fibers of $\mathbb{P}(V) \times V$ which restricts to $\mathcal{O}(-1)$. By tensoring and duality a hermitian metric ℓ_p is consistently defined along the fibers of $\mathcal{O}(p)$ for each $p \in \mathbb{Z}$. Here ℓ_1 is the quotient metric of the metric along the fibers of $\mathbb{P}(V) \times V^*$. The first Chern form computes as

$$c_1(\mathcal{O}(p), \ell_p) = p\Omega_0 \tag{1.79}$$

where Ω_0 is the Fubini Study Form on $\mathbb{P}(V)$.

§2 Meromorphic Maps on Parabolic Manifolds

A) Parabolic manifolds.

Let M be a connected, complex manifold of dimension m. Let $\tau \geqslant 0$ be a non-negative function of class C^∞ on M. For $0 \leqslant r \in \mathbb{R}$ and $A \subseteq M$ define

$$A(r) = \{x \in A \mid \tau(x) < r^2\} \tag{2.1}$$

$$A[r] = \{r \in A \mid \tau(x) \leqslant r^2\} \tag{2.2}$$

$$A<r> = \{x \in A \mid \tau(x) = r^2\} \tag{2.3}$$

$$A_* = A - A[0] = \{x \in A \mid \tau(x) > 0\} \tag{2.4}$$

$$\upsilon = dd^c\tau \qquad \omega = dd^c \log \tau \tag{2.5}$$

$$\sigma = d^c \log \tau \wedge (dd^c \log \tau)^{m-1} \tag{2.6}$$

The function τ is said to be an *exhaustion* if τ is unbounded and if $M[r]$ is compact for all $r \geqslant 0$. In this case, $M[r]$ is called the *closed pseudoball* and $M(r)$ the *open pseudoball* and $M<r>$ the *pseudosphere* of radius r. The function τ is said to be *parabolic* if

$$\omega \geqslant 0 \qquad \omega^m \equiv 0 \not\equiv \upsilon^m \quad \text{on } M_* . \tag{2.7}$$

Then $\upsilon \geqslant 0$ on M and $\upsilon > 0$ on some non-empty open subset of M.

If τ is a parabolic exhaustion, (M,τ) is said to be a **parabolic manifold**. A parabolic function is said to be **strict** if $\upsilon > 0$. A strictly parabolic exhaustion τ_0 of $\mathbb{C}^m$ is defined by $\tau_0(\mathfrak{z}) = \|\mathfrak{z}\|^2$ for all $\mathfrak{z} \in \mathbb{C}^m$. Up to biholomorphic isometry, $(\mathbb{C}^m, \tau_0)$ is the only strictly parabolic manifold of dimension m (Stoll [106], Burns [9], P. Wong [124]). Let M be a connected complex manifold of dimension m. Let $\pi : M \longrightarrow \mathbb{C}^m$ be a surjective, proper holomorphic map. Then $\tau = \tau_0 \circ \pi = \|\pi\|^2$ is a parabolic exhaustion of M and (M,τ) is called a **parabolic covering manifold of** $\mathbb{C}^m$. In particular any affine algebraic manifold of dimension m can be represented as parabolic covering manifold of $\mathbb{C}^m$. A Riemann surface is parabolic if every subharmonic function which is bounded above is constant. The cartesian product of parabolic manifolds is parabolic. For more details see Stoll [103], [105], [106], [107], [108].

Let (M,τ) be a parabolic manifold of dimension m. Then

(2.8) $$\tau^{p+1}\omega^p = \tau^p\upsilon^p - p\,d\tau \wedge d^c\tau \wedge \upsilon^{p-1} \quad \text{if } 0 < p \in \mathbb{Z}$$

(2.9) $$\tau\upsilon^m = m\,d\tau \wedge d^c\tau \wedge \upsilon^{m-1} = m\tau^m\, d\tau \wedge \sigma$$

(2.10) $$d\sigma = 0$$

Take $r > 0$. Then $\partial M(r)$ is the boundary of $M(r)$ and

(2.11) $$dM(r) = \{x \in \partial M(r) \mid d\tau(x) \neq 0\}$$

is open in the topology of $\partial M(r)$. Here $dM(r)$ can be regarded as a smooth boundary manifold of $M(r)$ oriented to the exterior of $M(r)$. The complement $M\langle r\rangle - dM(r)$ is compact. Then $M(r)$ is a **Stokes domain** in the sense of Tung [112], if for every point $p \in M\langle r\rangle$ there is an open neighborhood U of p and a biholomorphic map $\mathfrak{z} : U \longrightarrow U'$ onto an open subset U' of $\mathbb{C}^m$ such that $\mathfrak{z}(U \cap M\langle r\rangle)$ has finite $(2m-1)$-

dimensional Hausdorff measure and such that $\mathfrak{z}(U \cap (M\langle r\rangle - dM(r)))$ has zero $(2m-1)$-dimensional Hausdorff measure. Let

$$(2.12) \qquad \mathcal{E}_\tau = \{r \in \mathbb{R}^+ \mid M(r) \text{ is a Stokes domain}\}$$

$$(2.13) \qquad \hat{\mathcal{E}}_\tau = \{r \in \mathbb{R}^+ \mid d\tau(x) \neq 0 \text{ for all } x \in M\langle r\rangle\}$$

Then $\hat{\mathcal{E}}_\tau \subseteq \mathcal{E}_\tau$ and $\mathbb{R}^+ - \hat{\mathcal{E}}_\tau$ has measure zero. If $r \in \mathcal{E}_\tau$, then $M\langle r\rangle = \partial M(r) = \overline{dM(r)}$ and Stokes Theorem holds for $M(r)$. If $r \in \hat{\mathcal{E}}_\tau$, then $M\langle r\rangle = \partial M(r) = dM(r)$ is smooth. For most purposes $\hat{\mathcal{E}}_\tau$ would suffice, but in some proofs a proper modification $\pi : \tilde{M} \longrightarrow M$ along a thin analytic subset of M is used. If $r \in \mathcal{E}_\tau$, then $\pi^{-1}(M(r))$ is a Stokes domain in $\tilde{M}$, while the boundary of $\pi^{-1}(M(r))$ may not be smooth if $r \in \hat{\mathcal{E}}_\tau$. In particular, this holds if $\tilde{M}$ is not a manifold but a complex space. For $r \in \mathcal{E}_\tau$, the number

$$(2.14) \qquad \varsigma = \int_{M\langle r\rangle} \sigma > 0$$

is constant. If $r \geqslant 0$, then

$$(2.15) \qquad \int_{M[r]} \upsilon^m = \int_{M(r)} \upsilon^m = \varsigma r^{2m} > 0$$

In particular $M[r] \neq \emptyset$ for all $r \geqslant 0$.

B) <u>Divisors.</u>

Let M be a connected, complex manifold of dimension m. On M we can identify a divisor with its multiplicity function. Let $f \not\equiv 0$ be a holomorphic function on M. Take $x \in M$. Let $\mathcal{O}_x$ be the ring of germs of holomorphic functions at x. Let $\mathfrak{m}_x$ be the maximal ideal in $\mathcal{O}_x$. Then f defines a germ $0 \neq f_x \in \mathcal{O}_x$. One and only one integer $p \geq 0$ exists such that $f_x \in \mathfrak{m}_x^p - \mathfrak{m}_x^{p+1}$. Then $\mu_f^0(x) = p$ is said to be the **zero multiplicity** of f at x. The function $\nu : M \longrightarrow \mathbb{R}$ is said to be a **divisor** on M if and only if for every point $x \in M$ there exists an open, connected, neighborhood U of x and holomorphic functions $g \not\equiv 0$ and $h \not\equiv 0$ on U such that

$$\nu | U = \mu_g^0 - \mu_h^0 \tag{2.16}$$

The set $\mathfrak{D}_M$ of divisors on M is a module under function addition. A divisor is **non-negative** if $\nu(x) \geq 0$ for all $x \in M$, which is the case if and only if for every point $x \in M$ there is an open, connected neighborhood U of x and a holomorphic function $g \not\equiv 0$ on U such that $\nu | U = \mu_g^0$. The **support** $S = \operatorname{supp} \nu$ of the divisor ν is either empty or an analytic set of pure dimension $m - 1$. Inversely, if S is an analytic set of pure dimension $m - 1$, there exists one and only one divisor ν_S such that $\nu_S(x) = 1$ for every simple point $x \in S$. Let ν be a divisor with support S. Let $\mathfrak{B}$ be the set of branches of S. Let $\mathbb{R}(S)$ be the set of simple points of S. Then for each branch $B \in \mathfrak{B}$, there exists an integer $k_B \neq 0$ such that $\nu(x) = k_B$ for all $x \in B \cap \mathbb{R}(S)$. Moreover, we have the locally finite representation

$$\nu = \sum_{B \in \mathfrak{B}} k_B \nu_B \tag{2.17}$$

Here $\nu \geq 0$ if and only if $k_B > 0$ for all $B \in \mathfrak{B}$.

Let N and M be connected complex manifolds. Let $f : M \longrightarrow N$ be a holomorphic map. Let ν be a divisor on N with $f(M) \not\subseteq \operatorname{supp} \nu$. Then there exists one and only one **pullback divisor** $f^*(\nu)$ such that if $g \not\equiv 0$ and $h \not\equiv 0$ are holomorphic functions on an open, connected subset U of M with $\nu | U = \mu_g^0 - \mu_h^0$ then

$$f^*(\nu) | f^{-1}(U) = \mu_{g \circ f}^0 - \mu_{h \circ f}^0 \tag{2.18}$$

if $f^{-1}(U) \neq \emptyset$. Then $f^* : \vartheta_N \longrightarrow \vartheta_M$ is a homomorphism. If $\nu \geqslant 0$, then $f^*(\nu) \geqslant 0$.

Let $s \not\equiv 0$ be a holomorphic section of a holomorphic vector bundle W over M. Then s defines a **(zero)-divisor** μ_s. For every point $x \in M$, there is an open, connected neighborhood U of x, a holomorphic section t of W over U and a holomorphic function $h \not\equiv 0$ on U such that $s|U = h \cdot t$ and such that the zero set of t has at most dimension $m-2$. Here h and t are uniquely defined up to units and μ_s is well defined by $\mu_s|U = \mu_h^0$. A holomorphic vector function $\mathfrak{w} : M \longrightarrow V$ can be viewed as a holomorphic section in the trivial bundle $M \times V$. Hence the **(zero)-divisor** $\mu_{\mathfrak{w}}$ of $\mathfrak{w}$ is defined. Take $b \in \mathbb{P}_1$. Let f be a meromorphic function on M. If $f \not\equiv b$, the **b-divisor** μ_f^b of f is defined. For each point $x \in M$ there is an open, connected neighborhood U of x and there are coprime holomorphic functions g and h on U such that $hf = g$. Then $\mu_f^b|U = \mu_{g-bh}^0$ if $b \neq \infty$ and $\mu_f^\infty|U = \mu_h^0$. If $f \not\equiv 0$, then $\mu_f = \mu_f^0 - \mu_f^\infty$ is called the **divisor** of f. Also μ_f^∞ is called the **pole divisor** of f.

Let (M,τ) be a parabolic manifold of dimension m. Let ν be a divisor on M with support S. The **counting function** of ν is defined by

$$n_\nu(r) = r^{2-2m} \int_{S[r]} \nu v^{m-1} \qquad \text{if } m > 1 \tag{2.19}$$

$$n_\nu(r) = \sum_{z \in S[r]} \nu(z) \qquad \text{if } m = 1\,. \tag{2.20}$$

Then $n_\nu(r) \longrightarrow n_\nu(0)$ for $r \longrightarrow 0$. If $m > 1$, then

$$(2.21) \qquad n_\nu(r) = \int_{S[r]-S[0]} \nu\omega^{m-1} + n_\nu(0) \qquad \text{for} \quad r > 0.$$

For $0 < s < r$, the **valence function** N_ν of ν is defined by

$$(2.22) \qquad N_\nu(r,s) = \int_s^r n_\nu(t) \frac{dt}{t}.$$

If ν and μ are divisors on M, then $n_{-\mu}(r) = -n_\mu(r)$ and $N_{-\mu}(r,s) = -N_\mu(r,s)$ and $n_{\mu+\nu}(r) = n_\mu(r) + n_\nu(r)$ and $N_{\mu+\nu}(r,s) = N_\mu(r,s) + N_\nu(r,s)$. If $\nu \geqslant 0$, then $n_\nu(r) \geqslant 0$ and $N_\nu(r,s) \geqslant 0$ increase with r. If $f \not\equiv 0$ is a meromorphic function on M, define $N_f(r,s) = N_{\mu_f}(r,s)$. For $0 < s < r \in \mathcal{E}_\tau$ with $s \in \mathcal{E}_\tau$ we have the **Jensen Formula**

$$(2.23) \qquad N_f(r,s) = \int_{M\langle r\rangle} \log |f| \sigma - \int_{M\langle s\rangle} \log |f| \sigma.$$

(See Stoll [103] IV 11.)

C) **Meromorphic maps.**

Let M and N be connected complex manifolds. Define $m = \dim M$ and $n = \dim N$. Let S be a proper, analytic subset of M. Define $A = M - S$. Let $f : A \longrightarrow N$ be a holomorphic map. The closure Γ_f of the graph $\{(x,f(x)) \mid x \in A\}$ in $M \times N$ is called the **closed graph** of f. Then f is said to be **meromorphic** on M if the closed graph Γ_f is an analytic subset of $M \times N$ and if the projection $\pi : \Gamma_f \longrightarrow M$ is proper. If f is meromorphic, a maximal open subset $A(f)$ of M exists such that f continues onto $A(f)$ as a holomorphic map into N. Define $\tilde{A}(f) = \pi^{-1}(A(f)) \subseteq \Gamma_f$. Then $A(f)$ is the maximal open subset of M such that $\pi : \tilde{A}(f) \longrightarrow A(f)$ is biholomorphic.

The complement $I_f = M - A(f)$ is analytic with $I_f \subseteq S$ and is called the **indeterminacy** of f. Since M is a manifold, $\dim I_f \leq m - 2$ and

$$(2.24) \qquad I_f = \{x \in M \mid \#\pi^{-1}(x) > 1\}$$

in fact, $\dim_y \pi^{-1}(x) > 0$ for all $y \in \pi^{-1}(x)$ and $x \in I_f$.

Let $\psi : \Gamma_f \longrightarrow N$ be the projection. If $B \subseteq M$, define $f(B) = \psi(\pi^{-1}(B))$. If $C \subseteq N$, define $f^{-1}(C) = \pi(\psi^{-1}(C))$. Let ν be a divisor on N with support S. Assume that $f(A(f)) \not\subseteq C$. Hence the pullback divisor $f^*(\nu)$ is defined on A(f). Since $\dim I_f \leq m - 2$, the divisor $f^*(\nu)$ continues uniquely to a divisor on M again denoted by $f^*(\nu)$ and called the **pullback divisor** of ν under the meromorphic map f.

Let V be a complex vector space of dimension $n + 1 > 1$. Let S be a proper, analytic subset of M. Put $A = M - S$. Let $f : A \longrightarrow \mathbb{P}(V)$ be a holomorphic map. Let U be a connected, open, non-empty subset of M. A holomorphic map $\mathfrak{v} : U \longrightarrow V$ is said to be a **representation** of f (at a point $p \in M$ if $p \in U$) if $\mathfrak{v} \not\equiv 0$ and if $f(x) = \mathbb{P}(\mathfrak{v}(x))$ for all $x \in U \cap A$ with $\mathfrak{v}(x) \neq 0$. The map f is meromorphic on M if and only if there is a representation at every point of M. Assume that f is meromorphic. A representation $\mathfrak{v} : U \longrightarrow V$ is said to be **reduced** if $\dim \mathfrak{v}^{-1}(0) \leq m - 2$ which is the case if and only if $\mathfrak{v}^{-1}(0) = U \cap I_f$. In other words, a representation $\mathfrak{v} : U \longrightarrow V$ is reduced if and only if $\mu_{\mathfrak{v}} \equiv 0$. A meromorphic map admits a reduced representation at every point of M and on every Cousin II open subset of M. Let $\mathfrak{v}_j : U_j \longrightarrow V$ be representations of f for $j = 1,2$ with $U_1 \cap U_2 \neq \emptyset$. Then there is an unique meromorphic function v on $U_1 \cap U_2$ with $\mathfrak{v}_1 = v\mathfrak{v}_2$ on $U_1 \cap U_2$. If $\mathfrak{v}_2$ is reduced, then

v is holomorphic, if also $\mathfrak{v}_1$ is reduced, the holomorphic function v has no zeros.

Let $\{U_\lambda\}_{\lambda\in\Lambda} = \mathfrak{U}$ be a family of subsets of M. If $\lambda = (\lambda_0, \dots, \lambda_p) \in \Lambda^{p+1}$, define

(2.25) $$U_\lambda = U_{\lambda_0,\dots,\lambda_p} = U_{\lambda_0} \cap \dots \cap U_{\lambda_p}$$

(2.26) $$\Lambda[p] = \{\lambda \in \Lambda^{p+1} \mid U_\lambda \neq \emptyset\}.$$

Here $\Lambda[p] = \Lambda[p,\mathfrak{U}]$ is called the $\underline{p}^{\text{th}}$ **nerve** of $\mathfrak{U}$. The family $\mathfrak{U}$ is said to be **open** if U_λ is open for each $\lambda \in \Lambda$. The family $\mathfrak{U}$ is said to be a **covering** of M if and only if $M = \bigcup_{\lambda\in\Lambda} U_\lambda$. If $U \subseteq M$, define

(2.27) $$\Lambda_U = \{\lambda \in \Lambda \mid U_\lambda \cap U \neq \emptyset\}$$

The trivial bundle over M with general fiber V is denoted by $V_M = M \times V$. If $U \neq \emptyset$ is a subset of M and $\mathfrak{v} : U \longrightarrow V$ a function, then $\mathfrak{v}$ defines a section $\tilde{\mathfrak{v}} : U \longrightarrow V_M$ by setting $\tilde{\mathfrak{v}}(x) = (x, \mathfrak{v}(x))$ for all $x \in U$. If U is open, then $\tilde{\mathfrak{v}}$ is holomorphic if and only if $\mathfrak{v}$ is holomorphic. Sometimes we write $\tilde{\mathfrak{v}} = \mathfrak{v}$ and identify $(x, \mathfrak{v}(x)) = \mathfrak{v}(x)$.

PROPOSITION 2.1. *Let $f : M \longrightarrow \mathbb{P}(V)$ be a meromorphic map. Then there exists, uniquely up to isomorphism, a holomorphic line bundle L_f, called the* **hyperplane section bundle** *of f, over M and a holomorphic section F of $V_M \otimes L_f$ over M such that for each reduced representation $\mathfrak{v} : U \longrightarrow V$ there exists a holomorphic frame $\mathfrak{v}^\Delta$ of L_f over U such that $F \mid U = \tilde{\mathfrak{v}} \otimes \mathfrak{v}^\Delta$. Given $\mathfrak{v}$ and F, the frame $\mathfrak{v}^\Delta$ is unique. Moreover $I_f = Z(F)$ is the zero set of the*

<u>section</u> F. <u>Over</u> A(f), <u>the bundle</u> L_f <u>is isomorphic to the pullback</u> $f^*(\mathcal{O}(1))$ <u>of the hyperplane section bundle</u> $\mathcal{O}(1)$ <u>on</u> $\mathbb{P}(V)$. <u>If</u> $\mathfrak{w} : U \longrightarrow V$ <u>is a reduced representation with</u> $U \subseteq A(f)$, <u>then the section</u> $\tilde{\mathfrak{w}}$ <u>in</u> V_M <u>is a section in the subbundle</u> $f^*(\mathcal{O}(-1))$ <u>and is a frame of</u> $f^*(\mathcal{O}(-1))$ <u>over</u> U. <u>Then</u> $\mathfrak{w}^\Delta = \tilde{\mathfrak{w}}^*$ <u>is the dual frame of</u> $\tilde{\mathfrak{w}}$ <u>if we identify</u> $L_f | A(f) = f^*(\mathcal{O}(1))$.

<u>PROOF</u>. There exists a family $(\mathfrak{w}_\lambda)_{\lambda \in \Lambda}$ of reduced representations $\mathfrak{w}_\lambda : U_\lambda \longrightarrow V$ of f such that $\mathfrak{U} = \{U_\lambda\}_{\lambda \in \Lambda}$ is an open covering of M. For each pair $(\lambda,\mu) \in \Lambda[1]$ one and only one holomorphic function $v_{\lambda\mu} : U_{\lambda\mu} \longrightarrow \mathbb{C} - \{0\}$ exists such that

(2.28) $$\mathfrak{w}_\lambda = v_{\lambda\mu}\mathfrak{w}_\mu \quad \text{on } U_{\lambda\mu} .$$

Then $v_{\lambda\mu} \cdot v_{\mu\lambda} = 1$ and $v_{\lambda\lambda} = 1$. If $(\lambda,\mu,\rho) \in \Lambda[2]$, then

(2.29) $$v_{\lambda\mu}v_{\mu\rho}v_{\rho\lambda} = 1 \quad \text{on } U_{\lambda\mu\rho} .$$

The cocycle $\{v_{\lambda\mu}\}_{(\lambda,\mu)\in\Lambda[1]}$ determines a holomorphic line bundle L_f on M and a family $(\mathfrak{w}^\Delta_\lambda)_{\lambda\in\Lambda}$ of holomorphic frames $\mathfrak{w}^\Delta_\lambda$ of L_f over U_λ such that

(2.30) $$\mathfrak{w}^\Delta_\lambda = v_{\mu\lambda}\mathfrak{w}^\Delta_\mu \quad \text{on } U_{\lambda\mu}$$

for all pairs $(\lambda,\mu) \in \Lambda[1]$. The transition formulas (2.28) and (2.30) show that there exists a global holomorphic section F of $V_M \otimes L_f$ such that

(2.31) $$F | U_\lambda = \tilde{\mathfrak{w}}_\lambda \otimes \mathfrak{w}^\Delta_\lambda .$$

Let $\mathfrak{w} : U \longrightarrow V$ be a reduced representation. For each $\lambda \in \Lambda_U$, there is a holomorphic function $v_\lambda : U \cap U_\lambda \longrightarrow \mathbb{C} - \{0\}$ such that $\mathfrak{w}_\lambda = v_\lambda\mathfrak{w}$ on $U \cap U_\lambda$. If $(\lambda,\mu) \in \Lambda_U[1]$, then $v_\lambda = v_{\lambda\mu}v_\mu$ on

$U \cap U_{\lambda\mu}$. Hence $v_\lambda \mathfrak{w}_\lambda^\Delta = v_\mu \mathfrak{w}_\mu^\Delta$ on $U \cap U_{\lambda\mu}$. A holomorphic frame $\mathfrak{w}^\Delta$ of L_f over U is defined by $\mathfrak{w}^\Delta | (U \cap U_\lambda) = v_\lambda \mathfrak{w}_\lambda^\Delta$ for $\lambda \in \Lambda_U$. Then $F|U = \tilde{\mathfrak{w}} \otimes \mathfrak{w}^\Delta$. Let φ be any holomorphic frame of L_f over U such that $F|U = \tilde{\mathfrak{w}} \otimes \varphi$. A holomorphic function without zeros exists on U such that $\varphi = h\mathfrak{w}^\Delta$. Hence

$$\tilde{\mathfrak{w}} \otimes \mathfrak{w}^\Delta = F|U = \tilde{\mathfrak{w}} \otimes \varphi = h\tilde{\mathfrak{w}} \otimes \mathfrak{w}^\Delta.$$

Thus $h \equiv 1$ and $\varphi = \mathfrak{w}^\Delta$.

Let $\hat{L}_f$, $\hat{F}$, $\hat{\mathfrak{w}}^\Delta$ be another solution. Let $\mathfrak{w} : U \longrightarrow V$ be a reduced representation of f. A bundle isomorphism $\alpha_{\mathfrak{w}} : L_f | U \longrightarrow \hat{L}_f | U$ is uniquely defined by $\alpha_{\mathfrak{w}} \circ \mathfrak{w}^\Delta = \hat{\mathfrak{w}}^\Delta$. Let $\mathfrak{w} : W \longrightarrow V$ be another reduced representation of f with $U \cap W \neq \emptyset$. A holomorphic function h without zeros exists on $U \cap W$ such that $\mathfrak{w} = h\mathfrak{w}$ on $U \cap W$. Then $\mathfrak{w}^\Delta = h\mathfrak{w}^\Delta$ and $\hat{\mathfrak{w}}^\Delta = h\hat{\mathfrak{w}}^\Delta$. On $U \cap W$ we have

$$\alpha_{\mathfrak{w}} \circ \mathfrak{w}^\Delta = h\alpha_{\mathfrak{w}} \circ \mathfrak{w}^\Delta = h\hat{\mathfrak{w}}^\Delta = \hat{\mathfrak{w}}^\Delta = \alpha_{\mathfrak{w}} \circ \mathfrak{w}^\Delta.$$

Hence $\alpha_{\mathfrak{w}} = \alpha_{\mathfrak{w}}$ over $U \cap W$. Therefore a global bundle isomorphism $\alpha : L_f \longrightarrow \hat{L}_f$ is defined by $\alpha \circ \mathfrak{w}^\Delta = \hat{\mathfrak{w}}^\Delta$ for all reduced representations $\mathfrak{w} : U \longrightarrow V$. Obviously $(\mathrm{Id} \otimes \alpha) \circ F = \hat{F}$. Hence the construction is unique up to an isomorphism.

The holomorphic map $f : A(f) \longrightarrow \mathbb{P}(V)$ pulls back $\mathcal{O}(-1)$ to

$$(2.32) \qquad f^*(\mathcal{O}(-1)) = \{(x,\mathfrak{x}) \in M \times V \mid \mathfrak{x} \in E[f(x)]\} \subset M \times V = V_M .$$

Let $\mathfrak{v} : U \longrightarrow V_*$ be a reduced representation of f on $U \subseteq A(f)$. Then $\tilde{\mathfrak{v}}(x) = (x, \mathfrak{v}(x)) \in \{x\} \times E(x) = f^*(\mathcal{O}(-1))_x$. Hence $\tilde{\mathfrak{v}}$ is a holomorphic frame of $f^*(\mathcal{O}(-1))$ over U. Let $\tilde{\mathfrak{v}}^*$ be the dual frame of $f^*(\mathcal{O}(1))$ over U. Let $\mathfrak{w} : W \longrightarrow V_*$ be another reduced representation of f with $W \subseteq A(f)$ and $W \cap U \neq \emptyset$. Then there is a holomorphic function h without zeros on $W \cap U$ such that $\mathfrak{v} = h\mathfrak{w}$ on $W \cap U$. Then $h\tilde{\mathfrak{v}}^* = \tilde{\mathfrak{w}}^*$ on $U \cap W$. Therefore $\tilde{\mathfrak{v}} \otimes \tilde{\mathfrak{v}}^* = \tilde{\mathfrak{w}} \otimes \tilde{\mathfrak{w}}^*$ on $U \cap W$. Consequently, a global holomorphic section $\tilde{F}$ of $f^*(\mathcal{O}(1))$ exists such that $\tilde{F}|U = \tilde{\mathfrak{v}} \otimes \tilde{\mathfrak{v}}^*$ for each reduced representation $\mathfrak{v} : U \longrightarrow V_*$ of f with $U \subseteq A(f)$. Trivially $L_{f|A(f)} = L_f|A(f)$ with representation section $F|A(f)$. Therefore a bundle isomorphism $\alpha : L_f|A(f) \longrightarrow f^*(\mathcal{O}(1))$ exists such that $\alpha \circ \mathfrak{v}^\Delta = \tilde{\mathfrak{v}}^*$ for each reduced representation $\mathfrak{v} : U \longrightarrow V_*$ with $U \subseteq A(f)$. Hence $(\mathrm{Id} \otimes \alpha) \circ F|A(f) = \tilde{F}$. q.e.d.

The section $F = F_f$ is called the <u>representation section</u> of f in L_f.

Let $V_1, \ldots, V_k$ and W be hermitian vector spaces. Let

$$\Theta : V_1 \times \ldots \times V_k \longrightarrow W$$

be an operation homogeneous of degree $(q_1, \ldots, q_k) \in \mathbb{Z}^k$. For $j = 1, \ldots, k$, let $f_j : M \longrightarrow \mathbb{P}(V_j)$ be meromorphic maps. The operation Θ extends to a fiber preserving, holomorphic map

$$\Theta : (V_{1M} \otimes L_{f_1}) \oplus \ldots \oplus (V_{kM} \otimes L_{f_k}) \longrightarrow W_M \otimes L_{f_1}^{q_1} \otimes \ldots \otimes L_{f_k}^{q_k} .$$

A holomorphic section

$$F_{f_1} \ominus \ldots \ominus F_{f_k} \in \Gamma(M, W_M \otimes L_{f_1}^{q_1} \otimes \ldots \otimes L_{f_k}^{q_k})$$

is defined. If $\mathfrak{v}_j : U \longrightarrow V_j$ is a reduced representation of f_j for $j = 1, \ldots, k$, then

$$F_{f_1} \ominus \ldots \ominus F_{f_k} | U = (\mathfrak{v}_1 \ominus \ldots \ominus \mathfrak{v}_k) \otimes (\mathfrak{v}_1^{\Delta})^{q_1} \otimes \ldots \otimes (\mathfrak{v}_k^{\Delta})^{q_k} .$$

Then $(f_1, \ldots, f_k)$ is said to be free for $\ominus$, if $F_{f_1} \ominus \ldots \ominus F_{f_k} \not\equiv 0$ which is the case if $\mathfrak{v}_1 \ominus \ldots \ominus \mathfrak{v}_k \not\equiv 0$ on U for at least one (and hence all) possible choices of $\mathfrak{v}_1, \ldots, \mathfrak{v}_k$. If so, we write

$$\mu_{f_1 \dot{\ominus} \ldots \dot{\ominus} f_k} = \mu_{F_{f_1} \ominus \ldots \ominus F_{f_k}}$$

for the divisor of this section and, on a parabolic manifold, denote the counting function and valence function of this divisor by

$$n_{f_1 \dot{\ominus} \ldots \dot{\ominus} f_k} \quad \text{and} \quad N_{f_1 \dot{\ominus} \ldots \dot{\ominus} f_k} .$$

Now we will consider examples of such operations.

Take $p \in \mathbb{Z}[0,n]$ and $q \in \mathbb{Z}[0,n]$. Define $\mu = \mu(p,q) = \mathrm{Min}(p+1, q+1)$. Take $\rho \in \mathbb{Z}[0,\mu]$. Let L be a holomorphic line bundle over M. Abbreviate

$$V(p,q,L) = (\textstyle\bigwedge_{p+1} V_M) \otimes (\bigwedge_{q+1} V_M^*) \otimes L .$$

A contraction

$$\boxplus^{\rho} = \boxplus^{\rho} \times \mathrm{Id} : V(p,q,L) \longrightarrow V(p-\rho, q-\rho, L) \tag{2.33}$$

is defined where $\boxplus^0$ is the identity, $\boxplus^1 = \boxplus$ and $\boxplus^{\rho} = \boxplus^{\sigma} \circ \boxplus^{\rho-\sigma}$ if $\sigma \in \mathbb{Z}[0,\rho]$. If $\rho = \mu = q + 1$, then $\llcorner = \boxplus^{q+1}$ is the interior

product mapping into $\bigwedge_{p-q} V_M \otimes L$, if $\rho = \mu = p+1$, then $\llcorner = \boxplus^{p+1}$ is the **interior product** mapping into $\bigwedge_{q-p} V_M^* \otimes L$.

Let $f : M \longrightarrow \mathbb{P}(\bigwedge_{p+1} V)$ and $g : M \longrightarrow \mathbb{P}(\bigwedge_{q+1} V^*)$ be meromorphic maps. Abbreviate $F = F_f$ and $G = F_g$. Then $F \otimes G$ is a holomorphic section of the holomorphic vector bundle.

(2.34) $$\bigwedge_{p+1} V_M \otimes \bigwedge_{q+1} V_M^* \otimes L_f \otimes L_g$$

with the proper commutation of terms. Hence $F \boxplus^{\rho} G = \boxplus^{\rho}(F \otimes G)$ is a holomorphic section in

(2.35) $$\bigwedge_{p+1-\rho} V_M \otimes \bigwedge_{q+1-\rho} V_M^* \otimes L_f \otimes L_g$$

If $\rho = 0$, then $F \boxplus^0 G = F \otimes G$; if $\rho = 1$, abbreviate $F \boxplus^1 G = F \boxplus G$. If $\rho = \mu = q+1$, we write $F \boxplus^{q+1} G = F \llcorner G$. If $\rho = \mu = p+1$, we write $F \boxplus^{p+1} G = G \llcorner F$. If $\rho = p+1 = q+1$, we write $F \boxplus^{p+1} G = \langle F, G \rangle$, which is a holomorphic section in the line bundle $L_f \otimes L_g$.

Let $\mathfrak{v} : U \longrightarrow \bigwedge_{p+1} V$ be a reduced representation of f and let $\mathfrak{w} : U \longrightarrow \bigwedge_{q+1} V^*$ be a reduced representation of g. Then

(2.36) $$F \boxplus^{\rho} G | U = (\mathfrak{v} \boxplus^{\rho} \mathfrak{w}) \otimes \mathfrak{v}^{\Delta} \otimes \mathfrak{w}^{\Delta}$$

The pair (f,g) is said to be **free of order** ρ, if $F \boxplus^{\rho} G \not\equiv 0$, which is the case if $\mathfrak{v} \boxplus^{\rho} \mathfrak{w} \not\equiv 0$ on U for at least one (and consequently all) possible choices of $\mathfrak{v}$ and $\mathfrak{w}$. The pair f,g is said to be **free** if it is free of order 1, and it is said to be **strictly free** if it is free of order μ. If $\rho = \mu = p+1$, we have

(2.37) $$(G \llcorner F) | U = (\mathfrak{w} \llcorner \mathfrak{v}) \otimes \mathfrak{v}^{\Delta} \otimes \mathfrak{w}^{\Delta}$$

If $\rho = \mu = q + 1$, we have

(2.38) $$(F \llcorner G) | U = (\mathfrak{v} \llcorner \mathfrak{w}) \otimes \mathfrak{v}^{\Delta} \otimes \mathfrak{w}^{\Delta}$$

If (f,g) is free of order ρ, the divisor $\mu_{F \boxplus^{\rho} G}$ is defined on M. If $\mathfrak{v}$ and $\mathfrak{w}$ are reduced representations of f respectively g on U, then

(2.39) $$\mu_{F \boxplus^{\rho} G} | U = \mu_{\mathfrak{v} \boxplus^{\rho} \mathfrak{w}}$$

(2.40) $$\mu_{F \boxplus G} | U = \mu_{\mathfrak{v} \boxplus \mathfrak{w}} \qquad \text{if } \rho = 1$$

(2.41) $$\mu_{G \llcorner F} | U = \mu_{\mathfrak{w} \llcorner \mathfrak{v}} \qquad \text{if } \rho = \mu = p + 1$$

(2.42) $$\mu_{F \llcorner G} | U = \mu_{\mathfrak{v} \llcorner \mathfrak{w}} \qquad \text{if } \rho = \mu = q + 1$$

(2.43) $$\mu_{<F,G>} | U = \mu_{<\mathfrak{v}, \mathfrak{w}>} \qquad \text{if } \rho = \mu = p + 1 = q + 1$$

If (M,τ) is a parabolic manifold, we abbreviate the counting functions

(2.44) $$n_{\mu_{F \boxplus^{\rho} G}} = n_{f \dot{\boxplus}^{\rho} g}$$

(2.45) $$n_{\mu_{F \boxplus G}} = n_{f \dot{\boxplus} g} \qquad \text{if } \rho = 1$$

(2.46) $$n_{\mu_{G \llcorner F}} = n_{g \dot{\llcorner} f} \qquad \text{if } \rho = p + 1 = \mu$$

(2.47) $$n_{\mu_{F \llcorner G}} = n_{f \dot{\llcorner} g} \qquad \text{if } \rho = q + 1 = \mu$$

(2.48) $$n_{\mu_{<F,G>}} = n_{f;g} \qquad \text{if } \rho = \mu = p + 1 = q + 1$$

and in full analogy for the valence functions.

LEMMA 2.2. *Let* $f : M \longrightarrow G_p(V)$ *and* $g : M \longrightarrow G_q(V^*)$ *be meromorphic maps. Take* $x \in M - (I_f \cup I_g)$. *Then* $(F \boxplus G)(x) = 0$ *if and only if* $E(f(x)) \subseteq E[g(x)]$. *Also* $(F \boxplus^{\rho} G)(x) = 0$ *if and only if* $\dim E(f(x)) \cap E[g(x)] > p + 1 - \rho$. *If* $q \leqslant p$, *then* $(F \llcorner G)(x) = 0$ *if and only if* $f(x) \in \ddot{E}_p[g(x)]$.

The proof follows immediately from (2.36), (2.37), Lemma 1.3, Lemma 1.8, and (1.8). If $p = q$, we can assume as well that $p = q = 0$. Then $\mu_{f,g}$ permits another interpretation. Let

$$\text{(2.49)} \qquad \pi_1 : \mathbb{P}(V) \times \mathbb{P}(V^*) \longrightarrow \mathbb{P}(V) \qquad \pi_2 : \mathbb{P}(V) \times \mathbb{P}(V^*) \longrightarrow \mathbb{P}(V^*)$$

be the projections. They lift the line bundles and we obtain

$$\text{(2.50)} \qquad \mathcal{O}(a,b) = \pi_1^*\mathcal{O}(a) \otimes \pi_2^*\mathcal{O}(b) \qquad \text{for} \quad (a,b) \in \mathbb{Z}^2$$

where $\mathcal{O}(a,b)^* = \mathcal{O}(-a,-b)$. If $x = (v,w) \in \mathbb{P}(V) \times \mathbb{P}(V^*)$ and $\mathfrak{v} \in V_*$ and $\mathfrak{w} \in V_*^*$ with $v = \mathbb{P}(\mathfrak{v})$ and $w = \mathbb{P}(\mathfrak{w})$, then $(\mathfrak{v}, \mathfrak{w}) \in \mathcal{O}(-1,-1)_x$. A global holomorphic section η of $\mathcal{O}(1,1)$ is defined by

$$\text{(2.51)} \qquad <(\mathfrak{v}, \mathfrak{w}), \eta(v,w)> = <\mathfrak{v}, \mathfrak{w}> \in \mathbb{C}.$$

If $a = \mathbb{P}(\mathfrak{a}) \in \mathbb{P}(V^*)$ and $b = \mathbb{P}(\mathfrak{b}) \in \mathbb{P}(V^*)$ and if $\mathfrak{v} \in V - E[a]$ and $\mathfrak{w} \in V^* - E[\mathfrak{b}]$, then

$$\text{(2.52)} \qquad \eta(v,w) = <\mathfrak{v}, \mathfrak{w}> \frac{\mathfrak{a} \mid E(v)}{<\mathfrak{v}, \mathfrak{a}>} \otimes \frac{\mathfrak{b} \mid E(w)}{<\mathfrak{b}, \mathfrak{w}>}$$

The divisor μ_η has multiplicity one on its support S, i.e. $\mu_\eta = \mu_S$ where

$$\text{(2.53)} \qquad S = \operatorname{supp} \eta = \{(v,w) \in \mathbb{P}(V) \times \mathbb{P}(V^*) \mid v \in \ddot{E}[w]\}$$

is smooth, compact, complex submanifold of dimension $2n-1$ called the <u>incidence</u> manifold. With these notations we have

LEMMA 2.3. *Let $f : M \longrightarrow \mathbb{P}(V)$ and $g : M \longrightarrow \mathbb{P}(V^*)$ be meromorphic maps. A meromorphic map $h : M \longrightarrow \mathbb{P}(V) \times \mathbb{P}(V^*)$ is defined by $h(x) = (f(x), g(x))$ for all $x \in M - (I_f \cup I_g)$. Then the pair f, g is free if and only if $h(M) \not\subseteq S$. If so, then $\mu_{f,g} = h^*(\mu_\eta)$.*

PROOF. $h(M) \not\subseteq S$, if and only if $x \in M - (I_f \cup I_g)$ exists such that $f(x) \notin \ddot{E}[g(x)]$ which is the case if and only if the pair f, g is free. Assume that the pair f, g is free. Take $x \in M - (I_f \cup I_g)$. Take $a \in \mathbb{P}(\mathfrak{a}) \in \mathbb{P}(V^*)$ and $b = \mathbb{P}(\mathfrak{b}) \in \mathbb{P}(V)$ such that $f(x) \notin \ddot{E}[a]$ and $g(x) \notin \ddot{E}[b]$. Take reduced representations $\mathfrak{v} : U \longrightarrow V_*$ of f and $\mathfrak{w} : U \longrightarrow V_*^*$ of g with $x \in U$ such that $\mathfrak{v}(z) \notin E[\mathfrak{a}]$ and $\mathfrak{w}(z) \notin E[\mathfrak{b}]$ for all $z \in U$. By (2.52) we have

$$h^*(\mu_\eta)|U = \mu^0_{<\mathfrak{v}, \mathfrak{w}>} = \mu_{<F,G>}|U = \mu_{f,g}|U \qquad \text{q.e.d.}$$

Again let $f : M \longrightarrow \mathbb{P}(\bigwedge_{p+1} V)$ and $g : M \longrightarrow \mathbb{P}(\bigwedge_{q+1} V^*)$ be meromorphic maps. Define $\mu = \operatorname{Min}(p+1, q+1)$. Take $\rho \in \mathbb{Z}[0, \mu]$. Assume that the pair f, g is free of order ρ. Take reduced representations $\mathfrak{v} : U \longrightarrow \bigwedge_{p+1} V$ of f and $\mathfrak{w} : U \longrightarrow \bigwedge_{q+1} V^*$ of g, then $\mathfrak{v} \boxplus^\rho \mathfrak{w} \not\equiv 0$ on U. If $\tilde{\mathfrak{v}} : \tilde{U} \longrightarrow \bigwedge_{p+1} V$ and $\tilde{\mathfrak{w}} : \tilde{U} \longrightarrow \bigwedge_{q+1} V^*$ are other reduced representations of f respectively g with $U \cap \tilde{U} \neq \emptyset$, then there are holomorphic functions φ and ψ on $U \cap \tilde{U}$ without zeros such that $\tilde{\mathfrak{v}} = \varphi \mathfrak{v}$ and $\tilde{\mathfrak{w}} = \psi \mathfrak{w}$ on $U \cap \tilde{U}$. Therefore $\tilde{\mathfrak{v}} \boxplus^\rho \tilde{\mathfrak{w}} = \varphi\psi\, \mathfrak{v} \boxplus^\rho \mathfrak{w}$ on $U \cap \tilde{U}$. Hence one and only one meromorphic map

$$f \boxplus^\rho g : M \longrightarrow \mathbb{P}\Big(\bigwedge_{p+1-\rho} V \otimes \bigwedge_{q+1-\rho} V^*\Big) \tag{2.54}$$

is defined such that $\mathfrak{v} \boxplus^{\rho} \mathfrak{w}$ is a representation of $f \boxplus^{\rho} g$ on U if $\mathfrak{v} : U \longrightarrow V^*$ is a reduced representations of f and if $\mathfrak{w} = U \longrightarrow V^*$ is a reduced representation of g. However $\mathfrak{v} \boxplus^{\rho} \mathfrak{w}$ may not be reduced.

If $\rho = 0$, we write $f \boxplus^0 g = f \otimes g$. If $\rho = 1$, we write $f \boxplus^1 g = f \boxplus g$. If $\rho = p + 1 = \mu$ we write $f \boxplus^{\rho} g = g \llcorner f$ if $\rho = q + 1 = \mu$ we write $f \boxplus^{\rho} g = f \llcorner g$. If $\rho = \mu = p + 1 = q + 1$ we write $f \boxplus^{\rho} g = \langle f,g\rangle = \infty \in \mathbb{P}_0$ which is the constant map.

Let $V_1, \ldots, V_p$ and W hermitian vector spaces and let Θ be a p-fold unitary projective operation of degree $(q_1, \ldots, q_p)$. Let $f_j : M \longrightarrow \mathbb{P}(V_j)$ be meromorphic maps. Then

$$\square f_1 \dot{\Theta} \ldots \dot{\Theta} f_p \square : (M - (I_{f_1} \cup \ldots \cup I_{f_p})) \longrightarrow \mathbb{R}_+ \tag{2.55}$$

is a non-negative real valued function. The operation $\dot{\Theta}$ can be specified following (1.42) – (1.55).

Let $f : M \longrightarrow \mathbb{P}(\underset{p+1}{\wedge} V)$ and $g : M \longrightarrow \mathbb{P}(\underset{q+1}{\wedge} V)$ be meromorphic maps with $p + q < n$. Abbreviate $F = F_f$ and $G = F_g$. An exterior product

(2.56)
$$\wedge : ((\underset{p+1}{\wedge} V_M) \otimes L_f) \oplus ((\underset{q+1}{\wedge} V_M) \otimes L_g) \longrightarrow (\underset{p+q+2}{\wedge} V_M) \otimes L_g \otimes L_g$$

is defined by

$$(\mathfrak{v} \otimes \varphi) \wedge (\mathfrak{w} \otimes \psi) = (\mathfrak{v} \wedge \mathfrak{w}) \otimes \varphi \otimes \psi \tag{2.57}$$

where $\mathfrak{v} \in \underset{p+1}{\wedge} V$, $\mathfrak{w} \in \underset{q+1}{\wedge} V$, $\varphi \in L_{f,x}$, $\psi \in L_{g,x}$ and $x \in M$. A holomorphic section $F \wedge G$ of $(\underset{p+q+2}{\wedge} V_M) \otimes L_f \otimes L_g$ over M is defined. If $\mathfrak{v} : U \longrightarrow \underset{p+1}{\wedge} V$ and $\mathfrak{w} : U \longrightarrow \underset{q+1}{\wedge} V$ are reduced representations of f and g respectively, then

$$F \wedge G = (\tilde{\mathfrak{v}} \wedge \tilde{\mathfrak{w}}) \otimes \mathfrak{v}^{\Delta} \otimes \mathfrak{w}^{\Delta} \quad \text{over } U. \tag{2.58}$$

The pair f,g is said to be free for $\wedge$, if $F \wedge G \not\equiv 0$, in which case the exterior divisor $\mu_{f \dot{\wedge} g} = \mu_{F \wedge G}$ exists such that

$$\mu_{f \dot{\wedge} g} | U = \mu_{F \wedge G} | U = \mu_{\mathfrak{v} \wedge \mathfrak{w}} . \tag{2.59}$$

On a parabolic manifold abbreviate the counting and valence functions by

$$n_{\mu_{f \dot{\wedge} g}} = n_{f \dot{\wedge} g} \qquad N_{\mu_{f \dot{\wedge} g}} = N_{f \dot{\wedge} g} . \tag{2.60}$$

If the pair f,g is free for $\wedge$, a meromorphic map

$$f \wedge g : M \longrightarrow \mathbb{P}(\underset{p+q+2}{\wedge} V) \tag{2.60}$$

is defined such that $\mathfrak{v} \wedge \mathfrak{w}$ is a representation of $f \wedge g$ on U if $\mathfrak{v} : U \longrightarrow \underset{p+1}{\wedge} V$ and $\mathfrak{w} : U \longrightarrow \underset{q+1}{\wedge} V$ are reduced representations of f and g respectively. If f maps into $G_p(V)$ and if g maps into $G_q(V)$, then $f \wedge g$ maps into $G_{p+q+1}(V)$ and if both maps are holomorphic at x, then $(F \wedge G)(x) = 0$ if and only if $f(x) \in \ddot{E}_p(g(x))$, see (1.6) and (2.36).

<u>**LEMMA**</u> **2.4.** <u>Let</u> p,q <u>and</u> k <u>be integers with</u> $0 \leqslant q < p \leqslant n$ <u>and with</u> $0 \leqslant k \leqslant p - q - 1$. <u>Let</u> $f : M \longrightarrow \mathbb{P}(\underset{p+1}{\wedge} V)$, <u>and</u> $g : M \longrightarrow \mathbb{P}(\underset{q+1}{\wedge} V^*)$, <u>and</u> $h : M \longrightarrow \mathbb{P}(\underset{k+1}{\wedge} V^*)$ <u>be meromorphic maps.</u> <u>Assume that the pair</u> (f,g) <u>is strictly free and that the pair</u> (g,h) <u>is free</u>

for $\wedge$. *Then the pair* $(f, g \wedge h)$ *is strictly free if and only if the pair* $(f \llcorner g, h)$ *is strictly free. If so then*

$$(2.62) \qquad f \llcorner (g \wedge h) = (f \llcorner g) \llcorner h : M \longrightarrow \mathbb{P}(\underset{p-q-k-1}{\wedge} V)$$

$$(2.63) \qquad \mu_{f \dot{\llcorner} (g \wedge h)} + \mu_{g \dot{\wedge} h} = \mu_{f \dot{\llcorner} g} + \mu_{(f \llcorner g) \dot{\llcorner} h} \,.$$

PROOF. Take reduced representations $\mathfrak{w} : U \longrightarrow \underset{p+1}{\wedge} V$ of f and $\mathfrak{v} : U \longrightarrow \underset{q+1}{\wedge} V^*$ of g and $\mathfrak{h} : U \longrightarrow \underset{k+1}{\wedge} V^*$ of h and $\mathfrak{g} : U \longrightarrow \underset{p-q}{\wedge} V$ of $f \llcorner g$ and $\mathfrak{q} : U \longrightarrow \underset{q+k+2}{\wedge} V^*$ of $g \wedge h$. Then $(f, g \wedge h)$ is strictly free if and only if $\mathfrak{w} \llcorner (\mathfrak{v} \wedge \mathfrak{h}) \not\equiv 0$ and $(f \llcorner g, h)$ is strictly free if and only if $(\mathfrak{w} \llcorner \mathfrak{v}) \llcorner \mathfrak{h} \not\equiv 0$ where $\mathfrak{w} \llcorner (\mathfrak{v} \wedge \mathfrak{h}) = (\mathfrak{w} \llcorner \mathfrak{v}) \llcorner \mathfrak{h}$. Hence $(f, g \wedge h)$ is strictly free if and only if $(f \llcorner g, h)$ is strictly free. Also $f \llcorner (g \wedge h) = (f \llcorner g) \llcorner h$ follows. Assume that $(f, g \wedge h)$ is strictly free. Holomorphic functions $P \not\equiv 0$ and $Q \not\equiv 0$ exists on U such that $\mathfrak{w} \llcorner \mathfrak{v} = P \mathfrak{g}$ and $\mathfrak{v} \wedge \mathfrak{h} = Q \mathfrak{q}$. Then

$$\mu_{f \dot{\llcorner} g} | U = \mu_{\mathfrak{w} \llcorner \mathfrak{v}} = \mu_P^0 \qquad \mu_{g \dot{\wedge} h} | U = \mu_{\mathfrak{v} \wedge \mathfrak{h}} = \mu_Q^0$$

$$\mu_{f \dot{\llcorner} (g \wedge h)} | U = \mu_{\mathfrak{w} \llcorner \mathfrak{q}} \qquad \mu_{(f \llcorner g) \dot{\llcorner} h} | U = \mu_{\mathfrak{g} \llcorner \mathfrak{h}}$$

$$P \mathfrak{g} \llcorner \mathfrak{h} = (\mathfrak{w} \llcorner \mathfrak{v}) \llcorner \mathfrak{h} = \mathfrak{w} \llcorner (\mathfrak{v} \wedge \mathfrak{h}) = Q\, \mathfrak{w} \llcorner \mathfrak{q} \,.$$

Hence

$$\mu_{f \dot{\llcorner} g} + \mu_{(f \llcorner g) \dot{\llcorner} h} = \mu_{g \dot{\wedge} h} + \mu_{f \dot{\llcorner} (g \wedge h)}$$

on U. q.e.d.

§3 The First Main Theorem

Let (M,τ) be a parabolic manifold of dimension m. Let W be a holomorphic vector bundle over a connected, complex manifold N. Let α be a holomorphic section of W over N. Let $f : M \longrightarrow N$ be a meromorphic map with indeterminancy I_f. Then W pulls back to a holomorphic vector bundle $f^*(W)$ over $M - I_f$ and there is a holomorphic bundle map $\tilde{f} : f^*(W) \longrightarrow W$ over $f : M - I_f \longrightarrow N$ and a holomorphic section α_f of $f^*(W)$ over $M - I_f$ such that $\tilde{f} \circ \alpha_f = \alpha \circ f$. If we use the standard model, we have

$$f^*(W) = \{(x,w) \in (M - I_f) \times W \mid w \in W_{f(x)}\}$$

$$\tilde{f}(x,w) = w \qquad \alpha_f(x) = (x, \alpha(f(x))) .$$

If $f(M - I_f) \not\subseteq Z(\alpha)$, then $\alpha \not\equiv 0$ and $\alpha_f \not\equiv 0$. Hence a non-negative divisor $\mu_f^\alpha = \mu_{\alpha_f}$ is defined on $M - I_f$. Since $\dim I_f \leqq m - 2$, the divisor μ_f^α continues uniquely to a divisor on M again denoted by μ_f^α.

THEOREM 3.1. **Green Residue Theorem**. Let κ be a hermitian metric along the fibers of W. Assume that $f(M - I_f) \not\subseteq Z(\alpha)$. Abbreviate $\nu = \mu_f^\alpha$. Take $s \in \mathfrak{L}_\tau$ and $r \in \mathfrak{L}_\tau$ with $0 < s < r$. Then we have

$$(3.1)\qquad \int_s^r \int_{M[t]} dd^c \log \|\alpha \circ f\|_\kappa^2 \wedge \upsilon^{m-1} \frac{dt}{t^{2m-1}} + N_\nu(r,s)$$

$$= \int_{M\langle r\rangle} \log \|\alpha \circ f\|_\kappa\, \sigma - \int_{M\langle s\rangle} \log \|\alpha \circ f\|_\kappa\, \sigma .$$

PROOF. If f is holomorphic, the theorem follows from Stoll [103], [108], (6.43). Assume that f is meromorphic. Let Γ_f be the closed graph of f with projections $\pi : \Gamma_f \longrightarrow M$ and $\psi : \Gamma_f \longrightarrow N$. Let $\pi_0 : \hat{M} \longrightarrow \Gamma_f$ be a resolution of singularities of Γ_f. Then $\hat{\pi} = \pi \circ \pi_0 : \hat{M} \longrightarrow M$ is a surjective proper, holomorphic map and $\hat{\tau} = \tau \circ \hat{\pi}$ is a parabolic exhaustion of the m-dimensional, connected, complex manifold $\hat{M}$. The map $\hat{f} = \psi \circ \pi_0 : M \longrightarrow N$ is holomorphic with $\hat{f}(M) \not\subseteq Z(\alpha)$. Hence $\hat{\nu} = \mu^{\alpha}_{\hat{f}}$ exists. Also $s \in \mathcal{E}_{\hat{\tau}}$ and $r \in \mathcal{E}_{\hat{\tau}}$.

Therefore (3.1) holds with $M, \upsilon, f, \sigma, \nu$ replaced by $\hat{M}, \hat{\upsilon}, \hat{f}, \hat{\sigma}, \hat{\nu}$. Abbreviate $\hat{I}_f = \hat{\pi}^{-1}(I_f)$. Then $\hat{I}_f$ is analytic with $\dim \hat{I}_f \leqslant m - 1$. The restriction $\hat{\pi} : \hat{M} - \hat{I}_f \longrightarrow M - I_f$ is biholomorphic. Hence $\hat{\pi}^*(\upsilon) = \hat{\upsilon}$ and $\hat{\pi}^*(\sigma) = \hat{\sigma}$ and $\hat{M}[t] = \hat{\pi}^{-1}(M[t])$ and $\hat{M}\langle r\rangle = \hat{\pi}^{-1}(M\langle r\rangle)$. Hence we obtain (3.1) if we can show that $N_{\hat{\nu}}(r,s) = N_\nu(r,s)$.

Define $S = \operatorname{supp} \nu$ and $\hat{S} = \operatorname{supp} \hat{\nu}$. Let $\hat{S}_0$ be the union of all branches of $\hat{S}$ which are contained in $\hat{I}_f$. Let $\hat{S}_1$ be the union of all other branches of $\hat{S}$. There are divisors $\hat{\nu}_j \geqslant 0$ with $\operatorname{supp} \hat{\nu}_j = \hat{S}_j$ for $j = 0,1$ such that $\hat{\nu} = \hat{\nu}_0 + \hat{\nu}_1$. Hence $N_{\hat{\nu}}(r,s) = N_{\hat{\nu}_0}(r,s) + N_{\hat{\nu}_1}(r,s)$. Consider the standard models

$$\hat{f}^*(W)|(\hat{M} - \hat{I}_f) = \{(x,w) \in (\hat{M} - \hat{I}_f) \times W \mid w \in W_{\hat{f}(x)}\}$$

$$f^*(W) = \{(y,w) \in (M - I_f) \times W \mid w \in W_{f(y)}\}$$

where

$$f \circ \hat{\pi} = f \circ \pi \circ \pi_0 = \psi \circ \pi_0 = \hat{f} \quad \text{on} \quad \hat{M} - \hat{I}_f .$$

Hence a biholomorphic bundle isomorphism $\tilde{\pi} : \hat{f}^*(W)|(\hat{M} - \hat{I}_f) \longrightarrow f^*(W)$ is defined by $\tilde{\pi}(x,w) = (\hat{\pi}(x),w)$. For $x \in \hat{M} - \hat{I}_f$ we have

$$\tilde{\pi}(\alpha_{\hat{f}}(x)) = (\hat{\pi}(x), \alpha(\hat{f}(x))) = (\hat{\pi}(x), \alpha(f(\hat{\pi}(x)))) = \alpha_f(\hat{\pi}(x))$$

$$\tilde{\pi} \circ \alpha_{\hat{f}} = \alpha_f \circ \hat{\pi} \quad \text{on} \quad \hat{M} - \hat{I}_f .$$

Take $x \in \hat{M} - \hat{I}_f$. Then $\vartheta_0(x) = 0$ and we have

$$\vartheta_1(x) = \vartheta(x) = \mu_{\alpha_{\hat{f}}}(x) = \mu_{\alpha_f}(\hat{\pi}(x)) = \nu(\hat{\pi}(x)) = \hat{\pi}^*(\nu)(x) .$$

Since $\dim(\hat{I}_f \cap \operatorname{supp} \vartheta_1) < m - 1$ and $\dim(I_f \cap \operatorname{supp} \nu) < m - 1$, we obtain

$$N_{\vartheta_1}(r,s) = N_\nu(r,s) .$$

Let $j : I_f \longrightarrow M$ and $\hat{j} : \hat{I}_f \longrightarrow \hat{M}$ be the inclusion maps. Since $\dim I_f \leqslant m - 2$, we have

$$\hat{j}^*(\hat{\upsilon}^{m-1}) = \hat{j}^*\hat{\pi}^*(\upsilon^{m-1}) = (\hat{\pi} \circ \hat{j})^*(\upsilon^{m-1}) = (j \circ \hat{\pi})^*(\upsilon^{m-1})$$

$$= \hat{\pi}^*((j^*(\upsilon))^{m-1}) = \hat{\pi}^*(0) = 0$$

since $\dim I_f \leqslant m - 2$. Therefore $N_{\mathfrak{D}_0}(r,s) = 0$. We obtain $N_{\mathfrak{D}}(r,s) = N_\nu(r,s)$. q.e.d.

If $W = L$ is a holomorphic line bundle, then

$$c(L,\kappa) = -\, dd^c \log \|\alpha\|_\kappa^2 \tag{3.2}$$

is the <u>Chern form</u> of L for κ on $N - Z(\alpha)$. Of course, $c(L,\kappa)$ is defined and of class C^∞ on N and does not depend on the section α. On $M - \operatorname{supp} \nu$, we obtain

$$f^*(c(L,\kappa)) = -\, dd^c \log \|\alpha \circ f\|_\kappa^2 \tag{3.3}$$

For $t > 0$, the <u>spherical image</u> is defined by

$$A_f(t,L,\kappa) = \frac{1}{t^{2m-2}} \int_{M[t]} f^*(c(L,\kappa)) \wedge \upsilon^m . \tag{3.4}$$

For $0 < s < r$, the <u>characteristic</u> is defined by

$$T_f(r,s,L,\kappa) = \int_s^r A_f(t) \frac{dt}{t} . \tag{3.5}$$

For $t > 0$, the <u>counting function</u> $n_f(t,\alpha,L) = n_\nu(t)$ is defined. For $0 < s < r$, the <u>valence function</u> $N_f(r,s,\alpha,L) = N_\nu(r,s)$ is defined. For $r \in \mathfrak{L}_\tau$, the <u>compensation function</u> is defined by

$$m_f(r,\alpha,L,\kappa) = \int_{M\langle r\rangle} \log \frac{1}{\|\alpha \circ f\|_\kappa} \sigma . \tag{3.6}$$

Then Theorem 3.1 becomes the <u>First Main Theorem</u> for <u>line bundles</u>

$$T_f(r,s,L,\kappa) = N_f(r,s,\alpha,L) + m_f(r,\alpha,L,\kappa) - m_f(s,\alpha,L,\kappa) \tag{3.7}$$

for $0 < s < r \in \mathfrak{L}_\tau$ with $s \in \mathfrak{L}_\tau$.

REMARK 1. The compensation function $m_f(r,a,L,\kappa)$ extends to a continuous function of $r > 0$ such that (3.7) holds for all $0 < s < r$.

REMARK 2. $N_f(r,s,a,L) \geq 0$ since $\mu_a \geq 0$. If $c(L,\kappa) \geq 0$, then $A_f(t,L,\kappa) \geq 0$ increases and $T_f(r,s,L,\kappa) \geq 0$ increases in r. If N is compact, κ can be taken such that $\|a\|_\kappa \leq 1$, hence $m_f(r,a,L,\kappa) \geq 0$. For details see Stoll [103].

REMARK 3. If $M = N$ and if f is the identity map, we omit f as an index.

REMARK 4. Let V be a hermitian vector space of dimension $n + 1 \geq 1$. Let $f : M \longrightarrow \mathbb{P}(V)$ be a meromorphic map. Let L_f be the hyperplane section bundle of f on M. Take a hermitian metric κ along the fibers of L_f. Then the characteristic function $T(r,s,L_f,\kappa_f)$ is defined. Assume that there is a global reduced representation $\mathfrak{w} : M \longrightarrow V$ of f. Then there is a holomorphic frame $\mathfrak{w}^\Delta$ of L_f over M, in particular, L_f is trivial. The Green Residue Theorem implies

$$(3.8) \qquad - T(r,s,L_f,\kappa_f) = \int_{M\langle r\rangle} \log \|\mathfrak{w}^\Delta\|_\kappa \, \sigma - \int_{M\langle s\rangle} \log \|\mathfrak{w}^\Delta\|_\kappa \, \sigma$$

for $0 < s < r \in \mathfrak{L}_\tau$ with $s \in \mathfrak{L}_\tau$.

Now we will introduce the value distribution functions for a meromorphic map into projective space. Again let V be a hermitian vector space of dimension $n + 1 > 1$. The hermitian metric on V induces a hermitian metric along the fibers of the hyperplane section bundle $\mathcal{O}(1)$ whose Chern form is the Fubini-Study form Ω_0. Let $f : M \longrightarrow \mathbb{P}(V)$ be a meromorphic map. For $t > 0$, the **spherical image** is defined by

$$(3.9) \qquad A_f(t) = \frac{1}{t^{2m-2}} \int_{M[t]} f^*(\Omega_0) \wedge \upsilon^{m-1} \geqq 0 .$$

The function A_f increases. Define

$$(3.10) \qquad A_f(0) = \lim_{0<t\to 0} A_f(0) \qquad A_f(\infty) = \lim_{t\to\infty} A_f(t) .$$

Observe that $M_*[t] = M[t] - M[0]$. Then

$$A_f(t) = \int_{M_*[t]} f^*(\Omega_0) \wedge \omega^{m-1} + A_f(0) .$$

For $0 < s < r$, the <u>characteristic</u> of f is defined by

$$(3.11) \qquad T_f(r,s) = \int_s^r A_f(t) \frac{dt}{t} \geqq 0 .$$

Then

$$(3.12) \qquad \frac{T_f(r,s)}{\log r} \longrightarrow A_f(\infty) \qquad \text{for } r \longrightarrow \infty$$

If f is constant, then $A_f(\infty) = A_f(t) = 0$ for $t > 0$ and $T_f(r,s) = 0$ for $0 < s < r$. If f is not constant, then $A_f(\infty) > A_f(t) > 0$ for $t > 0$ and $T_f(r,s) > 0$ for $0 < s < r$. Also $T_f(r,s) \longrightarrow \infty$ for $r \longrightarrow \infty$. (Stoll [103] Theorem 12.8 and Stoll [108] Lemma 10.4.)

<u>PROPOSITION</u> 3.2. Let L_f be the hyperplane section bundle of f on M. Let κ be a hermitian metric along the fibers of L_f. Let $F = F_f$ be the representation section of f in $V_M \otimes L_f$. Let ℓ be the hermitian metric along the fibers of $V_M = M \times V$, defined by the hermitian metric on V. Abbreviate $\lambda = \ell \otimes \kappa$. Take $s \in \mathfrak{L}_\tau$ and $r \in \mathfrak{L}_\tau$ with $0 < s < r$. Then we have

$$\text{(3.13)}\qquad T_f(r,s) - T(r,s,L_f,\kappa_f) = \int_{M\langle r\rangle} \log \|F\|_\lambda \sigma - \int_{M\langle s\rangle} \log \|F\|_\lambda \sigma .$$

If $\mathfrak{v} : M \longrightarrow V$ is a global reduced representation of f, we obtain

$$\text{(3.14)}\qquad T_f(r,s) = \int_{M\langle r\rangle} \log \|\mathfrak{v}\|\sigma - \int_{M\langle s\rangle} \log \|\mathfrak{v}\|\sigma$$

PROOF. Let $\mathfrak{v} : U \longrightarrow V$ be a reduced representation of f. Then we have $F|U = \mathfrak{v} \otimes \mathfrak{v}^\Delta$ and

$$\text{(3.15)}\qquad dd^c \log \|F\|_\lambda^2 | U = dd^c \log \|\mathfrak{v}\|^2 + dd^c \log \|\mathfrak{v}^\Delta\|_\kappa^2$$

$$= (f^*(\Omega_0) - c(L_f,\kappa))|U$$

Thus the Green Residue Formula, Theorem 3.1 implies (3.13).

If $\mathfrak{v} : M \longrightarrow V$ is a global, reduced representation of f, then $\|F\|_\lambda = \|\mathfrak{v}\| \, \|\mathfrak{v}^\Delta\|_\kappa$. Hence (3.13) and (3.8) imply (3.14). q.e.d.

PROPOSITION 3.3. Let $\mathfrak{v} : M \longrightarrow V$ be a representation of f on M. Let $N_\mathfrak{v}$ be the valence function of the divisor of $\mathfrak{v}$. Assume that the Theorem of Cousin II holds on M. Then

$$\text{(3.16)}\qquad T_f(r,s) = \int_{M\langle r\rangle} \log \|\mathfrak{v}\|\sigma - \int_{M\langle s\rangle} \log \|\mathfrak{v}\|\sigma - N_\mathfrak{v}(r,s)$$

for $0 < s < r \in \mathcal{E}_\tau$ with $s \in \mathcal{E}_\tau$.

PROOF. A reduced representation $\mathfrak{w} : M \longrightarrow V$ exists. A holomorphic function $h : M \longrightarrow \mathbb{C}$ exists such that $\mathfrak{v} = h\mathfrak{w}$. Then $\mu_\mathfrak{v} = \mu_h^0$. We have

$$T_f(r,s) = \int_{M\langle r\rangle} \log \|\mathfrak{w}\| \sigma - \int_{M\langle s\rangle} \log \|\mathfrak{w}\| \sigma$$

$$N_{\mathfrak{w}}(r,s) = \int_{M\langle r\rangle} \log |h| \sigma - \int_{M\langle s\rangle} \log |h| \sigma .$$

Addition implies (3.16). q.e.d.

Let $V_1, \ldots, V_k$ and W be hermitian vector spaces. Let

$$\Theta : V_1 \times \ldots \times V_k \longrightarrow W$$

be an operation homogeneous of degree $(q_1, \ldots, q_k) \in \mathbb{Z}^k$. For $j = 1, \ldots, k$ let $f_j : M \longrightarrow \mathbb{P}(V_j)$ be a meromorphic map such that $(f_1, \ldots, f_k)$ is free for Θ. Let $N_{f_1\dot{\Theta}\ldots\dot{\Theta}f_k}$ be the valence of the divisor $\mu_{f_1\dot{\Theta}\ldots\dot{\Theta}f_k}$ and define the <u>compensation</u> <u>function</u> by

(3.17) $$m_{f_1\dot{\Theta}\ldots\dot{\Theta}f_k}(r) = \int_{M\langle r\rangle} \log \frac{1}{\square f_1 \dot{\Theta} \ldots \dot{\Theta} f_k \square} \sigma$$

for $r \in \mathfrak{L}_\tau$.

<u>THEOREM</u> 3.4. <u>The General First Main Theorem</u>. If $s \in \mathfrak{L}_\tau$ and $r \in \mathfrak{L}_\tau$ with $0 < s < r$, then

(3.18) $$\sum_{j=1}^{q} q_j T_{f_j}(r,s) - T_{f_1\Theta\ldots\Theta f_k}(r,s)$$

$$= N_{f_1\dot{\Theta}\ldots\dot{\Theta}f_k}(r,s) + m_{f_1\dot{\Theta}\ldots\dot{\Theta}f_k}(r) - m_{f_1\dot{\Theta}\ldots\dot{\Theta}f}(s)$$

<u>PROOF</u>. Abbreviate $h = f_1 \Theta \ldots \Theta f_k : M \longrightarrow \mathbb{P}(W)$. Let $L_j = L_{f_j}$ be the hyperplane section bundle of f_j and take a hermitian metric κ_j

along the fibers of L_j. Let ℓ_j be the hermitian metric of V_j and the induced metric along the fibers of $V_{jM} = M \times V_j$. Abbreviate $\lambda_j = \ell_j \otimes \kappa_j$. Let L_h be the hyperplane section bundle of h. Take a hermitian metric κ_h along the fibers of L_h. Let κ_h^* be the metric along L_h^* dual to κ_h. Then

$$\kappa = (\kappa_1)^{q_1} \otimes \ldots \otimes (\kappa_k)^{q_k} \otimes \kappa_h^* \tag{3.19}$$

is a hermitian metric along the fibers of the line bundle

$$L = (L_1)^{q_1} \otimes \ldots \otimes (L_k)^{q_k} \otimes L_h^* . \tag{3.20}$$

We have $c(L_h^*, \kappa_h^*) = - c(L_h, \kappa_h)$ and

$$c(L, \kappa) = \sum_{j=1}^{k} q_j c(L_j, \kappa_j) - c(L_h, \kappa_h) \tag{3.21}$$

which implies

$$T(r,s,L,\kappa) = \sum_{j=1}^{k} q_j T(r,s,L_j,\kappa_j) - T(r,s,L_h,\kappa_h) \tag{3.22}$$

Let $F_j = F_{f_j}$ be the representation section of f_j for $j = 1, \ldots ,k$ and let F_h be the representation section of H. Let $\lambda_j = \ell_j \otimes \kappa_j$ and $\lambda_h = \ell_h \otimes \kappa_h$ be the respective hermitian metrics.

We shall construct a section in L. Let $(U_\lambda)_{\lambda \in \Lambda}$ be an open covering of M by open, connected Stein subsets $U_\lambda \neq \emptyset$ with $H^2(U_\lambda, \mathbb{Z}) = 0$. For each $\lambda \in \Lambda$ we select reduced representations $\mathfrak{v}_{\lambda j} : U_\lambda \longrightarrow V_j$ of f_j and $\mathfrak{w}_\lambda : U_\lambda \longrightarrow W$ of h on U_λ. Since $\mathfrak{v}_{\lambda 1} \odot \ldots \odot \mathfrak{v}_{\lambda k}$ is a representation of h on U_λ, a holomorphic function $W_\lambda \not\equiv 0$ exists on U_λ such that

(3.23) $$\mathfrak{w}_{\lambda 1} \odot \ldots \odot \mathfrak{w}_{\lambda k} = W_\lambda \mathfrak{W}_\lambda \qquad \text{on } U_\lambda .$$

We have

(3.24) $$F_j | U_\lambda = \mathfrak{w}_{\lambda j} \otimes \mathfrak{w}^{\Delta}_{\lambda j} \qquad F_h | U_\lambda = \mathfrak{W}_\lambda \otimes \mathfrak{W}^{\Delta}_{\lambda}$$

(3.25) $$F_1 \odot \ldots \odot F_k | U_\lambda$$
$$= (\mathfrak{w}_{\lambda 1} \odot \ldots \odot \mathfrak{w}_{\lambda k}) \otimes (\mathfrak{w}^{\Delta}_{\lambda 1})^{q_1} \otimes \ldots \otimes (\mathfrak{w}^{\Delta}_{\lambda k})^{q_k}$$

Let $\mathfrak{W}^{\Delta *}_{\lambda}$ be the holomorphic frame of L^*_h over U_λ which is dual to $\mathfrak{W}^{\Delta}_{\lambda}$. Then

(3.26) $$\mathfrak{f}_\lambda = (\mathfrak{w}^{\Delta}_{\lambda 1})^{q_1} \otimes \ldots \otimes (\mathfrak{w}^{\Delta}_{\lambda k})^{q_k} \otimes \mathfrak{W}^{\Delta *}_{\lambda}$$

is a holomorphic frame of L over U_λ. For each pair $(\lambda,\mu) \in \Lambda[1]$ there are zero free holomorphic transition functions $v_{\lambda\mu j}$ and $w_{\lambda\mu}$ on $U_{\lambda\mu} = U_\lambda \cap U_\mu$ such that

(3.27) $$\mathfrak{w}_{\lambda j} = v_{\lambda\mu j} \mathfrak{w}_{\mu j} \qquad \mathfrak{W}_\lambda = w_{\lambda\mu} \mathfrak{W}_\mu$$

(3.28) $$\mathfrak{w}^{\Delta}_{\mu j} = v_{\lambda\mu j} \mathfrak{w}^{\Delta}_{\mu j} \qquad \mathfrak{W}^{\Delta}_{\mu} = w_{\lambda\mu} \mathfrak{W}^{\delta}_{\mu}$$

(3.29) $$\mathfrak{W}^{\Delta *}_{\lambda} = w_{\lambda\mu} \mathfrak{W}^{\Delta *}_{\mu}$$

(3.30) $$\mathfrak{f}_\mu = (v_{\lambda\mu 1})^{q_1} \ldots (v_{\lambda\mu k})^{q_k} \, w_{\mu\lambda} \mathfrak{f}_\lambda$$

on $U_{\lambda\mu}$. Therefore we obtain

$$W_\lambda \mathfrak{w}_\lambda = \mathfrak{w}_{\lambda 1} \odot \ldots \odot \mathfrak{w}_{\lambda k}$$

$$= (v_{\lambda\mu 1})^{q_1} \ldots (v_{\lambda\mu k})^{q_k} \mathfrak{w}_{\mu 1} \odot \ldots \odot \mathfrak{w}_{\mu k}$$

$$= (v_{\lambda\mu 1})^{q_1} \ldots (v_{\lambda\mu k})^{q_k} W_\mu \mathfrak{w}_\mu$$

$$= (v_{\lambda\mu 1})^{q_1} \ldots (v_{\lambda\mu k})^{q_k} w_{\mu\lambda} W_\mu \mathfrak{w}_\lambda$$

or

$$W_\lambda = (v_{\lambda\mu 1})^{q_1} \ldots (v_{\lambda\mu k})^{q_k} w_{\mu\lambda} W_\mu$$

$$W_\lambda \varphi_\lambda = (v_{\lambda\mu 1})^{q_1} \ldots (v_{\lambda\mu k})^{q_k} w_{\mu\lambda} \varphi_\lambda W_\mu = W_\mu \varphi_\mu$$

on $U_{\lambda\mu}$. Therefore one and only one holomorphic section W of L over M exists such that $W | U_\lambda = W_\lambda \varphi_\lambda$ for all $\lambda \in \Lambda$. Observe that

$$\mu_{f_1 \dot{\odot} \ldots \dot{\odot} f_k} = \mu_{F_1 \odot \ldots \odot F_k} = \mu_{\mathfrak{w}_1 \odot \ldots \odot \mathfrak{w}_k} = \mu_{W_\lambda} = \mu_W$$

on U_λ, hence $\mu_{f_1 \dot{\odot} \ldots \dot{\odot} f_k} = \mu_W$ on M, which implies

$$N_W(r,s) = N_{f_1 \dot{\odot} \ldots \dot{\odot} f_k}.$$

Therefore (3.7) and (3.22) imply

$$\text{(3.31)} \qquad \sum_{j=1}^{k} q_j T(r,s,L_j,\kappa_j) - T(r,s,L_h,\kappa_h)$$

$$= N_{f_1 \dot{\odot} \ldots \dot{\odot} f_k}(r,s) + \int_{M\langle r\rangle} \log \frac{1}{\|W\|_\kappa} \sigma - \int_{M\langle s\rangle} \log \frac{1}{\|W\|_\kappa} \sigma .$$

Now, Proposition 3.2 implies

(3.32) $$T_{f_j}(r,s) \mathcal{L} T(r,s,L_j,\kappa_j) = \int_{M\langle r\rangle} \log \|F_j\|_{\lambda_j}\, \sigma \mathcal{L} \int_{M\langle s\rangle} \log \|F_j\|_{\lambda_j}\, \sigma$$

(3.33) $$T_h(r,s) - T(r,s,L_h,\kappa_h) = \int_{M\langle r\rangle} \log \|F_h\|_{\lambda_h}\, \sigma - \int_{M\langle s\rangle} \log \|F_h\|_{\lambda_h}\, \sigma$$

Define

(3.34) $$u = \frac{\|F_1\|_{\lambda_1}^{q_1} \cdots \|F_k\|_{\lambda_k}^{q_k}}{\|W\|_{\kappa}\ \|F_h\|_{\lambda_h}}$$

Then (3.31), (3.32) and (3.33) combine to

(3.35) $$\sum_{j=1}^{k} q_j T_{f_j}(r,s) - T_h(r,s)$$

$$= N_{f_1 \dot{\odot} \cdots \dot{\odot} f_k}(r,s) + \int_{M\langle r\rangle} \log u\sigma - \int_{M\langle s\rangle} \log u\sigma$$

For each $\lambda \in \Lambda$, we have $\|\mathfrak{v}^{\Delta}_{\lambda}\|_{\kappa_h}\ \|\mathfrak{w}^{\Delta *}_{\lambda}\|_{\kappa^*_h} = 1$ on U_λ. Hence

$$u = \frac{(\|\mathfrak{v}_{\lambda 1}\|\ \|\mathfrak{v}^{\Delta}_{\lambda 1}\|_{\kappa_1})^{q_1} \cdots (\|\mathfrak{v}_{\lambda k}\|\ \|\mathfrak{v}^{\Delta}_{\lambda k}\|_{\kappa_k})^{q_k}}{|W_\lambda|\,(\|\mathfrak{v}^{\Delta}_{\lambda 1}\|_{\kappa_1})^{q_1} \cdots (\|\mathfrak{v}^{\Delta}_{\lambda k}\|_{\kappa_k})^{q_k}\ \|\mathfrak{w}^{\Delta *}_{\lambda}\|_{\kappa^*_h}\ \|F_h\|_{\lambda_h}}$$

$$= \frac{\|\mathfrak{v}_{\lambda 1}\|^{q_1} \cdots \|\mathfrak{v}_{\lambda k}\|^{q_k}}{|W_\lambda|\ \|\mathfrak{w}^{\Delta *}_{\lambda}\|_{\kappa^*_h}\ \|\mathfrak{w}_{\lambda}\|\ \|\mathfrak{w}^{\Delta}_{\lambda}\|_{\kappa_h}}$$

$$= \frac{\|\mathfrak{v}_{\lambda 1}\|^{q_1} \cdots \|\mathfrak{v}_{\lambda k}\|^{q_k}}{\|\mathfrak{v}_{\lambda_1} \odot \cdots \odot \mathfrak{v}_{\lambda k}\|} = \frac{1}{\Box f_1 \dot{\odot} \cdots \dot{\odot} f_k \Box}$$

Hence (3.35) coincides with (3.18). q.e.d.

The compensation function $m_{f_1\dot{\odot}\ldots\dot{\odot}f_k}$ extends from ℓ_τ to a continuous function on $\mathbb{R}^+$ such that (3.18) holds for all $0 < s < r$.

In order to simplify the notation, we will also write

$$A_f(t) = A(t;f) \qquad T_f(r,s) = T(r,s;f)$$

$$m_f(r) = m(r;f) \qquad N_f(r,s) = N(r,s;f)$$

etc. Now, we will study the case of special operations.

1. <u>SPECIAL CASE</u>: <u>The First Main Theorem for the exterior product.</u> Take positive integers n_j with $n_1 + \ldots + n_k \leqslant n + 1$. Define $V_j = \bigwedge_{n_j} V$ and $q_j = 1$ for $j = 1, \ldots ,k$. Take the exterior product $\wedge = \odot$ as homogeneous operation of degree $(1, \ldots ,1)$. Let $f_j : M \longrightarrow \mathbb{P}(\bigwedge_{n_j} V)$ be meromorphic maps such that $(f_1, \ldots ,f_k)$ is free for $\wedge$. Put $p = n_1 + \ldots + n_k$ and $W = \bigwedge_p V$. Then we have

$$\text{(3.36)} \qquad \sum_{j=1}^{k} T_{f_j}(r,s) = T_{f_1\wedge\ldots\wedge f_k}(r,s) + N_{f_1\dot{\wedge}\ldots\dot{\wedge}f_k}(r,s)$$

$$+ m_{f_1\dot{\wedge}\ldots\dot{\wedge}f_k}(r) - m_{f_1\dot{\wedge}\ldots\dot{\wedge}f_k}(s)$$

If $n_1 + \ldots + n_k = n + 1$, then $T_{f_1\wedge\ldots\wedge f_k}(r,s) = 0$. If f_j maps into the Grassmannian cone: $f_j : M \longrightarrow G_{n_j-1}(V)$, then

$\Box\, f_1 \dot{\wedge} \ldots \dot{\wedge} f_k \,\Box \leqslant 1$ and the compensation function

(3.37) $$m_{f_1 \dot{\wedge} \ldots \dot{\wedge} f_k}(r) \geq 0$$

is non-negative. The terms in (3.36) are invariant under permutation of the maps $f_1, \ldots, f_k$.

If $n_1 = n_2 = \ldots = n_k = 1$, then $f_1, \ldots, f_k$ are said to be in general position, if and only if $(f_1, \ldots, f_k)$ is free for the operation $\wedge$. This situation will be explored in Section 7.

2. **SPECIAL CASE: The First Main Theorem for a contraction.** Take $k = 2$, $p \in \mathbb{Z}[0,n]$, $q \in \mathbb{Z}[0,n]$ and $V_1 = \bigwedge_{p+1} V$ and $V_2 = \bigwedge_{q+1} V^*$. Let $f = f_1 : M \longrightarrow \mathbb{P}(\bigwedge_{p+1} V)$ and $g = f_2 : M \longrightarrow \mathbb{P}(\bigwedge_{q+1} V^*)$ be meromorphic maps. Define $\mu = \text{Min}(p+1, q+1)$ and take $\rho \in \mathbb{N}[1,\mu]$. Then (f,g) is free of order ρ, if and only if (f,g) is free for the operation $\boxplus^\rho$, which we will assume now. Define

$$W = (\bigwedge_{p+1-\rho} V) \otimes (\bigwedge_{q+1-\rho} V^*)$$

Then $\boxplus^\rho$ is a homogeneous operation of degree $(1,1)$. We obtain

(3.38) $$T_f(r,s) + T_g(r,s) - T(r,s;f \boxplus^\rho g)$$

$$= N(r,s;f \dot{\boxplus}^\rho g) + m(r,f \dot{\boxplus}^\rho g) - m(s;f \dot{\boxplus}^\rho g)$$

If $f : M \longrightarrow G_p(V)$ and $g : M \longrightarrow G_q(V^*)$ map into Grassmann manifolds, then $\Box f \dot{\boxplus}^\rho g \Box \leq 1$ and $m(r;f \dot{\boxplus}^\rho g) \geq 0$. If $\rho = 1$, we abbreviate $\boxplus = \boxplus^1$.

3. **SPECIAL CASE: The First Main Theorem for the interior product.** Take $k = 2$, $p \in \mathbb{Z}[0,n]$, $q \in \mathbb{Z}[0,p]$ and $V_1 = \bigwedge_{p+1} V$ and $V_2 = \bigwedge_{q+1} V^*$. Let $f = f_1 : M \longrightarrow \mathbb{P}(\bigwedge_{p+1} V)$ and $g = f_2 : M \longrightarrow \mathbb{P}(\bigwedge_{q+1} V^*)$ be meromorphic

maps. Define $W = \bigwedge_{p-q} V$. Then (f,g) is strictly free, if and only if (f,g) is free for $\llcorner$, where we take the interior product $\llcorner$ as operation $\odot$ of degree $(1,1)$. We obtain

(3.39)
$$T_f(r,s) + T_g(r,s) - T_{f\llcorner g}(r,s)$$

$$= N_{f\dot{\llcorner} g}(r,s) + m_{f\dot{\llcorner} g}(r) - m_{f\dot{\llcorner} g}(s) .$$

which is also a subcase of Special Case 2 with $0 \leqslant \rho = \mu = q + 1 \leqslant p + 1$. If $f : M \longrightarrow G_p(V)$ and $g : M \longrightarrow G_q(V^*)$, then $\Box\, f\dot{\llcorner} g\, \Box \leqslant 1$ and $m_{f\dot{\llcorner} g}(r) \geqslant 0$. If g is constant, then $T_g(r,s) = 0$ and (3.39) is well known (see Wu [126], Theorem 4.9).

We assumed $0 \leqslant q \leqslant p$. The case $0 \leqslant p \leqslant q$ can be obtained by replacing V by V^* and V^* by $V = V^{**}$. The details are left to the reader.

4. <u>SPECIAL CASE</u>: <u>The First Main Theorem for the inner product</u>. Take $k = 2$, $V_1 = V$, $V_2 = V^*$, $W = \mathbb{C}$. Let $f : M \longrightarrow \mathbb{P}(V)$ and $g : M \longrightarrow \mathbb{P}(V^*)$ be meromorphic maps such that (f,g) is free. The inner product $<\, ,\, > : V \times V^* \longrightarrow \mathbb{C}$ is a homogeneous operation of degree $(1,1)$. We have

(3.40)
$$\boxed{T_f(r,s) + T_g(r,s) = N_{f,g}(r,s) + m_{f,g}(r) - m_{f,g}(s)}$$

where $m_{f,g}(r) \geqslant 0$. If $g \equiv a \in \mathbb{P}(V^*)$ is constant, we write

(3.41)
$$N_f(r,s;a) = N_{f,a}(r,s) \qquad m_f(r;a) = m_{f,a}(r)$$

and obtain the classical <u>First Main Theorem for Hyperplanes</u>

(3.42)
$$T_f(r,s) = N_f(r,s;a) + m_f(r;a) - m_f(s;a) .$$

If we integrate over $\mathbb{P}(V^*)$ we obtain

$$(3.43) \qquad \int_{a\in\mathbb{P}(V^*)} m_f(r,a)\Omega_0(a) = \frac{\varsigma}{2}\sum_{\nu=1}^{n}\frac{1}{\nu}$$

$$(3.44) \qquad \boxed{T_f(r,s) = \int_{a\in\mathbb{P}(V^*)} N_f(r,s;a)\Omega_0(a)}$$

__We will give an alternative proof of__ (3.40). Recall (2.49) – (2.53) and Lemma (2.3). A meromorphic map $h : M \longrightarrow \mathbb{P}(V) \times \mathbb{P}(V^*)$ is defined by $h(x) = (f(x),g(x))$ for all $x \in M - (I_f \cup I_g)$. The hermitian metric ℓ on V induces hermitian metrics along the fibers of the bundles $V^*_{\mathbb{P}(V)}$, $V_{\mathbb{P}(V^*)}$, $\mathcal{O}(1)$ and $H = \mathcal{O}(1,1)$. We have $h(M) \not\subset S$ and

$\mu_{f,g} = h^*(\mu_\eta) = \mu_h^\eta$.

__CLAIM__: Take $x = \mathbb{P}(\mathfrak{k}) \in \mathbb{P}(V)$ and $\alpha \in V^*$. Then $\alpha | E(x) \in \mathcal{O}(1)_x$ and

$$(3.45) \qquad \|\alpha | E(x)\|_\ell = \frac{|\langle \mathfrak{k},\alpha\rangle|}{\|\mathfrak{k}\|}$$

__PROOF OF THE CLAIM__: Let $\mathfrak{n}_0, \ldots, \mathfrak{n}_n$ be an orthonormal base of V with $\mathfrak{k} = x_0\mathfrak{n}_0$. Let $\imath_0, \ldots, \imath_n$ be the dual base. Then $\langle \mathfrak{k},\imath_0\rangle = x_0$ and $\langle \mathfrak{k},\imath_j\rangle = 0$ for $j = 1, \ldots, n$. Hence $\imath_j | E(x) = 0$ for $j = 1, \ldots, n$. Hence $\imath_1, \ldots, \imath_n$ spans the kernel of the evaluation map $V^* \longrightarrow \mathcal{O}(1)_x$. Therefore $\mathbb{C}\imath_0$ is isometric to $\mathcal{O}(1)_x$. We have $\alpha = a_0\imath_0 + \ldots + a_n\imath_n$ and

$$\|\alpha | E(x)\|_\ell = |a_0| = |\langle \mathfrak{n}_0,\alpha\rangle| = \frac{|\langle \mathfrak{k},\alpha\rangle|}{\|\mathfrak{k}\|} . \qquad \text{q.e.d.}$$

Take $x = \mathbb{P}(\mathfrak{z}) \in \mathbb{P}(V)$ and $y = \mathbb{P}(\mathfrak{w}) \in \mathbb{P}(V^*)$. Pick $\mathfrak{a} \in V_*^*$ with $\langle \mathfrak{z},\mathfrak{a} \rangle \neq 0$ and $\mathfrak{b} \in V_*$ with $\langle \mathfrak{b},\mathfrak{w} \rangle \neq 0$. Now (2.52) implies

$$\begin{aligned} \|\eta(x,y)\|_\ell &= \|\langle \mathfrak{z},\mathfrak{w} \rangle \frac{\mathfrak{a} \mid E(x)}{\langle \mathfrak{a},\mathfrak{z} \rangle} \otimes \frac{\mathfrak{b} \mid E(y)}{\langle \mathfrak{b},\mathfrak{w} \rangle}\|_\ell \\ &= |\langle \mathfrak{z},\mathfrak{w} \rangle| \frac{\|\mathfrak{a} \mid E(x)\|_\ell}{|\langle \mathfrak{a},\mathfrak{z} \rangle|} \frac{\|\mathfrak{b} \mid E(y)\|_\ell}{|\langle \mathfrak{b},\mathfrak{w} \rangle|} \\ &= \frac{|\langle \mathfrak{z},\mathfrak{w} \rangle|}{\|\mathfrak{z}\| \, \|\mathfrak{w}\|} = \Box\, x,y \,\Box \end{aligned}$$

If $0 < s < r \in \mathbb{R}_\tau$ and $s \in \mathbb{R}_\tau$, then (3.7) implies

$$T_h(r,s,H,\ell) = N_h(r,s,\eta,H) + m_h(r,\eta,H,\ell) - m_h(s,\eta,H,L)$$

where

$$\begin{aligned} m_h(r,\eta,H,\ell) &= \int_{M\langle r \rangle} \log \frac{1}{\|\eta \circ h\|} \sigma \\ &= \int_{M\langle r \rangle} \log \frac{1}{\Box\, f,g \,\Box} \sigma = m_{f,g}(r) \end{aligned}$$

$$m_h(s,\eta,H,\ell) = m_{f,g}(s)$$

Also Lemma 2.3 implies $N_h(r,s,\eta,H) = N_{f,g}(r,s)$. We have

$$c(H,\ell) = f^*(c(\mathcal{O}(1),\ell)) + g^*(c(\mathcal{O}(1),\ell)) = f^*(\Omega_0) + g^*(\Omega_0)$$

$$T_h(r,s,H,\ell) = T_f(r,s) + T_g(r,s)$$

and we obtain

$$T_f(r,s) + T_g(r,s) = N_{f,g}(r,s) + m_{f,g}(r) - m_{f,g}(s)$$

which proves (3.40).

This proof is already known, see Shiffman [84].

§4 Associated Maps

Let M be a connected, complex manifold of dimension m. A biholomorphic map

$$\mathfrak{z} = (z^1, \ldots, z^m) : U \longrightarrow U' \tag{4.1}$$

of an open subset $U \neq \emptyset$ of M onto an open subset U' of $\mathbb{C}^m$ is called a **chart**. Define

$$\chi = \frac{i}{2\pi} \sum_{\mu=1}^{m} dz^\mu \wedge d\bar{z}^\mu \qquad \zeta = dz^1 \wedge \ldots \wedge dz^m \tag{4.2}$$

For $0 \leqslant q \in \mathbb{Z}$ abbreviate

$$i_q = \left(\frac{i}{2\pi}\right)^q (-1)^{\frac{q(q-1)}{2}} q! \tag{4.3}$$

Then

$$\chi^m = i_m \zeta \wedge \bar{\zeta} > 0. \tag{4.4}$$

Dependence on $\mathfrak{z}$ is indicated by an index $\mathfrak{z}$ as needed $U = U_{\mathfrak{z}}$, $\zeta = \zeta_{\mathfrak{z}}$, $\chi = \chi_{\mathfrak{z}}$ etc. Let $\{\mathfrak{z}_\lambda\}_{\lambda \in \Lambda}$ be a family of charts $\mathfrak{z}_\lambda : U_{\mathfrak{z}_\lambda} \longrightarrow U'_{\mathfrak{z}_\lambda}$ of M. Abbreviate $U_\lambda = U_{\mathfrak{z}_\lambda}$, $U'_\lambda = U'_{\mathfrak{z}_\lambda}$, $\chi_\lambda = \chi_{\mathfrak{z}_\lambda}$, $\zeta_\lambda = \zeta_{\mathfrak{z}_\lambda}$ and use the notation (2.25) – (2.27) for the family $\mathfrak{U} = \{U_\lambda\}_{\lambda \in \Lambda}$ of open subsets of M. Then $\{\mathfrak{z}_\lambda\}_{\lambda \in \Lambda}$ is called a

<u>chart atlas</u> of M if and only if $\mathfrak{U}$ is a covering of M. If $(\lambda,\mu) \in \Lambda[1]$, then there is a holomorphic function $\Delta_{\lambda\mu}$ without zeros on $U_{\lambda\mu}$ such that

(4.5) $$\zeta_\lambda = \Delta_{\lambda\mu}\zeta_\mu \qquad \Delta_{\lambda\mu}\Delta_{\mu\lambda} = 1 \quad \text{on } U_{\lambda\mu} .$$

Then $\Delta_{\lambda\lambda} = 1$ on U_λ and if $(\lambda,\mu,\rho) \in \Lambda[2]$, we have

(4.6) $$\Delta_{\lambda\mu}\Delta_{\mu\rho}\Delta_{\rho\lambda} = 1 \quad \text{on } U_{\lambda\mu\rho} .$$

Each ζ_λ is a holomorphic frame of the canonical bundle K of M and $(\Delta_{\lambda\mu})_{(\lambda,\mu)\in\Lambda[1]}$ is a basic cocycle of K.

Let B be a holomorphic form of bidegree $(m - 1,0)$ on M. If $\mathfrak{z} : U \longrightarrow U'$ is a chart, holomorphic functions B_μ exist uniquely on U such that

(4.7) $$B = \sum_{\mu=1}^{m} (-1)^{\mu-1} B_\mu dz^1 \wedge \ldots \wedge dz^{\mu-1} \wedge dz^{\mu+1} \wedge \ldots \wedge dz^m .$$

Let V be a complex vector space. Let $\mathfrak{v} : U \longrightarrow V$ be a holomorphic vector function. A holomorphic vector function $\mathfrak{v}' : U \longrightarrow V$, called the <u>**B-derivative of**</u> $\mathfrak{v}$, is uniquely defined by

(4.8) $$d\mathfrak{v} \wedge B = \mathfrak{v}'\zeta \qquad \mathfrak{v}' = \sum_{\mu=1}^{m} B_\mu \frac{\partial \mathfrak{v}}{\partial z^\mu} .$$

The operation can be iterated: $\mathfrak{v}^{(p)} = (\mathfrak{v}^{(p-1)})'$. Put $\mathfrak{v}^{(0)} = \mathfrak{v}$. Abbreviate

(4.9) $$\mathfrak{v}_{\underline{p}} = \mathfrak{v} \wedge \mathfrak{v}' \wedge \ldots \wedge \mathfrak{v}^{(p)} : U \longrightarrow \tilde{G}_p(V).$$

Dependence on $\mathfrak{z}$ is denoted by an index $\mathfrak{z}$ as $\mathfrak{v}' = \mathfrak{v}'_{\mathfrak{z}}$, $\mathfrak{v}_{\underline{p}} = \mathfrak{v}_{\mathfrak{z}\underline{p}}$ etc.

Let V be a complex vector space of dimension $n + 1 > 1$. Let $f : M \longrightarrow \mathbb{P}(V)$ be a meromorphic map. Then $\{\mathfrak{a}_\lambda, \mathfrak{w}_\lambda\}_{\lambda\in\Lambda}$ is said to be a **representation family** of f, if $\mathfrak{a}_\lambda : U_\lambda \longrightarrow U'_\lambda$ is a chart and $\mathfrak{w}_\lambda : U_\lambda \longrightarrow V$ is a reduced representation of f. If nothing else is said, we assume that each U_λ is connected, Stein and $H^2(U_\lambda, \mathbb{Z}) = 0$. Then U_λ is a Cousin II domain, each divisor on U_λ is principal and each meromorphic map has a reduced representation on U_λ. The representation family is said to be a **representation atlas** if $\mathfrak{U} = \{U_\lambda\}_{\lambda\in\Lambda}$ is a covering of M, which we will assume now. If $(\lambda,\mu) \in \Lambda[1]$, a holomorphic function $v_{\lambda\mu}$ without zeros is uniquely defined on $U_{\lambda\mu}$ by

$$(4.10) \qquad \mathfrak{w}_\lambda = v_{\lambda\mu}\mathfrak{w}_\mu \qquad v_{\lambda\mu}v_{\mu\lambda} = 1$$

on $U_{\lambda\mu}$. Then $v_{\lambda\lambda} = 1$ on U_λ for all $\lambda \in \Lambda$. If $(\lambda,\mu,\rho) \in \Lambda[2]$, then

$$(4.11) \qquad v_{\lambda\mu}v_{\mu\rho}v_{\rho\lambda} = 1 \quad \text{on } U_{\lambda\mu\rho}.$$

If $0 \leqslant p \in \mathbb{Z}$ and $(\lambda,\mu) \in \Lambda[1]$, then

$$(4.12) \qquad \mathfrak{w}_{\lambda\underline{p}} = (v_{\lambda\mu})^{p+1}(\Delta_{\mu\lambda})^{\frac{p(p+1)}{2}}\mathfrak{w}_{\mu\underline{p}}$$

on $U_{\lambda\mu}$. Let L_f be the hyperplane section bundle of f. Then $\mathfrak{w}^{\Delta}_\lambda$ is a holomorphic frame of L_f over U_λ. If $(\lambda,\mu) \in \Lambda[1]$, then

$$(4.13) \qquad \mathfrak{w}^{\Delta}_\mu = v_{\lambda\mu}\mathfrak{w}^{\Delta}_\lambda \quad \text{on } U_{\lambda\mu}.$$

These transformation formulas imply the existence and uniqueness of a holomorphic section $F_p = F_{pf}$, called the pth **representation section of** f, of the holomorphic vector bundle

(4.14) $$L_f[p] = (\bigwedge_{p+1} V_M) \otimes (L_f)^{p+1} \otimes K^{\frac{p(p+1)}{2}}$$

such that

(4.15) $$SF_p | U_\lambda = \mathfrak{v}_{\lambda p} \otimes (\mathfrak{v}_\lambda^\Delta)^{p+1} \otimes \zeta_\lambda^{\frac{p(p+1)}{2}}$$

for all $\lambda \in \Lambda$. Here $F_0 = F \not\equiv 0$ but $F_p \equiv 0$ if $p > n$. If $F_p \equiv 0$, then $F_{p+1} \equiv 0$. Hence $\ell_f \in \mathbb{Z}[0,n]$ exists uniquely such that $F_p \not\equiv 0$ if $0 \leqslant p \leqslant \ell_f$ and $F_p \equiv 0$ if $p > \ell_f$. We call ℓ_f the **generality index** of f for B and f is said to be **general of order** p **for** B if $0 \leqslant p \leqslant \ell_f$. If M admits m analytically independent holomorphic functions, then for any finite sets of meromorphic maps defined on M, there exists a holomorphic form B of degree $m - 1$ on M such that the generality index of each of these maps f for B equals the dimension of the smallest projective plane containing the image of f. (See Stoll [100], Theorem 7.11.) The map f is said to be **general for** B if $\ell_f = n$.

For each $p \in \mathbb{Z}[0,\ell_f]$, the pth **associated map** $f_p = \mathbb{P} \circ F_p : M \longrightarrow G_p(V)$ of f is defined. Here f_p is meromorphic and $\mathfrak{v}_{\lambda p} : U_\lambda \longrightarrow \tilde{G}_p(V)$ is a representation of f_p for each $\lambda \in \Lambda$ and (4.15) implies

(4.16) $$\mu_{F_p} | U_\lambda = \mu_{\mathfrak{v}_{\lambda p}} \qquad \text{for all } \lambda \in \Lambda .$$

In particular, $\mathfrak{v}_{\lambda p}$ may not be reduced. Define $L_f[-1] = M \times \mathbb{C} = \mathbb{C}_M$ and let F_{-1} be the trivial section defined by $F_{-1}(z) = (z,1)$. Then $\mu_{F_{-1}} = 0$. For $0 \leqslant p \leqslant \ell_f$, the pth **stationary divisor**

(4.17) $$\ell_{f_p} = \mu_{F_{p-1}} - 2\mu_{F_p} + \mu_{F_{p+1}} \geq 0$$

is non-negative (Stoll [93], [100]).

Take $p \in \mathbb{Z}[0,n]$. Let $\varphi : M \longrightarrow G_p(V)$ be a meromorphic map. A non-negative form of class C^∞ is given on $M - I_\varphi$ by

(4.18) $$\mathbb{H}(\varphi) = m i_{m-1}\, \varphi^*(\Omega_p) \wedge B \wedge \bar{B} \geq 0\ .$$

If $\mathfrak{z} : U_{\mathfrak{z}} \longrightarrow U'_{\mathfrak{z}}$ is a chart, a non-negative function $H_{\mathfrak{z}}(\varphi)$ is defined on $U_{\mathfrak{z}} - I_\varphi$ by

(4.19) $$\mathbb{H}(\varphi) = H_{\mathfrak{z}}(\varphi)^2\, \chi_{\mathfrak{z}}^m$$

where $H_{\mathfrak{z}}(\varphi)^2$ is of class C^∞ on $U_{\mathfrak{z}} - I_f$. If $\lambda \in \Lambda$ abbreviate $H_\lambda(\varphi) = H_{\mathfrak{z}_\lambda}(\varphi)$. Returning to our given meromorphic map $f : M \longrightarrow \mathbb{P}(V)$ and its associated maps $f_p : M \longrightarrow G_p(V)$ for $p \in \mathbb{Z}[0, \ell_f]$ we abbreviate $\mathbb{H}_p = \mathbb{H}(f_p)$, $H_{p\mathfrak{z}} = H_{\mathfrak{z}}(f_p)$, $H_{p\lambda} = H_{\mathfrak{z}_\lambda}(f_p)$. Then

(4.20) $$\mathbb{H}_p = m i_{m-1}\, f_p^*(\Omega_p) \wedge B \wedge \bar{B}$$

(4.21) $$\mathbb{H}_p | U_\lambda = H_{p\lambda}^2\, \chi_\lambda^m$$

(4.22) $$H_{p\lambda} = \frac{\| \mathfrak{w}_{\lambda \underline{p-1}} \|\ \| \mathfrak{w}_{\lambda \underline{p+1}} \|}{\| \mathfrak{w}_{\lambda \underline{p}} \|^2} \geq 0$$

where $H_{p\lambda} > 0$ on the complement of the thin analytic subset

(4.23) $\quad \tilde{I}_p = \tilde{I}_p(f) = I_{f_{p-1}} \cup I_{f_p} \cup I_{f_{p+1}} \cup \operatorname{supp}(\mu_{F_{p-1}} + \mu_{F_p} + \mu_{F_{p+1}})$

Define $\mathbb{H}_p = 0$ if $p < 0$ or if $p > \ell_f$. Also (4.22) implies $\mathbb{H}_p = 0$ if $p = \ell_f$. For $0 \leq p < \ell_f$ we obtain on $M - \tilde{I}_f$ the identity

(4.24) $$\operatorname{Ric} \mathbb{H}_p = f^*_{p-1}(\Omega_{p-1}) - 2f^*_p(\Omega_p) + f^*_{p+1}(\Omega_{p+1})$$

(4.25) $$m i_{m-1} \operatorname{Ric} \mathbb{H}_p \wedge B \wedge \bar{B} = \mathbb{H}_{p-1} - 2\mathbb{H}_p + \mathbb{H}_{p+1} .$$

Let (M,τ) be a parabolic manifold of dimension m. The open set

(4.26) $$M^+ = \{x \in M \mid \upsilon(x) > 0\} = \{x \in M \mid \upsilon(x)^m > 0\}$$

is not empty. Let Ψ be a positive form of degree $2m$ and class C^∞ on M. The <u>Ricci function</u> of Ψ is defined by

(4.27) $$\operatorname{Ric}(r,s,\Psi) = \int_s^r \int_{M[t]} (\operatorname{Ric} \Psi) \wedge \upsilon^{m-1} \frac{dt}{t^{2m-1}}$$

for $0 < s < r$. On M, a non-negative function v of class C^∞ is defined by

(4.28) $$\upsilon^m = v\Psi .$$

Define

(4.29) $$\mathfrak{L}^0_\tau = \{r \in \mathfrak{L}_\tau \mid (\log v)\sigma \text{ is integrable over } M\langle r\rangle\} .$$

Then $\mathfrak{L}^0_\tau$ does not depend on the choice of Ψ and $\mathbb{R}^+ - \mathfrak{L}^0_\tau$ has measure zero. For $0 < s < r \in \mathfrak{L}^0_\tau$ with $s \in \mathfrak{L}^0_\tau$ the <u>Ricci function of</u> τ is defined by

$$(4.30)\qquad \mathrm{Ric}_\tau(r,s) = \frac{1}{2}\int_{M\langle r\rangle} \log v\sigma - \frac{1}{2}\int_{M\langle s\rangle} \log v\sigma + \mathrm{Ric}(r,s,\Psi)$$

does not depend on the choice of Ψ. In particular if (M,τ) is a parabolic covering manifold of $\mathbb{C}^m$, if $\pi : M \longrightarrow \mathbb{C}^m$ is a proper, surjective holomorphic map such that $\tau = \|\pi\|^2$ and if β is the **branching divisor** of π, then $\mathrm{Ric}_\tau(r,s) = N_\beta(r,s) \geq 0$. If $\pi = (\pi_1, \ldots, \pi_m)$, then β is the divisor of the holomorphic form $d\pi_1 \wedge \ldots \wedge d\pi_m \not\equiv 0$ on M. (See Stoll [103].)

Take $p \in \mathbb{Z}[0,n]$. Let $\varphi : M \longrightarrow G_p(V)$ be a meromorphic map. On M^+, a non-negative function $h(\varphi)$ is defined by

$$(4.31)\qquad \mathbb{H}(\varphi) = h(\varphi)^2\, \upsilon^m .$$

The function $h(\varphi)^2$ is of class C^∞ on M^+. For $r \in \mathfrak{L}_\tau$ define

$$(4.32)\qquad S_\varphi(r) = \int_{M\langle r\rangle} (\log h(\varphi))\sigma$$

provided the integral exists.

Returning to our given meromorphic map $f : M \longrightarrow \mathbb{P}(V)$ and its associated maps $f_p : M \longrightarrow G_p(V)$ for $p \in \mathbb{Z}[0,\ell_f]$, we abbreviate $h_p = h(f_p)$. Then $\mathbb{H}_p = h_p^2\, \upsilon^m$ on M^+. For $0 \leq p < \ell_f$ and $r \in \mathfrak{L}_\tau^0$ the integral $S_{f_p}(r)$ exists. For $0 < s < r \in \mathfrak{L}_\tau^0$ with $s \in \mathfrak{L}_\tau^0$ we have the **Plücker Difference Formula**

$$(4.33)\qquad N_{f_p}(r,s) + T_{f_{p-1}}(r,s) - 2T_{f_p}(r,s) + T_{f_{p+1}}(r,s)$$

$$= S_{f_p}(r) - S_{f_p}(s) + \mathrm{Ric}_\tau(r,s)$$

where $T_{f_q}(r,s) = 0$ if $q < 0$ or $q \geq \ell_f$. (See Stoll [93] Satz 15.3 and Stoll [108] Theorem 7.6.)

Let $f : M \longrightarrow \mathbb{P}(V)$ and $g : M \longrightarrow \mathbb{P}(V^*)$ be meromorphic maps. Then $(\mathfrak{z}_\lambda, \mathfrak{v}_\lambda, \mathfrak{w}_\lambda)_{\lambda\in\Lambda}$ is called a <u>representation atlas of the pair (f,g)</u> if $(\mathfrak{z}_\lambda, \mathfrak{v}_\lambda)_{\lambda\in\Lambda}$ is a representation atlas of f and $(\mathfrak{z}_\lambda, \mathfrak{w}_\lambda)_{\lambda\in\Lambda}$ is a representation atlas of g. For each $(\lambda,\mu) \in \Lambda[1]$ there are holomorphic functions $\Delta_{\lambda\mu}$, $v_{\lambda\mu}$ and $w_{\lambda\mu}$ without zeros on $U_{\lambda\mu}$ such that (4.5), (4.6). (4.10), (4.11), (4.12), (4.13), (4.14), (4.15) hold and such that

$$\mathfrak{w}_\lambda = w_{\lambda\mu}\mathfrak{w}_\mu \qquad \mathfrak{w}_\mu^\Delta = w_{\lambda\mu}\mathfrak{w}_\lambda^\Delta \qquad w_{\lambda\mu}w_{\mu\lambda} = 1 \tag{4.34}$$

$$\mathfrak{w}_{\lambda p} = (w_{\lambda\mu})^{p+1}(\Delta_{\mu\lambda})^{\frac{p(p+1)}{2}}\mathfrak{w}_{\mu p} \tag{4.35}$$

hold on $U_{\lambda\mu}$. If $\lambda \in \Lambda$, then $w_{\lambda\lambda} = 1$ on U_λ. If $(\lambda,\mu,\rho) \in \Lambda[2]$,

$$w_{\lambda\mu}w_{\mu\rho}w_{\rho\lambda} = 1 \quad \text{on } U_{\lambda\mu\rho}. \tag{4.36}$$

The pth representation section F_{pg} of g is abbreviated to $F_{pg} = G_p$ and is a section in

$$L_g[p] = \left(\bigwedge_{p+1} V_M^*\right) \otimes (L_g)^{p+1} \otimes K^{\frac{p(p+1)}{2}} \tag{4.37}$$

$$G_p | U = \mathfrak{w}_{\lambda p} \otimes (\mathfrak{w}_\lambda^\Delta)^{p+1} \otimes \zeta_\lambda^{\frac{p(p+1)}{2}} \tag{4.38}$$

if $\lambda \in \Lambda$. Also we abbreviate

$$\mathbb{K}_p = \mathbb{H}(g_p) \qquad K_{p\lambda} = H_\lambda(g_p) \qquad k_p = h(g_p). \tag{4.39}$$

Then $\mathbb{K}_p = k_p^2 \upsilon^m$ on M^+. For $p \in \mathbb{Z}[0,\ell_g]$ and $q \in \mathbb{Z}[0,\ell_f]$ a non-negative form $\mathbb{K}_{pq}$ of degree $2m$ is defined on $M - (I_{f_q} \cup I_{g_p})$ by

$$\mathbb{K}_{pq} \mid U_\lambda = H_{q\lambda} K_{p\lambda} \chi_\lambda^m \tag{4.40}$$

for all $\lambda \in \Lambda$. Then

$$0 \leqslant \mathbb{K}_{pq} = h_q k_p \upsilon^m \leqslant \tfrac{1}{2}(\mathbb{H}_q + \mathbb{K}_p) \tag{4.41}$$

Take p and q in $\mathbb{Z}[0,n]$. Define $\mu = \mathrm{Min}(p+1,q+1)$ and $t = \frac{1}{2}(p(p+1) + q(q+1))$. Take $\rho \in \mathbb{Z}[0,\mu]$. Then $a = p + 1 - \rho \geqslant 0$ and $b = q + 1 - \rho \geqslant 0$. Define the holomorphic vector bundle

$$L^\rho_{f,g}[q,p] = (\textstyle\bigwedge_b V_M) \otimes (\bigwedge_a V^*_M) \otimes L_f^{q+1} \otimes L_g^{p+1} \otimes K^t \tag{4.42}$$

Considering (4.14) and (4.37) a contraction

$$\boxplus^\rho : L_f[q] \otimes L_g[p] \longrightarrow L^\rho_{f,g}[q,p] \tag{4.43}$$

is defined by $\boxplus^\rho = \boxplus^\rho \otimes \mathrm{Id}$ and the proper commutation of terms. Hence we have a holomorphic section $F_q \boxplus^\rho G_p = \boxplus^\rho F_q \otimes G_p$ of $L^\rho_{f,g}[q,p]$. The pair f,g is said to be <u>**free of order**</u> (q,p,ρ) if $F_q \boxplus^\rho G_p \not\equiv 0$. If $0 \leqslant q \leqslant \ell_f$ and $0 \leqslant p \leqslant \ell_g$, then f,g is free of order $(q,p,0)$. The pair f,g is said to be <u>**free of order**</u> (q,p) if it is free of order $(q,p,1)$. The pair f,g is said to be <u>**strictly free of order**</u> (q,p) if it is free of order (q,p,μ). We have $F_q \boxplus^0 G_p = F_q \otimes G_p$. We write $\boxplus^1 = \boxplus$; that is, $F_q \boxplus^1 G_p = F_q \boxplus G_p$. If $\rho = q + 1 \leqslant p + 1$ we write $F_q \boxplus^{q+1} G_p = G_p \llcorner F_q$. If $\rho = p + 1 \leqslant q + 1$, we write

$F_q \boxplus^{p+1} G_p = F_q \llcorner G_p$. If $\rho = q+1 = p+1$, we write $F_q \boxplus^{p+1} G_p = \langle F_q, G_p \rangle$, which is a section in a line bundle. The sections F_q, G_p and $F_q \boxplus^\rho G_p$ are not to be mistaken for the sections F_{f_q}, F_{g_p} and $F_{f_g} \boxplus^\rho G_{g_p}$. The following lemma clarifies the situation.

LEMMA 4.1. _Take_ $q \in \mathbb{Z}[0,\ell_f]$ _and_ $p \in \mathbb{Z}[0,\ell_g]$. _Let_ ρ _be an integer with_ $0 \leq \rho \leq \mathrm{Min}(p+1,q+1)$. _Then the pair_ (f,g) _is free of order_ (p,q,ρ) _if and only if the pair_ (f_q,g_p) _is free of order_ ρ. _If so, then_

$$\mu_{F_q \boxplus^\rho G_p} = \mu_{F_q} + \mu_{G_q} + \mu_{f_q \dot{\boxplus}^\rho g_p} . \tag{4.44}$$

PROOF. Take $\lambda \in \Lambda$. Since U_λ is a Cousin II domain, there are reduced representations $\mathfrak{g}_\lambda$ of f_q on U_λ and $\mathfrak{h}_\lambda$ of g_p on U_λ such that $\mathfrak{w}_{\lambda q} = P_\lambda \mathfrak{g}_\lambda$ and $\mathfrak{v}_{\lambda p} = Q_\lambda \mathfrak{h}_\lambda$. Then $\mu_{F_q} | U_\lambda = \mu^0_{P_\lambda}$ and $\mu_{G_p} | U_\lambda = \mu^0_{Q_\lambda}$. Put $t = \frac{1}{2}(p(p+1) + q(q+1))$. We have

$$F_q \boxplus^\rho G_p = (\mathfrak{w}_{\lambda \underline{q}} \boxplus^\rho \mathfrak{v}_{\lambda \underline{p}}) \otimes (\mathfrak{w}^\Delta_\lambda)^{q+1} \otimes (\mathfrak{v}^\Delta_\lambda)^{p+1} \otimes \zeta^t_\lambda$$

$$= P_\lambda Q_\lambda (\mathfrak{g}_\lambda \boxplus^\rho \mathfrak{h}_\lambda) \otimes (\mathfrak{w}^\Delta_\lambda)^{q+1} \otimes (\mathfrak{v}^\Delta_\lambda)^{p+1} \otimes \zeta^t_\lambda$$

$$F_{f_q} \boxplus^\rho G_{g_p} = (\mathfrak{g}_\lambda \boxplus^\rho \mathfrak{h}_\lambda) \otimes \mathfrak{g}^\Delta_\lambda \otimes \mathfrak{h}^\Delta_\lambda$$

Hence $F_q \boxplus^\rho G_p \not\equiv 0$ if and only if $F_{f_q} \boxplus^\rho G_{g_p} \not\equiv 0$. If so, we have

$$\mu_{F_q \boxplus^\rho G_p} = \mu^0_{P_\lambda} + \mu^0_{Q_\lambda} + \mu_{\mathfrak{g}_\lambda \boxplus^\rho \mathfrak{h}_\lambda} = \mu_{F_q} + \mu_{G_p} + \mu_{f_q \dot{\boxplus}^\rho g_p}$$

on U_λ . q.e.d.

LEMMA 4.2. Take $p \in \mathbb{Z}[0,\ell_g]$. Then the pair (f,g) is free of order $(0,p)$ if and only if the pair (f,g) is free of order (q,p) for all $q \in \mathbb{Z}[0,\ell_f]$.

PROOF. Let $\mathfrak{w} : U \longrightarrow V$ and $\mathfrak{v} : U \longrightarrow V^*$ be reduced representations of f and g respectively. Then (f,g) is free of order (q,p) if $\mathfrak{w}_q \boxplus \mathfrak{v}_p \not\equiv 0$. By Lemma 1.2 we have

$$\mu^2 \| \mathfrak{w}_q \boxplus \mathfrak{v}_p \|^2 = \| \mathfrak{w}_q \|^2 \, \| \mathfrak{v}_p \|^2 \sum_{j=0}^{q} \| \mathfrak{v}_p \llcorner \mathfrak{w}^{(j)} \|^2$$

Hence if (f,g) is free of order $(0,p)$, then $\| \mathfrak{v}_p \llcorner \mathfrak{w} \| > 0$ on an open subset of U and $\mathfrak{w}_q \boxplus \mathfrak{v}_p \not\equiv 0$. Hence (f,g) is free of order (q,p). If (f,g) is free of order (q,p) for all $q \in \mathbb{Z}[0,\ell_f]$, then (f,g) is free of order $(0,p)$. q.e.d

Let Ω_{pq} be the Fubini Study Kähler form on $\mathbb{P}(\bigwedge_{q+1} V \otimes \bigwedge_{p+1} V^*)$. Then $\Omega_q = \Omega_{q,(-1)}$ on $\mathbb{P}(\bigwedge_{q+1} V)$ and $\Omega_p^* = \Omega_{(-1),p}$ on $\mathbb{P}(\bigwedge_{p+1} V^*)$.

Define

$$(4.45) \qquad \mathbb{H}^{\rho}_{pq} = \mathbb{H}(f_q \boxplus^{\rho} g_p) = m i_{m-1} (f_q \boxplus^{\rho} g_p)^*(\Omega_{q-\rho,p-\rho}) \wedge B \wedge \bar{B}$$

Define $H^{\rho}_{pq\lambda} = H_{\lambda}(f_q \boxplus^{\rho} g_p)$ and $h^{\rho}_{pq} = h(f_q \boxplus^{\rho} g_p)$. Then

$$(4.46) \qquad \mathbb{H}^{\rho}_{pq} | U_{\lambda} = (H^{\rho}_{pq\lambda})^2 \, \chi^m_{\lambda} \qquad \mathbb{H}^{\rho}_{pq} = (h^{\rho}_{pq})^2 \, \upsilon^m$$

For $\rho = 1$, write $\mathbb{H}^1_{pq} = \mathbb{H}_{pq}$, $\mathbb{H}^1_{pq\lambda} = H_{pq\lambda}$, $h^1_{pq} = h_{pq}$.
If $p \leqslant q$, then

(4.47) $$\mathbb{H}_{pq}^{p+1} = mi_{m-1}(f_q \mathsf{L} g_p)^*(\Omega_{q-p-1}) \wedge B \wedge \bar{B}$$

If $q \leqslant p$, then

(4.48) $$\mathbb{H}_{pq}^{q+1} = mi_{m-1}(g_p \mathsf{L} f_q)^*(\Omega_{p-q-1}) \wedge B \wedge \bar{B}$$

Take $q \in \mathbb{Z}[0,\ell_f]$ and $p \in \mathbb{Z}[0,\ell_g]$. Define $\mu = \mathrm{Min}(p+1,q+1)$. Take $\rho \in \mathbb{Z}[0,\mu]$. Define

(4.49) $$\Phi_{pq}^{(\rho)} = \binom{\mu}{\rho}^2 \,\square\, f_q \,\dot{\boxplus}^{\rho}\, g_p \,\square^2$$

(4.50) $$\Phi_{pq} = \Phi_{pq}^{(1)} = \mu^2 \,\square\, f_q \,\dot{\boxplus}\, g_p \,\square^2$$

(4.51) $$\Phi_q = \Phi_{0q} = \square\, f_q \,\dot{\boxplus}\, g \,\square^2 = \square\, f_q \dot{\mathsf{L}} g \,\square^2$$

(4.52) $$\Psi_p = \Phi_{p0} = \square\, f \,\dot{\boxplus}\, g_p \,\square^2 = \square\, g_p \dot{\mathsf{L}} f \,\square^2 .$$

Then Lemma 1.8 implies

(4.53) $$0 \leqslant \Phi_{pq}^{(\rho)} \leqslant \binom{\mu}{\rho} \qquad 0 \leqslant \Phi_{pq} \leqslant \mu$$

(4.54) $$0 \leqslant \Phi_q \leqslant 1 \qquad 0 \leqslant \Psi_p \leqslant 1$$

(4.55) $$\Phi_{pn} = p + 1 \qquad \Phi_{nq} = q + 1$$

If $p \leqslant q$, then $\Phi_{pq}^{(p+1)} = \square\, f_q \dot{\mathsf{L}} g_p \,\square^2$. If $q \leqslant p$, then $\Phi_{pq}^{(q+1)} = \square\, g_p \dot{\mathsf{L}} f_q \,\square^2$.

§5 Frenet Frames

Our theory could be carried out without Frenet frames. However Frenet frames provide a deeper geometric understanding of the theory and lead to new global invariants which describe the relative movement of two Frenet frames. Some of these invariants will appear in the Second Main Theorem for moving targets. Frenet frames and Frenet formulas on manifolds were introduced in [100]. For completeness sake the facts and proofs shall be given here.

Let M be a connected, complex manifold of dimension m. Let $B \not\equiv 0$ be a holomorphic form of bidegree $(m-1,0)$ on M. Let V be a hermitian vector space of dimension $n+1>1$. Let $f : M \longrightarrow \mathbb{P}(V)$ be a meromorphic map. Abbreviate $\ell_f = s$. Then $0 \leqslant s \leqslant n$. Let $I_p = Z(F_p)$ be the zero set of the pth representation section F_p of f. Then $I_0 = I_f$ and $I_{f_p} \subseteq I_p \subseteq I_{p+1}$ for $0 \leqslant p \leqslant s$. If $0 \leqslant p \leqslant s$, then $I_p \neq M$; if $p > s$, then $I_p = M$.

Let $(\mathfrak{v}_\lambda, \mathfrak{w}_\lambda)_{\lambda \in \Lambda}$ be a representation atlas of f. On occasion, if no confusion is to be feared, we will omit the index λ and write $\mathfrak{w}_{\lambda \underline{p}} = \mathfrak{w}_{\underline{p}}$ etc.

LEMMA 5.1. *If* $p \in \mathbb{Z}[1,s]$ *and* $z \in M - I_p$, *then*

$$E(f_{p-1}(z)) \subset E(f_p(z)) . \tag{5.1}$$

PROOF. Take $\lambda \in \Lambda$ with $z \in U_\lambda$. The identity

$\mathfrak{w}_{\lambda,\underline{p-1}}(z) \wedge \mathfrak{w}_\lambda^{(p)}(z) = \mathfrak{w}_{\lambda\underline{p}}(z) \neq 0$ implies

$$E(f_{p-1}(z)) = \{\mathfrak{x} \in V \mid \mathfrak{x} \wedge \mathfrak{w}_{\lambda\underline{p-1}}(z) = 0\}$$

$$\subseteq \{\mathfrak{x} \in V \mid \mathfrak{x} \wedge \mathfrak{w}_{\lambda\underline{p}}(z) = 0\} = E(f_p(z)) . \qquad \text{q.e.d.}$$

For $z \in M - I_s$ we have the **associated flag**

$$E(f(z)) \subset E(f_1(z)) \subset \dots \subset E(F_s(z)). \tag{5.2}$$

LEMMA 5.2. *Take $\lambda \in \Lambda$. For $p = 0,1, \dots ,s$ vector functions*

$$\mathfrak{e}_p^\lambda : U_\lambda - I_p \longrightarrow V \tag{5.3}$$

of class C^∞ are uniquely determined by

$$\|\mathfrak{e}_p^\lambda\| = 1 \qquad (\mathfrak{e}_p^\lambda \mid \mathfrak{e}_q^\lambda) = 0 \qquad \text{if} \quad 0 \leqslant q < p \leqslant s \tag{5.4}$$

$$\mathfrak{w}_{\lambda\underline{p}} = \|\mathfrak{w}_{\lambda\underline{p}}\| \, \mathfrak{e}_0^\lambda \wedge \dots \wedge \mathfrak{e}_p^\lambda . \tag{5.5}$$

PROOF. Define $\mathfrak{e}_0 = \mathfrak{w}/\|\mathfrak{w}\|$ on $U_\lambda - I_0$. Assume that $\mathfrak{e}_0, \dots ,\mathfrak{e}_{p-1}$ are already constructed where $p \leqslant s$. On $U_\lambda - I_p$ we have

$$\mathfrak{w}_{\underline{p}} = \|\mathfrak{w}_{\underline{p-1}}\| \mathfrak{e} \wedge \dots \wedge \mathfrak{e}_{p-1} \wedge \mathfrak{w}^{(p)} \neq 0 .$$

Hence

$$\tilde{\mathfrak{e}}_p = \mathfrak{w}^{(p)} - \sum_{j=0}^{p-1} (\mathfrak{w}^{(p)} \mid \mathfrak{e}_j)\mathfrak{e}_j \neq 0$$

on $U - I_p$. Then $\mathfrak{e}_p = \tilde{\mathfrak{e}}_p/\|\tilde{\mathfrak{e}}_p\|$ is of class C^∞ on $U - I_p$ with $\|\mathfrak{e}_p\| = 1$. If $0 \leq q < p$, then $(\mathfrak{e}_p | \mathfrak{e}_q) = 0$. Also

$$\mathfrak{w}_{\underline{p}} = \|\mathfrak{w}_{\underline{p-1}}\|\mathfrak{e}_0 \wedge \ldots \wedge \mathfrak{e}_{p-1} \wedge \mathfrak{w}^{(p)}$$

$$= \|\mathfrak{w}_{p-1}\| \; \|\tilde{\mathfrak{e}}_p\|\mathfrak{e}_0 \wedge \ldots \wedge \mathfrak{e}_p$$

where $\|\mathfrak{e}_0 \wedge \ldots \wedge \mathfrak{e}_p\| = \|\mathfrak{e}_0\| \ldots \|\mathfrak{e}_p\| = 1$. Hence $\|\mathfrak{w}_{\underline{p}}\| = \|\mathfrak{w}_{\underline{p-1}}\| \; \|\tilde{\mathfrak{e}}_p\|$. We have $\mathfrak{w}_{\underline{p}} = \|\mathfrak{w}_{\underline{p}}\|\mathfrak{e}_0 \wedge \ldots \wedge \mathfrak{e}_p$.

Thus a solution is constructed.

Suppose that $\hat{\mathfrak{e}}_0, \ldots, \hat{\mathfrak{e}}_p$ is another solution. By (5.5) $\hat{\mathfrak{e}}_0 = \mathfrak{e}_0$. Assume that $\hat{\mathfrak{e}}_j = \mathfrak{e}_j$ for $j = 0,1, \ldots, p - 1 < s$ is already proved. Take $z \in U - I_p$. Then $\hat{\mathfrak{e}}_p(z) \in E(f_p(z))$ is perpendicular to $E(f_{p-1}(z))$ which is spanned by $\hat{\mathfrak{e}}_0(z) = \mathfrak{e}_0(z), \ldots, \hat{\mathfrak{e}}_{p-1}(z) = \mathfrak{e}_{p-1}(z)$ and the same is true for $\mathfrak{e}_p(z)$. Therefore $\hat{\mathfrak{e}}_p(z) = a\mathfrak{e}_p(z)$ with $|a| = 1$. Also

$$a\|\mathfrak{w}_{\underline{p}}(z)\|\mathfrak{e}_0 \wedge \ldots \wedge \mathfrak{e}_p(z) = \|\mathfrak{w}_{\underline{p}}(z)\|\hat{\mathfrak{e}}_0(z) \wedge \ldots \wedge \hat{\mathfrak{e}}_p(z)$$

$$= \mathfrak{w}_{\underline{p}}(z) = \|\mathfrak{w}_{\underline{p}}(z)\|\mathfrak{e}_0(z) \wedge \ldots \wedge \mathfrak{e}_p(z)$$

Hence $a = 1$ and $\hat{\mathfrak{e}}_p(z) = \mathfrak{e}_p(z)$. q.e.d

The orthonormal frame $\mathfrak{e}_0^\lambda, \ldots, \mathfrak{e}_s^\lambda$ is called the <u>**Frenet Frame**</u> of f for $\lambda \in \Lambda$. For $z \in U_\lambda - I_s$, the system $\mathfrak{e}_0^\lambda(z), \ldots, \mathfrak{e}_s^\lambda(z)$ is an orthonormal flag base of the associated flag (5.2), but this condition does not define the flag base uniquely. We shall use the notations of Chapter 4.

<u>LEMMA</u> 5.3. <u>If</u> $(\lambda,\mu) \in \Lambda[1]$, <u>then</u>

$$\mathfrak{e}_p^\lambda = \frac{v_{\lambda\mu}}{|v_{\lambda\mu}|}\left[\frac{\Delta_{\mu\lambda}}{|\Delta_{\mu\lambda}|}\right]^p \mathfrak{e}_p^\mu \quad \text{on } U_{\lambda\mu} - I_p\,. \tag{5.6}$$

<u>PROOF</u>. For each $z \in U_{\lambda\mu} - I_p$ both $\mathfrak{e}_0^\lambda(z), \ldots, \mathfrak{e}_p^\lambda(z)$ and $\mathfrak{e}_0^\mu(z), \ldots, \mathfrak{e}_p^\mu(z)$ are orthonormal flag bases of the same flag. Therefore functions A_q of class C^∞ exist on $U_{\lambda\mu} - I_p$ such that $\mathfrak{e}_q^\lambda = A_q \mathfrak{e}_q^\mu$. Now (5.5) and (4.12) imply

$$A_0 \cdots A_p\, \mathfrak{e}_0^\mu \wedge \cdots \wedge \mathfrak{e}_p^\mu = \mathfrak{e}_0^\lambda \wedge \cdots \wedge \mathfrak{e}_p^\lambda = \frac{\mathfrak{w}_{\lambda\underline{p}}}{\|\mathfrak{w}_{\lambda\underline{p}}\|}$$

$$= \left[\frac{v_{\lambda\mu}}{|v_{\lambda\mu}|}\right]^{p+1} \left[\frac{\Delta_{\mu\lambda}}{|\Delta_{\mu\lambda}|}\right]^{\frac{p(p+1)}{2}} \frac{\mathfrak{w}_{\mu\underline{p}}}{\|\mathfrak{w}_{\mu p}\|}$$

$$= \left[\frac{v_{\lambda\mu}}{|v_{\lambda\mu}|}\right]^{p+1} \left[\frac{\Delta_{\mu\lambda}}{|\Delta_{\mu\lambda}|}\right]^{\frac{p(p+1)}{2}} \mathfrak{e}_0^\mu \wedge \cdots \wedge \mathfrak{e}_p^\mu$$

or

$$A_0 \cdots A_p = \left[\frac{v_{\lambda\mu}}{|v_{\lambda\mu}|}\right]^{p+1} \left[\frac{\Delta_{\mu\lambda}}{|\Delta_{\mu\lambda}|}\right]^{\frac{p(p+1)}{2}}$$

If we replace p by $p - 1$ and divide, we obtain (5.6). q.e.d.

Our operations on vector spaces extend easily to vector valued forms. Where confusion may occur, we indicate operations between forms by a dot.

Take $\lambda \in \Lambda$ and take integers p,q in $\mathbb{Z}[0,s]$. A form θ_{pq}^λ of degree 1 and class C^∞ is defined on $U_\lambda - (I_p \cup I_p)$ by

$$\theta_{pq}^\lambda = (d\mathfrak{e}_p^\lambda \,|\, \mathfrak{e}_q^\lambda) \tag{5.7}$$

If $p \in \mathbb{Z} - \mathbb{Z}[0,s]$ or $q \in \mathbb{Z} - \mathbb{Z}[0,s]$ define $\theta^\lambda_{pq} = 0$. On $U_\lambda - I_s$ a vector valued form of degree 1 and class C^∞ is defined by

(5.8) $$\mathfrak{v}^\lambda_p = d\mathfrak{t}^\lambda_p - \sum_{q=0}^{s} \theta^\lambda_{pq}\, \mathfrak{t}^\lambda_q$$

A number of relations between these forms will be proved. Our identities will be formulated and proved on $U_\lambda - I_s$ only, although they may hold on larger sets. If ω is a form of degree 1, let $\dot{\omega}$ be the component of bidegree (1,0) and let $\ddot{\omega}$ be the component of bidegree (0,1). If α is a form of bidegree (p,0) and if β is a form of bidegree (0,q) such that $\alpha \wedge \beta = 0$, then $\alpha = 0$ or $\beta = 0$.

THEOREM 5.4. *Frenet Identities. Take $\lambda \in \Lambda$. Then we have*

(5.9) $$\theta^\lambda_{pq} + \bar{\theta}^\lambda_{qp} = 0$$

(5.10) $$\dot{\theta}^\lambda_{pq} = -\overline{\ddot{\theta}^\lambda_{qp}} \qquad \ddot{\theta}^\lambda_{pq} = -\overline{\dot{\theta}^\lambda_{qp}}$$

(5.11) $$\theta^\lambda_{pq} \wedge B = 0 \qquad \text{if } q > p+1$$

(5.12) $$\theta^\lambda_{pq} \wedge \bar{B} = 0 \qquad \text{if } q < p-1$$

(5.13) $$\dot{\theta}^\lambda_{pq} = 0 \qquad \text{if } q \leqslant p-1$$

(5.14) $$\ddot{\theta}^\lambda_{pq} = 0 \qquad \text{if } q \geqslant p+1$$

(5.15) $$\theta^\lambda_{pq} \wedge B \wedge \bar{B} = 0 \qquad \text{if } |p-q| \geqslant 2$$

(5.16) $$(\mathfrak{v}^\lambda_p \mid \mathfrak{t}^\lambda_q) = 0 \qquad \text{if } 0 \leqslant p \leqslant s \text{ and } 0 \leqslant q \leqslant s$$

(5.17) $$\mathfrak{v}_p^\lambda = 0 \qquad \text{for } p = 0, \ldots, n \text{ if } s = n$$

(5.18) $$\mathfrak{v}_p^\lambda \wedge B = 0 \qquad \text{if } 0 \leqslant p \leqslant s$$

(5.19) $$\dot{\mathfrak{v}}_p^\lambda \wedge B = 0 = \ddot{\mathfrak{v}}_p^\lambda \qquad \text{if } 0 \leqslant p \leqslant s$$

(5.20) $$\frac{i}{4\pi}\,\theta_{pp}^\lambda = d^c \log \frac{\|\mathfrak{w}_{\lambda\underline{p-1}}\|}{\|\mathfrak{w}_{\lambda\underline{p}}\|} \qquad \text{if } 0 < p \leqslant s$$

(5.21) $$\dot{\theta}_{pp}^\lambda = \partial \log \frac{\|\mathfrak{w}_{\lambda\underline{p}}\|}{\|\mathfrak{w}_{\lambda\underline{p-1}}\|} \qquad \text{if } 0 < p \leqslant s$$

(5.22) $$\ddot{\theta}_{pp}^\lambda = \bar{\partial} \log \frac{\|\mathfrak{w}_{\lambda\underline{p-1}}\|}{\|\mathfrak{w}_{\lambda\underline{p}}\|} \qquad \text{if } 0 < p \leqslant s$$

(5.23) $$\frac{i}{4\pi}\,\theta_{00}^\lambda = d^c \log \frac{1}{\|\mathfrak{w}_\lambda\|}$$

(5.24) $$\dot{\theta}_{00}^\lambda = \bar{\partial} \log \frac{1}{\|\mathfrak{w}_\lambda\|} \qquad \ddot{\theta}_{00}^\lambda = \partial \log \|\mathfrak{w}_\lambda\|$$

(5.25) $$\theta_{p,p+1}^\lambda \wedge B = \dot{\theta}_{p,p+1}^\lambda \wedge B = H_{p\lambda}\zeta_\lambda \qquad \text{if } 0 \leqslant p \leqslant s$$

(5.26) $$\theta_{p+1,p}^\lambda \wedge \bar{B} = \ddot{\theta}_{p+1,p}^\lambda \wedge \bar{B} = -H_{p\lambda}\bar{\zeta}_\lambda \qquad \text{if } 0 \leqslant p \leqslant s$$

(5.27) $$\mathbb{H}_p = -i_m \theta_{p,p+1}^\lambda \wedge B \wedge \theta_{p+1,p}^\lambda \wedge \bar{B} \qquad \text{if } 0 \leqslant p \leqslant s$$

(5.28) $$d\mathfrak{z}_p^\lambda = \sum_{q=0}^{s} \theta_{pq}^\lambda \mathfrak{z}_q^\lambda + \mathfrak{v}_p^\lambda \qquad \text{if } 0 \leqslant p \leqslant s$$

(5.29) $$\partial\mathfrak{z}_p^\lambda = \sum_{q=p}^{s} \dot{\theta}_{pq}^\lambda \mathfrak{z}_q^\lambda + \dot{\mathfrak{v}}_p^\lambda \qquad \text{if } 0 \leqslant p \leqslant s$$

(5.30) $$\bar{\partial}\mathfrak{r}_p^\lambda = \sum_{q=0}^{p} \ddot{\theta}_{pq}^\lambda \mathfrak{r}_q^\lambda \qquad \text{if } 0 \leqslant p \leqslant s$$

(5.31) $$\partial\mathfrak{r}_p^\lambda \wedge B = (\dot{\theta}_{pp}^\lambda \mathfrak{r}_p^\lambda + \dot{\theta}_{p,p+1}^\lambda \mathfrak{r}_{p+1}^\lambda) \wedge B \qquad \text{if } 0 \leqslant p \leqslant s$$

(5.32) $$\bar{\partial}\mathfrak{r}_p^\lambda \wedge \bar{B} = (\ddot{\theta}_{p,p-1}^\lambda \mathfrak{r}_{p-1}^\lambda + \ddot{\theta}_{pp}\mathfrak{r}_p^\lambda) \wedge \bar{B} \qquad \text{if } 0 \leqslant p \leqslant s$$

(5.33) $$d\mathfrak{r}_p^\lambda \wedge B \wedge \bar{B}$$
$$= (\theta_{p,p-1}^\lambda \mathfrak{r}_{p-1}^\lambda + \theta_{pp}^\lambda \mathfrak{r}_p^\lambda + \theta_{p,p+1}^\lambda \mathfrak{r}_{p+1}^\lambda) \wedge B \wedge \bar{B}$$

__REMARK__. The equations (5.31) – (5.33) are called the __Frenet__ __Formulas__. The extensive list of identities is not yet complete (see Theorem 5.5). The list will be helpful and provides more information than given on previous occasions (Stoll [100], [108]).

__PROOF__. If $p \in \mathbb{Z}[0,s]$ and $q \in \mathbb{Z}[0,s]$, then

$$\theta_{pq} = (d\mathfrak{r}_p \mid \mathfrak{r}_p) = d(\mathfrak{r}_p \mid \mathfrak{r}_q) - (\mathfrak{r}_p \mid d\mathfrak{r}_q) = -(\mathfrak{r}_p \mid d\mathfrak{r}_q) = -\bar{\theta}_{qp}$$

which proves (5.9) and implies (5.10). Definitions (5.7) and (5.8) imply (5.16), (5.17), (5.28) and (5.29) immediately.

For $z \in U_\lambda - I_s$ we select vectors $\mathfrak{r}_{s+1}(z), \ldots, \mathfrak{r}_n(z)$ such that $\mathfrak{r}_0(z), \ldots, \mathfrak{r}_n(z)$ is an orthonormal base of V if $s < n$. Then

(5.34) $$\mathfrak{v}_p = \sum_{q=s+1}^{n} R_{pq}\mathfrak{r}_q \qquad \text{on } U_\lambda - I_s$$

where $R_{pq} = (\mathfrak{v}_p \mid \mathfrak{r}_q)$ are forms of degree 1. We have

(5.35) $$d\mathfrak{r}_p = \sum_{q=0}^{s} \theta_{pq}\mathfrak{r}_q + \sum_{q=s+1}^{n} R_{pq}\mathfrak{r}_q .$$

If $0 \leq a < b \leq n$, put $\mathfrak{e}_{a,b} = \mathfrak{e}_a \wedge \dots \wedge \mathfrak{e}_b$. If $0 \leq p \leq s$, then $\mathfrak{w}_{\underline{p}} = \|\mathfrak{w}_{\underline{p}}\|\mathfrak{e}_0 \wedge \dots \wedge \mathfrak{e}_p$. We obtain

(5.36)

$$\mathfrak{w}_{\underline{p-1}} \wedge \mathfrak{w}^{(p+1)}\zeta = \mathfrak{w}_{\underline{p}}'\zeta = d\mathfrak{w}_{\underline{p}} \wedge B$$

$$= (d\|\mathfrak{w}_{\underline{p}}\| \wedge B)\mathfrak{e}_0 \wedge \dots \wedge \mathfrak{e}_p$$

$$+ \|\mathfrak{w}_{\underline{p}}\| \sum_{j=0}^{p} \mathfrak{e}_{0,j-1} \wedge (d\mathfrak{e}_j \wedge B) \wedge \mathfrak{e}_{j+1,p}$$

$$= \|\mathfrak{w}_{\underline{p}}\|((d \log \|\mathfrak{w}_{\underline{p}}\| + \sum_{j=0}^{p} \theta_{jj}) \wedge B)\mathfrak{e}_0 \wedge \dots \wedge \mathfrak{e}_p$$

$$+ \|\mathfrak{w}_{\underline{p}}\| \sum_{j=0}^{p} \sum_{k=p+1}^{s} (-1)^{p-j}(\theta_{jk} \wedge B)\mathfrak{e}_{0,j-1} \wedge \mathfrak{e}_{j+1,p} \wedge \mathfrak{e}_k$$

$$+ \|\mathfrak{w}_{\underline{p}}\| \sum_{j=0}^{p} \sum_{k=s+1}^{n} (-1)^{p-j}(R_{jk} \wedge B)\mathfrak{e}_{0,j-1} \wedge \mathfrak{e}_{j+1,p} \wedge \mathfrak{e}_k$$

If $0 \leq r \leq p-1$, then $\mathfrak{e}_r \wedge \mathfrak{w}_{\underline{p-1}} = 0$. Hence

$$0 = \sum_{k=p+1}^{s} (-1)^p \theta_{rk} \wedge B\, \mathfrak{e}_0 \wedge \dots \wedge \mathfrak{e}_p \wedge \mathfrak{e}_k$$

$$+ \sum_{k=s+1}^{n} (-1)^n (R_{rk} \wedge B)\mathfrak{e}_0 \wedge \dots \wedge \mathfrak{e}_p \wedge \mathfrak{e}_p$$

which implies $\theta_{rk} \wedge B = 0$ for $0 \leq r \leq p-1$ and $p+1 \leq k \leq s$. Also we have $R_{rk} \wedge B = 0$ for $0 \leq r \leq p-1$ and $s+1 \leq k \leq n$. Therefore $\theta_{rk} \wedge B = 0$ if $k > r+1$ which proves (5.11) and implies

(5.12) and (5.15). Also we have $\mathfrak{v}_r \wedge B = 0$ for $0 \leqslant r \leqslant s-1$. If $s = n$, then $\mathfrak{v}_p = 0$ for $p = 0,1, \dots ,n$. Assume that $s < n$. Then $\mathfrak{w}'_{\underline{s}} = \mathfrak{w}_{\underline{s-1}} \wedge \mathfrak{w}^{(s+1)}$ and $\mathfrak{w}_{\underline{s}} \wedge \mathfrak{w}^{(s+1)} = \mathfrak{w}_{\underline{s+1}} = 0$. Hence

$$\mathfrak{w}^{(s+1)} = A_0 \mathfrak{w} + A_s \mathfrak{w}' + \dots + A_s \mathfrak{w}^{(s)} .$$

Hence $\mathfrak{w}'_{\underline{s}} = A_s \mathfrak{w}_{\underline{s}}$ and $\mathfrak{e}_s \wedge \mathfrak{w}'_{\underline{s}} = \mathfrak{e}_s \wedge A_s \mathfrak{w}_{\underline{x}} = 0$. Consequently, (5.36) for $p = s$ implies

$$\sum_{k=s+1}^{n} (-1)^s (R_{sk} \wedge B) \mathfrak{e}_0 \wedge \dots \wedge \mathfrak{e}_s \wedge \mathfrak{e}_k = 0$$

Consequently $R_{sk} \wedge B = 0$ for $k = s + 1, \dots ,n$. Hence $\mathfrak{v}_s \wedge B = 0$. We proved (5.18) which implies (5.19).

In (5.36) we take the hermitian product with $\mathfrak{w}_{\underline{p}}$ and obtain

$$(\mathfrak{w}'_{\underline{p}} | \mathfrak{w}_{\underline{p}}) = (d\mathfrak{w}_{\underline{p}} \wedge B | \mathfrak{w}_{\underline{p}})$$

$$= \| \mathfrak{w}_{\underline{p}} \|^2 (d \log \| \mathfrak{w}_{\underline{p}} \| + \sum_{j=0}^{p} \theta_{jj}) \wedge B$$

or

$$\sum_{j=0}^{p} \theta_{jj} \wedge B = (2\partial \log \| \mathfrak{w}_{\underline{p}} \| - d \log \| \mathfrak{w}_{\underline{p}} \|) \wedge B$$

$$= 2\pi i (d^c \log \| \mathfrak{w}_{\underline{p}} \|^2) \wedge B$$

$$\frac{i}{2\pi} \sum_{j=0}^{p} \theta_{jj} \wedge B = d^c \log \frac{1}{\| \mathfrak{w}_{\underline{p}} \|^2} \wedge B$$

If $p > 0$, we obtain

$$\frac{i}{4\pi}\,\theta_{pp} \wedge B = d^c \log \frac{\|\mathfrak{w}_{\underline{p-1}}\|}{\|\mathfrak{w}_{\underline{p}}\|} \wedge B$$

$$\frac{i}{4\pi}\,\theta_{00} \wedge B = d^c \log \frac{1}{\|\mathfrak{w}\|} \wedge B.$$

Degree considerations imply

$$\ddot{\theta}_{pp} \wedge B = \bar{\partial} \log \frac{\|\mathfrak{w}_{\underline{p-1}}\|}{\|\mathfrak{w}_{\underline{p}}\|} \wedge B$$

$$\ddot{\theta}_{00} \wedge B = \bar{\partial} \log \frac{1}{\|\mathfrak{w}\|} \wedge B$$

Since $B \not\equiv 0$ we obtain (5.22) and by conjugation (5.21) and by the same method (5.24). Also (5.21) and (5.22) imply (5.20).

If $0 < p < s$, then (5.36) implies

$$\|\mathfrak{w}_{\underline{p-1}}\|\;\|\mathfrak{w}_{\underline{p+1}}\|\;\mathfrak{w}_{\underline{p+1}}\zeta$$

$$= \|\mathfrak{w}_{\underline{p-1}}\|\;\|\mathfrak{w}_{\underline{p+1}}\|\;\mathfrak{w}_{\underline{p}} \wedge \mathfrak{w}^{(p+1)}\zeta$$

$$= \|\mathfrak{w}_{\underline{p-1}}\|\;\|\mathfrak{w}_{\underline{p+1}}\|\;\|\mathfrak{w}_{\underline{p}}\|(-1)^p\,\mathfrak{e}_p \wedge \mathfrak{e}_{0,p-1} \wedge \mathfrak{w}^{(p+1)}\zeta$$

$$= \|\mathfrak{w}_{\underline{p+1}}\|\;\|\mathfrak{w}_{\underline{p}}\|(-1)^p\,\mathfrak{e}_p \wedge \mathfrak{w}_{\underline{p-1}} \wedge \mathfrak{w}^{(p+1)}\zeta$$

$$= \|\mathfrak{w}_{\underline{p+1}}\|\;\|\mathfrak{w}_{\underline{p}}\|^2(\theta_{p,p+1} \wedge B)\mathfrak{e}_0 \wedge \cdots \wedge \mathfrak{e}_p \wedge \mathfrak{e}_{p+1}$$

$$= \|\mathfrak{w}_{\underline{p}}\|^2(\theta_{p,p+1} \wedge B)\,\mathfrak{w}_{\underline{p+1}}\,.$$

We have

$$\theta_{p,p+1} \wedge B = \frac{\|\mathfrak{w}_{\underline{p-1}}\| \; \|\mathfrak{w}_{\underline{p-1}}\|}{\|\mathfrak{w}_{\underline{p}}\|^2} \mathfrak{z} = H_p \mathfrak{z} \; .$$

If $p = s$, then $\theta_{s,s+1} = 0$ and $H_s = 0$. Hence (5.25) holds for $0 < p \leqslant s$. If $p = 0$, then

$$\|\mathfrak{w}_{\underline{1}}\| \mathfrak{w}_1 \mathfrak{z} = \|\mathfrak{w}_{\underline{1}}\| \mathfrak{w} \wedge \mathfrak{w}' \mathfrak{z} = \|\mathfrak{w}_{\underline{1}}\| \; \|\mathfrak{w}\| \mathfrak{e}_0 \wedge \mathfrak{w}' \mathfrak{z}$$

$$= \|\mathfrak{w}_{\underline{1}}\| \; \|\mathfrak{w}\|^2 \mathfrak{e}_0 \wedge (\theta_{01} \wedge B) \mathfrak{e}_1 = \|\mathfrak{w}\|^2 \theta_{01} \wedge B \; \mathfrak{w}_{\underline{1}}$$

We have

$$\theta_{01} \wedge B = \frac{\|\mathfrak{w}_{\underline{1}}\|}{\|\mathfrak{w}\|^2} \mathfrak{z} = H_0 \mathfrak{z}$$

which proves (5.25) for $0 \leqslant p \leqslant s$. Conjugation proves (5.26). For degree reasons, we have $\ddot{\theta}_{p,p+1} \wedge B = 0$, hence $\ddot{\theta}_{p,p+1} = 0$. By (5.11), we have $\ddot{\theta}_{pq} \wedge B = 0$ for $q \geqslant p + 1$ which implies (5.14) and by conjugation (5.13). We have

$$\mathbb{H}_p = H_p^2 \chi^m = H_p^2 i_m \mathfrak{z} \wedge \bar{\mathfrak{z}} = - i_m \theta^\lambda_{p,p+1} \wedge B \wedge \theta^\lambda_{p+1,p} \wedge \bar{B}$$

which is (5.27). We have proved (5.9) – (5.29) which imply (5.30) – (5.33) trivially. q.e.d.

<u>THEOREM</u> 5.5. <u>Frenet Curvature Formulas</u>. Take $\lambda \in \Lambda$. Then

(5.37) $$d\theta^\lambda_{pq} = \sum_{j=0}^{s} \theta^\lambda_{pj} \wedge \theta^\lambda_{jq} - (\mathfrak{w}^\lambda_p | \mathfrak{w}^\lambda_q)$$

(5.38) $$\frac{i}{2\pi}\, d\theta^{\lambda}_{pp} = f^*(\Omega_{p-1}) - f^*(\Omega_p) \qquad \text{on } U_\lambda - I_s$$

(5.39) $$m i_{m-1} \frac{i}{2\pi}\, d\theta^{\lambda}_{pp} \wedge B \wedge \bar{B} = \mathbb{H}_{p-1} - \mathbb{H}_p \qquad \text{on } U_\lambda - I_s$$

(5.40) $$d\theta^{\lambda}_{pq} \wedge B \wedge \bar{B} = 0 \qquad \text{if } |p-q| \geqslant 2$$

(5.41) $$d\theta^{\lambda}_{p,p+1} \wedge B \wedge \bar{B}$$
$$= (\theta^{\lambda}_{pp} \wedge \theta^{\lambda}_{p,p+1} + \theta^{\lambda}_{p,p+1} \wedge \theta^{\lambda}_{p+1,p+1}) \wedge B \wedge \bar{B}$$

(5.42) $$d\theta^{\lambda}_{p,p-1} \wedge B \wedge \bar{B}$$
$$= (\theta^{\lambda}_{p,p-1} \wedge \theta^{\lambda}_{p-1,p-1} + \theta^{\lambda}_{pp} \wedge \theta^{\lambda}_{p,p-1}) \wedge B \wedge \bar{B}$$

(5.53) $$d\theta^{\lambda}_{p,p+1} \wedge B \wedge \bar{B} = \bar{\partial}\, \dot{\theta}^{\lambda}_{p,p+1} \wedge B \wedge \bar{B}$$

(5.44) $$d\theta^{\lambda}_{p,p-1} \wedge B \wedge \bar{B} = \partial \ddot{\theta}^{\lambda}_{p,p-1} \wedge B \wedge \bar{B}.$$

<u>PROOF</u>. We have

$$d\theta_{pq} = d(d\mathfrak{e}_p | \mathfrak{e}_q) = -(d\mathfrak{e}_p | d\mathfrak{e}_q)$$

$$= -\sum_{j=0}^{s} (d\mathfrak{e}_p | \mathfrak{e}_j) \wedge \bar{\theta}_{qj} - (d\mathfrak{e}_p | \mathfrak{v}_q)$$

$$= \sum_{j=0}^{s} \theta_{pj} \wedge \theta_{jq} - \sum_{j=0}^{s} \theta_{pj} \wedge (\mathfrak{e}_j | \mathfrak{v}_q) - (\mathfrak{v}_p | \mathfrak{v}_q)$$

$$= \sum_{j=0}^{s} \theta_{pj} \wedge \theta_{jq} - (\mathfrak{v}_p | \mathfrak{v}_q)$$

which proves (5.37). We have

$$\frac{i}{2\pi}\, d\theta_{pp} = dd^c \log \frac{\| \mathfrak{w}_{p-1} \|^2}{\| \mathfrak{w}_p \|^2} = f^*(\Omega_{p-1}) - f^*(\Omega_p)$$

which proves (5.38) and implies (5.39).

If at least one term in $d\theta_{pq} \wedge B \wedge \bar{B}$ is not zero we have $|p - j| \leqslant 1$ and $|q - j| \leqslant 1$. Hence $|p - q| \leqslant 2$, which implies $d\theta_{pq} \wedge B \wedge \bar{B} = 0$ if $|p - q| \geqslant 3$. Here

$$d\theta_{p,p+2} \wedge B \wedge \bar{B} = (\theta_{p,p+1} \wedge \theta_{p-1,p+2} + \theta_{pp} \wedge \theta_{p,p+2}$$

$$+ \theta_{p,p+1} \wedge \theta_{p+1,p+2}) \wedge B \wedge \bar{B} = 0$$

since $\theta_{p-1,p+1} \wedge B \wedge \bar{B} = 0$ and $\theta_{p,p+2} \wedge B \wedge \bar{B} = 0$ and since $\theta_{p,p+1} \wedge \theta_{p+1,p+2}$ has bidegree (2,0). By conjugation we obtain $d\theta_{p,p-2} \wedge B \wedge \bar{B} = 0$. Hence we have proved (4.40). Now (5.37) implies (5.41) and (5.42) easily. Degree consideration give (5.43) and (5.44). q.e.d.

Assume that f is general for B. Hence we assume that $\ell_f = n = s$. Take $\lambda \in \Lambda$ and $z \in U_\lambda - I_n$. Then $\mathfrak{e}_0^\lambda(z), \ldots, \mathfrak{e}_n^\lambda(z)$ is an orthonormal base of V. Let $\mathfrak{e}_0^{\lambda *}(z), \ldots, \mathfrak{e}_n^{\lambda *}(z)$ be the dual base, which is an orthonormal base of V^*. The vector functions $\mathfrak{e}_p^{\lambda *} : U_\lambda - I_n \longrightarrow V^*$ are of class C^∞ and are called the <u>dual</u> <u>Frenet</u> <u>frame</u>. If $(\lambda,\mu) \in \Lambda[1]$, then Lemma 5.3 implies

$$\text{(5.45)} \qquad \mathfrak{e}_p^{\mu *} = \frac{v_{\lambda\mu}}{|v_{\lambda\mu}|} \left[\frac{\Delta_{\mu\lambda}}{|\Delta_{\mu\lambda}|}\right]^p \mathfrak{e}_p^{\lambda *} \quad \text{on } U_{\lambda\mu} - I_n$$

<u>**LEMMA**</u> **5.6.** <u>Take</u> $p \in \mathbb{Z}[0,n]$ <u>and</u> $\lambda \in \Lambda$. <u>Then</u>

$$\text{(5.46)} \qquad d\mathfrak{e}_p^{\lambda *} = - \sum_{q=0}^{n} \theta_{qp}^{\lambda} \mathfrak{e}_q^{\lambda *} \quad \text{on } U_\lambda - I_n .$$

<u>PROOF</u>. There are forms A_{pq} on $U_\lambda - I_n$ such that

$$d\mathfrak{e}_p^* = \sum_{j=0}^{n} A_{pq}\mathfrak{e}_q^* .$$

Then

$$A_{pq} = \langle d\mathfrak{e}_p^*, \mathfrak{e}_q\rangle = -\langle \mathfrak{e}_p^*, d\mathfrak{e}_q\rangle = -\theta_{qp} . \qquad \text{q.e.d.}$$

The Frenet Formulas for the dual frame follow immediately

$$(5.47) \qquad d\mathfrak{e}_p^{\lambda *} \wedge B = -\sum_{q=p-1}^{n} (\theta_{qp}^{\lambda} \wedge B)\mathfrak{e}_q^{\lambda *} \qquad \text{if } 0 < p \leqslant n$$

$$(5.48) \qquad d\mathfrak{e}_0^{\lambda *} \wedge B = -\sum_{q=0}^{n} (\theta_{q0}^{\lambda} \wedge B)\mathfrak{e}_q^{\lambda *}$$

$$(5.49) \qquad d\mathfrak{e}_p^{\lambda *} \wedge B \wedge \bar{B}$$

$$= -(\theta_{p-1,p}^{\lambda}\mathfrak{e}_{p-1}^{\lambda *} + \theta_{pp}^{\lambda}\mathfrak{e}_p^{\lambda *} + \theta_{p+1,p}^{\lambda}\mathfrak{e}_{p+1}^{\lambda *}) \wedge B \wedge \bar{B}.$$

Again, let $f : M \longrightarrow \mathbb{P}(V)$ be a meromorphic map general for B. In addition, we consider a meromorphic map $g : M \longrightarrow \mathbb{P}(V^*)$. Abbreviate $\ell_g = s$. Then $0 \leqslant s \leqslant n$. Let $(\mathfrak{z}_\lambda, \mathfrak{w}_\lambda, \mathfrak{w}_\lambda)_{\lambda\in\Lambda}$ be a representation atlas of the pair f,g. Adopt the notations of (3.34) to (3.44). Define $\mathfrak{J} = I_n(f) \cup I_s(g)$. The $\mathfrak{J} \cap U_\lambda$ is the union of the zero set of $\mathfrak{w}_{\lambda\underline{n}}$ and the zero set of $\mathfrak{w}_{\lambda\underline{s}}$. Let $\psi_0^\lambda, \ldots, \psi_s^\lambda$ be the Frenet frame of g. Define

$$(5.50) \qquad \eta_{pq}^{\lambda} = (d\psi_p^{\lambda} | \psi_q^{\lambda}) \qquad \mathfrak{v}_p^{\lambda} = d\psi_p^{\lambda} - \sum_{q=0}^{s} \eta_{pq}^{\lambda}\psi_q^{\lambda}$$

Then Theorems 5.4 and 5.5 hold in analogy. For each $\lambda \in \Lambda$ and $p \in \mathbb{Z}[0,s]$ and $q \in \mathbb{Z}[0,n]$ define the <u>stress</u> <u>coefficient</u>

$$\Lambda^{\lambda}_{pq} = (\varphi^{\lambda}_{p} | \mathfrak{e}^{\lambda *}_{q}) \tag{5.51}$$

which is a function of class C^{∞} on $U_{\lambda} - \mathfrak{Z}$. Obviously we have

$$\varphi^{\lambda}_{p} = \sum_{q=0}^{n} \Lambda^{\lambda}_{pq} \mathfrak{e}^{\lambda *}_{q} \quad \text{if} \quad 0 \leqslant p \leqslant s \tag{5.52}$$

Since $\varphi^{\lambda}_{0}, \ldots, \varphi^{\lambda}_{s}$ is an orthonormal system, we have

$$\sum_{j=0}^{n} \Lambda^{\lambda}_{pj} \bar{\Lambda}^{\lambda}_{qj} = \begin{cases} 1 & \text{if } p = q \\ 0 & \text{if } p \neq q \end{cases} \tag{5.53}$$

If $s = n$, the square matrix (Λ^{λ}_{pq}) is unitary and we have also

$$\sum_{j=0}^{n} \Lambda^{\lambda}_{jp} \bar{\Lambda}^{\lambda}_{jq} = \begin{cases} 1 & \text{if } p = q \\ 0 & \text{if } p \neq q \end{cases} \tag{5.54}$$

LEMMA 5.7. Take $(\lambda,\mu) \in \Lambda[1]$ and $p \in \mathbb{Z}[0,s]$ and $q \in \mathbb{Z}[0,n]$. On $U_{\lambda\mu} - \mathfrak{Z}$ we have

$$\Lambda^{\lambda}_{pq} = \frac{v_{\lambda\mu}}{|v_{\lambda\mu}|} \frac{w_{\lambda\mu}}{|w_{\lambda\mu}|} \left[\frac{\Delta_{\lambda\mu}}{|\Delta_{\lambda\mu}|}\right]^{p+q} \Lambda^{\mu}_{pq} \tag{5.55}$$

PROOF. If $\alpha \in \mathbb{C}$ with $|\alpha| = 1$, then $\bar{\alpha} = \alpha^{-1}$. Hence (5.6) for φ and (5.45) imply

$$\begin{aligned}
\Lambda^{\lambda}_{pq} &= (\varphi^{\lambda}_{p} | \mathfrak{e}^{\lambda *}_{q}) \\
&= \frac{\bar{v}_{\mu\lambda}}{|v_{\mu\lambda}|} \frac{w_{\lambda\mu}}{|w_{\lambda\mu}|} \left[\frac{\bar{\Delta}_{\lambda\mu}}{|\Delta_{\lambda\mu}|}\right]^{q} \left[\frac{\Delta_{\mu\lambda}}{|\Delta_{\mu\lambda}|}\right]^{p} (\varphi^{\mu}_{p} | \mathfrak{e}^{\mu *}_{q}) \\
&= \frac{v_{\lambda\mu}}{|v_{\lambda\mu}|} \frac{w_{\lambda\mu}}{|\Delta_{\mu\lambda}|} \left[\frac{\Delta_{\mu\lambda}}{|\Delta_{\mu\lambda}|}\right]^{p+q} \Lambda^{\mu}_{pq} .
\end{aligned}$$

q.e.d.

We will use this transition formula to introduce important invariants, but we will first derive some formulas which calculate the stress coefficients in terms of the representations $\mathfrak{w}$ and $\mathfrak{v}$.

<u>LEMMA</u> 5.8. <u>Take</u> $\lambda \in \Lambda$ <u>and</u> $p \in \mathbb{Z}[1,s]$ <u>and</u> $q \in \mathbb{Z}[1,n]$. <u>Then we have</u>

(5.56) $$\Lambda^{\lambda}_{pq} = \frac{\langle (\mathfrak{w}_{\lambda_{\underline{q}}} L^* \mathfrak{w}_{\lambda_{\underline{q-1}}}), (\mathfrak{v}_{\lambda_{\underline{p}}} L^* \mathfrak{v}_{\lambda_{\underline{p-1}}}) \rangle}{\| \mathfrak{w}_{\lambda_{\underline{q}}} \| \, \| \mathfrak{w}_{\lambda_{\underline{q-1}}} \| \, \| \mathfrak{v}_{\lambda_{\underline{p}}} \| \, \| \mathfrak{v}_{\lambda_{\underline{p-1}}} \|}$$

(5.57) $$\Lambda^{\lambda}_{pq} = (-1)^q \frac{(\mathfrak{w}_{\lambda_{\underline{q}}} L (\mathfrak{v}_{\lambda_{\underline{p}}} L^* \mathfrak{v}_{\lambda_{\underline{p-1}}}) \mid \mathfrak{w}_{\lambda_{\underline{q-1}}})}{\| \mathfrak{w}_{\lambda_{\underline{q}}} \| \, \| \mathfrak{w}_{\lambda_{\underline{q-1}}} \| \, \| \mathfrak{v}_{\lambda_{\underline{p}}} \| \, \| \mathfrak{v}_{\lambda_{\underline{p-1}}} \|}$$

(5.58) $$\Lambda^{\lambda}_{pq} = (-1)^p \frac{(\mathfrak{v}_{\lambda_{\underline{p}}} L (\mathfrak{w}_{\lambda_{\underline{q}}} L^* \mathfrak{w}_{\lambda_{\underline{q-1}}}) \mid \mathfrak{v}_{\lambda_{\underline{p-1}}})}{\| \mathfrak{v}_{\lambda_{\underline{p}}} \| \, \| \mathfrak{v}_{\lambda_{\underline{p-1}}} \| \, \| \mathfrak{w}_{\lambda_{\underline{q}}} \| \, \| \mathfrak{w}_{\lambda_{\underline{q-1}}} \|}$$

(5.59) $$\Lambda^{\lambda}_{0q} = \frac{\langle (\mathfrak{w}_{\lambda_{\underline{q}}} L^* \mathfrak{w}_{\lambda_{\underline{q-1}}}), \mathfrak{v}_{\lambda} \rangle}{\| \mathfrak{w}_{\lambda_{\underline{q}}} \| \, \| \mathfrak{w}_{\lambda_{\underline{q-1}}} \| \, \| \mathfrak{v}_{\lambda} \|}$$

$$= (-1)^q \frac{((\mathfrak{w}_{\lambda_{\underline{q}}} L \mathfrak{v}_{\lambda}) \mid \mathfrak{w}_{\lambda_{\underline{q-1}}})}{\| \mathfrak{w}_{\lambda_{\underline{q}}} \| \, \| \mathfrak{v}_{\lambda} \| \, \| \mathfrak{w}_{\lambda_{\underline{q-1}}} \|}$$

(5.60) $$\Lambda^{\lambda}_{p0} = \frac{\langle \mathfrak{w}_{\lambda}, (\mathfrak{v}_{\lambda_{\underline{p}}} L^* \mathfrak{v}_{\lambda_{\underline{p-1}}}) \rangle}{\| \mathfrak{w}_{\lambda} \| \, \| \mathfrak{v}_{\lambda_{\underline{p}}} \| \, \| \mathfrak{v}_{\lambda_{\underline{p-1}}} \|}$$

$$= (-1)^p \frac{((\mathfrak{v}_{\lambda_{\underline{p}}} L \mathfrak{w}_{\lambda}) \mid \mathfrak{v}_{\lambda_{\underline{p-1}}})}{\| \mathfrak{v}_{\lambda_{\underline{p}}} \| \, \| \mathfrak{w}_{\lambda} \| \, \| \mathfrak{v}_{\lambda_{\underline{p-1}}} \|}$$

(5.61) $$\Lambda^{\lambda}_{00} = \frac{\langle \mathfrak{w}_{\lambda}, \mathfrak{v}_{\lambda} \rangle}{\| \mathfrak{w}_{\lambda} \| \, \| \mathfrak{v}_{\lambda} \|}$$

(5.62) $$\sum_{j=0}^{q} |\Lambda^{\lambda}_{pj}|^2 = \frac{\| \mathfrak{w}_{\lambda_{\underline{q}}} L (\mathfrak{v}_{\lambda_{\underline{p}}} L^* \mathfrak{v}_{\lambda_{\underline{p-1}}}) \|^2}{\| \mathfrak{w}_{\lambda_{\underline{q}}} \|^2 \, \| \mathfrak{v}_{\lambda_{\underline{p}}} \|^2 \, \| \mathfrak{v}_{\lambda_{\underline{p-1}}} \|^2}$$

$$(5.63)\qquad \sum_{j=0}^{p} |\Lambda_{jq}^{\lambda}|^2 = \frac{\|\mathfrak{v}_{\lambda\underline{p}} \llcorner (\mathfrak{w}_{\lambda\underline{q}} \llcorner^* \mathfrak{w}_{\lambda\underline{q-1}})\|^2}{\|\mathfrak{v}_{\lambda\underline{p}}\|^2 \;\|\mathfrak{v}_{\lambda\underline{q}}\|^2 \;\|\mathfrak{v}_{\lambda\underline{p-1}}\|^2}$$

If $0 \leqslant p \leqslant s$ and $0 \leqslant q \leqslant n$, then

$$(5.64)\qquad \Phi_{pq}|U_\lambda = \sum_{j=0}^{p} \sum_{k=0}^{q} |\Lambda_{jk}^{\lambda}|^2$$

$$(5.65)\qquad \Phi_q|U_\lambda = \sum_{j=0}^{q} |\Lambda_{0j}|^2$$

$$(5.66)\qquad \Psi_p|U_\lambda = \sum_{j=0}^{p} |\Lambda_{j0}|^2$$

$$(5.67)\qquad \Phi_{00} = \Phi_0 = \Psi_0 = \Box\, f;g\, \Box^2 = |\Lambda_{00}^{\lambda}|^2 \qquad \text{on } U_\lambda$$

<u>**PROOF**</u>. We have

(5.66)

$$\mathfrak{w}_{\underline{q}} = \|\mathfrak{w}_{\underline{q}}\|\, \mathfrak{e}_0 \wedge \dots \wedge \mathfrak{e}_q \qquad \mathfrak{v}_{\underline{p}} = \|\mathfrak{v}_p\|\, \mathfrak{f}_0 \wedge \dots \wedge \mathfrak{f}_p$$

$$(5.67)\qquad \mathfrak{e}_q = \frac{\mathfrak{w}_{\underline{q}} \llcorner^* \mathfrak{w}_{\underline{q-1}}}{\|\mathfrak{w}_{\underline{q}}\|\;\|\mathfrak{w}_{\underline{q-1}}\|} \qquad \mathfrak{f}_p = \frac{\mathfrak{v}_{\underline{p}} \llcorner^* \mathfrak{v}_{\underline{p-1}}}{\|\mathfrak{v}_{\underline{p}}\|\;\|\mathfrak{v}_{\underline{p-1}}\|}$$

$$(5.68)\qquad \langle \mathfrak{e}_q, \mathfrak{f}_p \rangle = \sum_{j=0}^{n} \Lambda_{pj} \langle \mathfrak{e}_q, \mathfrak{e}_j^* \rangle = \Lambda_{pq}$$

which proves (5.56). Also we have

$$(5.69)\qquad \frac{\mathfrak{w}_{\underline{q}} \llcorner (\mathfrak{v}_{\underline{p}} \llcorner^* \mathfrak{v}_{\underline{p-1}})}{\|\mathfrak{w}_{\underline{q}}\|\;\|\mathfrak{v}_{\underline{p}}\|\;\|\mathfrak{v}_{\underline{p-1}}\|} = \mathfrak{e}_0 \wedge \dots \wedge \mathfrak{e}_q \llcorner \mathfrak{f}_p$$

$$= \sum_{j=0}^{q} \Lambda_{pj} (-1)^j\, \mathfrak{e}_0 \wedge \dots \wedge \mathfrak{e}_{j-1} \wedge \mathfrak{e}_{j+1} \wedge \dots \wedge \mathfrak{e}_q$$

Taking the norm proves (5.62). Taking the hermitian product with $\underline{\mathfrak{w}}_{q-1} = \|\underline{\mathfrak{w}}_{q-1}\| \mathfrak{e}_0 \wedge \cdots \wedge \mathfrak{e}_{q-1}$ proves (5.57). Also we have

$$\frac{\underline{\mathfrak{v}}_p \llcorner (\underline{\mathfrak{w}}_q \llcorner^* \underline{\mathfrak{w}}_{q-1})}{\|\underline{\mathfrak{v}}_p\| \, \|\underline{\mathfrak{w}}_q\| \, \|\underline{\mathfrak{w}}_{q-1}\|} = \mathfrak{a}_0 \wedge \cdots \wedge \mathfrak{a}_p \llcorner \mathfrak{e}_q \tag{5.70}$$

$$= \sum_{j=0}^{p} \Lambda_{jq}(-1)^j \, \mathfrak{a}_0 \wedge \cdots \wedge \mathfrak{a}_{j-1} \wedge \mathfrak{a}_{j+1} \wedge \cdots \wedge \mathfrak{a}_p$$

Taking the norm proves (5.63). Taking the hermitian product with $\underline{\mathfrak{v}}_{p-1} = \|\underline{\mathfrak{v}}_{p-1}\| \, \mathfrak{a}_0 \wedge \cdots \wedge \mathfrak{a}_{p-1}$ proves (5.58). Also we have

$$\frac{\langle (\underline{\mathfrak{w}}_q \llcorner^* \underline{\mathfrak{w}}_{q-1}), \mathfrak{v} \rangle}{\|\underline{\mathfrak{w}}_q\| \, \|\underline{\mathfrak{w}}_{q-1}\| \, \|\mathfrak{v}\|} = \langle \mathfrak{a}_q | \mathfrak{e}_0 \rangle = \Lambda_{0q}$$

$$\frac{(\underline{\mathfrak{w}}_q \llcorner \mathfrak{v} \,|\, \underline{\mathfrak{w}}_{q-1})}{\|\underline{\mathfrak{w}}_q\| \, \|\underline{\mathfrak{w}}_{q-1}\| \, \|\mathfrak{v}\|} = (\mathfrak{e}_0 \wedge \cdots \wedge \mathfrak{e}_q \llcorner \mathfrak{a}_0 \,|\, \mathfrak{e}_0 \wedge \cdots \wedge \mathfrak{e}_{q-1})$$

$$= \sum_{j=1}^{q} (-1)^j \Lambda_{0q} (\mathfrak{e}_0 \wedge \cdots \wedge \mathfrak{e}_{j-1} \wedge \mathfrak{e}_{j+1} \wedge \cdots \wedge \mathfrak{e}_q \,|\, \mathfrak{e}_0 \wedge \cdots \wedge \mathfrak{e}_{q-1})$$

$$= (-1)^q \Lambda_{0q}$$

$$\frac{\langle \mathfrak{w}, (\underline{\mathfrak{v}}_p \llcorner^* \underline{\mathfrak{v}}_{p-1}) \rangle}{\|\mathfrak{w}\| \, \|\underline{\mathfrak{v}}_p\| \, \|\underline{\mathfrak{v}}_{p-1}\|} = \langle \mathfrak{e}_0, \mathfrak{a}_p \rangle = \Lambda_{p0}$$

$$\frac{(\underline{\mathfrak{v}}_p \llcorner \mathfrak{w} \,|\, \underline{\mathfrak{v}}_{p-1})}{\|\underline{\mathfrak{v}}_p\| \, \|\mathfrak{w}\| \, \|\underline{\mathfrak{v}}_{p-1}\|}$$

$$= (\mathfrak{a}_0 \wedge \cdots \wedge \mathfrak{a}_p \llcorner \mathfrak{e}_0 \,|\, \mathfrak{a}_0 \wedge \cdots \wedge \mathfrak{a}_{p-1})$$

$$= \sum_{j=0}^{p} (-1)^j \Lambda_{j0} (\mathfrak{a}_{0,j-1} \wedge \mathfrak{a}_{j+1,p} \,|\, \mathfrak{a}_0 \wedge \cdots \wedge \mathfrak{a}_{p-1})$$

$$= (-1)^p \Lambda_{p0}$$

$$\Lambda_{00} = (\mathfrak{w}_0 | \mathfrak{z}_0^*) = \langle \mathfrak{w}_0, \mathfrak{z}_0 \rangle = \frac{\langle \mathfrak{z}, \mathfrak{w} \rangle}{\| \mathfrak{z} \| \ \| \mathfrak{w} \|}$$

$$|\Lambda_{00}|^2 = \frac{|\langle \mathfrak{z}, \mathfrak{w} \rangle|^2}{\| \mathfrak{z} \|^2 \ \| \mathfrak{w} \|^2} = \Box f;g \Box^2 = \Phi_0 = \Psi_0 = \Phi_{00} \qquad \text{on } U_\lambda$$

which proves (5.59) -- (5.63) and (5.67). If $0 \leqslant p \leqslant s$ and $0 \leqslant q \leqslant n$, define $\mu = \mathrm{Min}(p+1, q+1)$. Then (1.20) implies

$$\Phi_{pq} = \mu^2 \Box f_q \dot{\boxplus} g_p \Box^2 = \mu^2 \| \mathfrak{z}_0 \wedge \ldots \wedge \mathfrak{z}_q \boxplus \mathfrak{w}_0 \wedge \ldots \wedge \mathfrak{w}_p \Box^2$$

$$= \| \sum_{j=0}^{p} \sum_{p=0}^{q} \Lambda_{jk} \hat{\mathfrak{z}}_k \otimes \hat{\mathfrak{w}}_j \|^2$$

$$= \sum_{j=0}^{p} \sum_{p=0}^{q} \sum_{r=0}^{p} \sum_{t=0}^{q} \Lambda_{jk} \bar{\Lambda}_{rt} (\hat{\mathfrak{z}}_k \otimes \hat{\mathfrak{w}}_j | \hat{\mathfrak{z}}_t \otimes \hat{\mathfrak{w}}_r)$$

$$= \sum_{j=0}^{p} \sum_{k=0}^{q} \sum_{r=0}^{p} \sum_{t=0}^{q} \Lambda_{jk} \bar{\Lambda}_{rt} (\hat{\mathfrak{z}}_k | \hat{\mathfrak{z}}_t)(\hat{\mathfrak{w}}_j | \hat{\mathfrak{w}}_r)$$

$$= \sum_{j=0}^{p} \sum_{p=0}^{q} |\Lambda_{jk}|^2 \qquad \text{on } U_\lambda$$

which proves (5.64) and implies (5.65) and (5.66). q.e.d.

The identity (5.64) is remarkable. It is surprising that the rectangular norm sum can be expressed as the contraction norm of the two associated meromorphic maps.

Take integers a_j, b_j, c_j, d_j with $0 \leqslant a_j \leqslant s$, $0 \leqslant b_j \leqslant n$, $0 \leqslant c_j \leqslant s$, $0 \leqslant d_j \leqslant n$ for $j = 1, \ldots, k$. Abbreviate

(5.71)

$$a = a_1 + \ldots + a_k, \ b = b_1 + \ldots + b_k, \ c = c_1 + \ldots + c_k, \ d = d_1 + \ldots + d_k$$

and assume

(5.72) $$a + b = c + d$$

If $(\lambda,\mu) \in \Lambda[1]$, we have

(5.73) $$\prod_{j=1}^{k} \Lambda^{\lambda}_{a_j b_j} \bar{\Lambda}^{\lambda}_{c_j d_j}$$

$$= \left[\frac{\Delta_{\mu\lambda}}{|\Delta_{\mu\lambda}|}\right]^{a+b} \left[\frac{\bar{\Delta}_{\mu\lambda}}{|\Delta_{\mu\lambda}|}\right]^{c+d} \prod_{j=1}^{k} \Lambda^{\mu}_{a_j b_j} \bar{\Lambda}^{\mu}_{c_j d_j}$$

$$= \prod_{j=1}^{k} \Lambda^{\mu}_{a_j b_j} \bar{\Lambda}^{\mu}_{c_j d_j}$$

on $U_{\lambda\mu} - \mathfrak{Z}$. Hence a function $S^{c_1 d_1 \cdots c_k d_k}_{a_1 b_1 \cdots a_k b_k}$ of class C^{∞}, called a <u>**stress**</u> <u>**invariant**</u> of level k is uniquely defined on $M - \mathfrak{Z}$ such that

(5.74) $$S^{c_1 d_1 \cdots c_k d_k}_{a_1 b_1 \cdots a_k b_k}$$

$$= \frac{1}{2}\left[\prod_{j=1}^{k} \Lambda^{\lambda}_{a_j b_j} \bar{\Lambda}^{\lambda}_{c_j d_j} + \prod_{j=1}^{k} \Lambda^{\lambda}_{c_j d_j} \bar{\Lambda}^{\lambda}_{a_j b_j}\right].$$

Obviously we have

(5.75) $$-1 \leq S^{c_1 d_1 \cdots c_k d_k}_{a_1 b_1 \cdots a_k b_k} = S^{a_1 b_1 \cdots a_k b_k}_{c_1 d_1 \cdots c_k d_k} \leq 1$$

(5.76) $$-0 \leq S^{c_1 d_1 \cdots c_k d_k}_{a_1 b_1 \cdots a_k b_k} = \prod_{j=1}^{k} S^{a_j b_j}_{a_j b_j} \leq 1 .$$

We introduce the summation convention, which can be repeated

(5.77) $$\sum_{j=0}^{r} S^{\cdots j \cdots}_{\cdots j \cdots} = S^{\cdots r \cdots}_{\cdots r \cdots}{}^{\!\!|} .$$

Hence we have

$$\mu^2 \square f_g \dot{\boxplus} g_p \square^2 = \Phi_{pq} = S\underset{pq}{\overset{pq}{\|}} = \sum_{j=0}^{p}\sum_{k=0}^{q} S^{jk}_{jk} \tag{5.78}$$

$$\square f_g \dot{L} g \square^2 = \Phi_q = S\underset{0q}{\overset{0q}{|}} = \sum_{k=0}^{q} S^{0k}_{0k} \tag{5.79}$$

$$\square g_p \dot{L} f \square^2 = \Psi_p = S\underset{p0}{\overset{p0}{|}} = \sum_{j=0}^{p} S^{j0}_{j0} \tag{5.80}$$

$$S^{p}_{p+1}\underset{q}{\overset{q}{|}}{}^{p+1}_{p}\underset{q}{\overset{q}{|}} = \sum_{j=0}^{q}\sum_{k=0}^{q} S^{p\;\;j\;\;p+1\;\;k}_{p+1\;\;j\;\;p\;\;k} \geqq 0 \tag{5.81}$$

$$S\underset{p}{\overset{p}{|}}{}^{q}_{q+1}\underset{p}{\overset{p}{|}}{}^{q+1}_{q} = \sum_{j=0}^{p}\sum_{k=0}^{p} S^{j\;\;q\;\;k\;\;q+1}_{j\;\;q+1\;\;k\;\;q} \geqq 0 \tag{5.82}$$

where the non-negativity in (5.81) and (5.82) follow from the local identities

$$S^{p}_{p+1}\underset{q}{\overset{q}{|}}{}^{p+1}_{p}\underset{q}{\overset{q}{|}}\,|U_\lambda = \Big|\sum_{j=0}^{q} \Lambda^{\lambda}_{p+1,j}\,\bar{\Lambda}^{\lambda}_{p,j}\Big|^2 \tag{5.83}$$

$$S\underset{p}{\overset{p}{|}}{}^{q}_{q+1}\underset{p}{\overset{p}{|}}{}^{q+1}_{q}\,|U_\lambda = \Big|\sum_{j=0}^{p} \Lambda^{\lambda}_{j\,q+1}\,\bar{\Lambda}^{\lambda}_{jq}\Big|^2 \tag{5.84}$$

If g is constant, the function Φ_q are well known in value distribution theory. If g is not constant, the functions Φ_q acquire new properties. The functions Φ_{pq} are new. Now we will establish a curvature identity for the Φ_{pq} which is remarkable for its symmetry and simplicity. If an index in Λ_{ab} or S^{cd}_{ab} is outside its natural range, then define $\Lambda_{ab} = 0$ respectively $S^{cd}_{ab} = 0$.

THEOREM 5.9. **The Stress Curvature Formula.** Take $p \in \mathbb{Z}[0,s]$ and $q \in \mathbb{Z}[0,n]$. Then we have

(5.85)
$$\boxed{\begin{aligned} & m\, i_{m-1}\, dd^c \Phi_{pq} \wedge B \wedge \bar{B} \\ &= (\Phi_{p+1,q} - 2\Phi_{pq} + \Phi_{p-1,q})\mathbb{K}_p \\ &\quad + (\Phi_{p,q+1} - 2\Phi_{pq} + \Phi_{p,q-1})\mathbb{H}_q \\ &\quad + 2S^{p}_{p+1}{}^{q+1}_{q}\, \mathbb{K}_{pq} \end{aligned}}$$

Also we state the formula for the gradient of Φ_{pq} and we shall prove both theorems together.

THEOREM 5.10. **The Stress Gradient Formula.** Take $p \in \mathbb{Z}[0,s]$ and $q \in \mathbb{Z}[0,n]$. Then we have

(5.86)
$$\boxed{\begin{aligned} & m\, i_{m-1}\, d\Phi_{pq} \wedge d^c\Phi_{pq} \wedge B \wedge \bar{B} \\ &= S\begin{matrix} p & q & p+1 & q \\ & | & & | \\ p+1 & q & p & q \end{matrix}\, \mathbb{K}_p + S\begin{matrix} p & q & p & q+1 \\ | & & | & \\ p & q+1 & p & q \end{matrix}\, \mathbb{H}_q \\ &\quad + 2S\begin{matrix} p & q & p & q+1 \\ & | & | & \\ p+1 & q & p & q \end{matrix}\, \mathbb{K}_{pq} \end{aligned}}$$

PROOF. We have

(5.87)
$$\begin{aligned} d\Lambda_{pq} &= d(\mathfrak{y}_p \mid \mathfrak{e}_q^*) = (d\mathfrak{y}_p \mid \mathfrak{e}_q^*) + (\mathfrak{y}_p \mid d\mathfrak{e}_q^*) \\ &= \sum_{j=0}^{s} \eta_{pj}(\mathfrak{y}_j \mid \mathfrak{e}_q^*) - \sum_{j=0}^{n} (\mathfrak{y}_p \mid \mathfrak{e}_q^*)\bar{\theta}_{jq} + (\mathfrak{v}_p \mid \mathfrak{e}_q^*) \\ &= \sum_{j=0}^{s} \Lambda_{jq}\eta_{pj} + \sum_{j=0}^{n} \Lambda_{pj}\theta_{qj} + (\mathfrak{v}_p \mid \mathfrak{e}_q^*) \end{aligned}$$

(5.88)
$$\partial\Lambda_{pq} = \sum_{j=p}^{s} \Lambda_{jq}\dot{\eta}_{pj} + \sum_{j=q}^{n} \Lambda_{pj}\dot{\theta}_{qj} + (\dot{\mathfrak{v}}_p \mid \mathfrak{e}_q^*)$$

$$(5.89) \qquad \bar{\partial}\Lambda_{pq} = \sum_{j=0}^{p} \Lambda_{jq}\ddot{\eta}_{pj} + \sum_{j=0}^{q} \Lambda_{pj}\ddot{\theta}_{qj}$$

$$(5.90) \qquad \partial\bar{\Lambda}_{pq} = - \sum_{j=0}^{p} \bar{\Lambda}_{jq}\dot{\eta}_{jp} - \sum_{j=0}^{q} \bar{\Lambda}_{pj}\dot{\theta}_{jq}$$

$$(5.91) \qquad \bar{\partial}\bar{\Lambda}_{pq} = - \sum_{j=p}^{s} \bar{\Lambda}_{jq}\ddot{\eta}_{jp} - \sum_{j=q}^{n} \bar{\Lambda}_{pj}\ddot{\theta}_{jq} + (\mathfrak{k}_q^* \,|\, \dot{\mathfrak{v}}_p)$$

Theorem 5.4 implies

$$(5.92) \quad \partial\Lambda_{pq} \wedge B = (\Lambda_{pq}(\dot{\eta}_{pp} + \dot{\theta}_{qq}) + \Lambda_{p+1,q}\dot{\eta}_{p,p+1} + \Lambda_{p,q+1}\dot{\theta}_{q,q+1}) \wedge B$$

$$(5.93) \quad \bar{\partial}\Lambda_{pq} \wedge \bar{B} = (\Lambda_{pq}(\ddot{\eta}_{pp} + \ddot{\theta}_{qq}) + \Lambda_{p-1,q}\ddot{\eta}_{p,p-1} + \Lambda_{p,q-1}\ddot{\theta}_{q,q-1}) \wedge \bar{B}$$

$$(5.94) \quad \partial\bar{\Lambda}_{pq} \wedge B = - (\bar{\Lambda}_{pq}(\dot{\eta}_{pp} + \dot{\theta}_{qq}) + \bar{\Lambda}_{p-1,q}\dot{\eta}_{p,p-1} + \bar{\Lambda}_{p,q-1}\dot{\theta}_{q-1,q}) \wedge B$$

$$(5.95) \quad \bar{\partial}\bar{\Lambda}_{pq} \wedge \bar{B} = - (\bar{\Lambda}_{pq}(\ddot{\eta}_{pp} + \ddot{\theta}_{qq}) + \bar{\Lambda}_{p+1,q}\ddot{\eta}_{p+1,p} + \bar{\Lambda}_{p,q+1}\ddot{\theta}_{q+1,q}) \wedge \bar{B}$$

Now (5.74) implies

$$(5.96) \qquad \partial S^{pq}_{pq} \wedge B = \partial(\Lambda_{pq}\bar{\Lambda}_{pq}) \wedge B = \bar{\Lambda}_{pq}\partial\Lambda_{pq} \wedge B + \Lambda_{pq}\partial\bar{\Lambda}_{pq} \wedge B$$

$$= (\bar{\Lambda}_{pq}\Lambda_{p+1,q}\dot{\eta}_{p,p+1} - \Lambda_{pq}\bar{\Lambda}_{p-1,q}\dot{\eta}_{p-1,p}) \wedge B$$

$$+ (\bar{\Lambda}_{pq}\Lambda_{p,q+1}\dot{\theta}_{q,q+1} - \Lambda_{pq}\bar{\Lambda}_{p,q-1}\dot{\theta}_{q-1,q}) \wedge B \,.$$

Summation yields

(5.97) $\partial\Phi_{pq} \wedge B$

$$= \sum_{j=0}^{q} \bar{\Lambda}_{pj}\Lambda_{p+1,j}\dot{\eta}_{p,p+1} \wedge B + \sum_{k=0}^{p} \bar{\Lambda}_{kq}\Lambda_{k,q+1}\dot{\theta}_{q,q+1} \wedge B$$

(5.98) $\bar{\partial}\Phi_{pq} \wedge \bar{B}$

$$= - \sum_{j=0}^{q} \Lambda_{pj}\bar{\Lambda}_{p+1,j}\ddot{\eta}_{p+1,p} \wedge B - \sum_{k=0}^{p} \Lambda_{kq}\bar{\Lambda}_{k,q+1}\ddot{\theta}_{q+1,q} \wedge \bar{B}$$

(5.99)
$$\begin{aligned}(-1)^{m-1}\partial\bar{\partial}\Phi_{pq} \wedge B \wedge \bar{B} &= (-1)^m\bar{\partial}(\partial\Phi_{pq} \wedge B) \wedge \bar{B} \\
&= \sum_{j=0}^{q} \Lambda_{p+1,j}\dot{\eta}_{p,p+1} \wedge B \wedge \bar{\partial}\bar{\Lambda}_{pj} \wedge \bar{B} \\
&+ \sum_{j=0}^{q} \bar{\Lambda}_{pj}\dot{\eta}_{p,p+1} \wedge B \wedge \bar{\partial}\Lambda_{p+1,j} \wedge \bar{B} \\
&+ \sum_{j=0}^{q} \bar{\Lambda}_{pj}\Lambda_{p+1,j}(-1)^m\, \bar{\partial}\dot{\eta}_{p,p+1} \wedge B \wedge \bar{B} \\
&+ \sum_{k=0}^{p} \Lambda_{k,q+1}\dot{\theta}_{q,q+1} \wedge B \wedge \bar{\partial}\bar{\Lambda}_{kq} \wedge \bar{B} \\
&+ \sum_{k=0}^{p} \bar{\Lambda}_{kq}\dot{\theta}_{q,q+1} \wedge B \wedge \bar{\partial}\Lambda_{k,q+1} \wedge \bar{B} \\
&+ \sum_{k=0}^{p} \bar{\Lambda}_{kq}\Lambda_{k,q+1}(-1)^m\bar{\partial}\dot{\theta}_{q,q+1} \wedge B \wedge \bar{B}\end{aligned}$$

Since $i_m = mi_{m-1}(-1)^{m-1}\frac{i}{2\pi}$, we obtain

(5.100)
$$\begin{aligned}mi_{m-1}dd^c\Phi_{pq} \wedge B \wedge \bar{B} &= mi_{m-1}\frac{i}{2\pi}\,\partial\bar{\partial}\Phi_{pq} \wedge B \wedge \bar{B} \\
&= \sum_{x=1}^{8}\sum_{j=0}^{q} A_j(x) + \sum_{x=9}^{16}\sum_{k=0}^{p} A_k(x)\end{aligned}$$

where

$$A_j(1) = -\, i_m \Lambda_{p+1,j} \bar{\Lambda}_{pj} \dot{\eta}_{p,p+1} \wedge B \wedge (\ddot{\eta}_{pp} + \ddot{\theta}_{jj}) \wedge \bar{B}$$

$$A_j(2) = -\, i_m \Lambda_{p+1,j} \bar{\Lambda}_{p+1,j} \dot{\eta}_{p,p+1} \wedge B \wedge \ddot{\eta}_{p+1,p} \wedge \bar{B}$$

$$A_j(3) = -\, i_m \Lambda_{p+1,j} \bar{\Lambda}_{p,j+1} \dot{\eta}_{p,p+1} \wedge B \wedge \ddot{\theta}_{j+1,j} \wedge \bar{B}$$

$$A_j(4) = i_m \bar{\Lambda}_{pj} \Lambda_{p+1,j} \dot{\eta}_{p,p+1} \wedge B \wedge (\ddot{\eta}_{p+1,p+1} + \ddot{\theta}_{j,j}) \wedge \bar{B}$$

$$A_j(5) = i_m \bar{\Lambda}_{pj} \Lambda_{p,j} \dot{\eta}_{p,p+1} \wedge B \wedge \ddot{\eta}_{p+1,p} \wedge \bar{B}$$

$$A_j(6) = i_m \bar{\Lambda}_{pj} \Lambda_{p+1,j-1} \dot{\eta}_{p,p+1} \wedge B \wedge \ddot{\theta}_{j,j-1} \wedge \bar{B}$$

$$A_j(7) = i_m \bar{\Lambda}_{pj} \Lambda_{p+1,j} \dot{\eta}_{p,p+1} \wedge B \wedge \ddot{\eta}_{pp} \wedge \bar{B}$$

$$A_j(8) = -\, i_m \bar{\Lambda}_{pj} \Lambda_{p+1,j} \dot{\eta}_{p,p+1} \wedge B \wedge \ddot{\eta}_{p+1,p+1} \wedge \bar{B}$$

$$A_k(9) = -\, i_m \Lambda_{k,q+1} \bar{\Lambda}_{kq} \dot{\theta}_{q,q+1} \wedge B \wedge (\ddot{\eta}_{k,k} + \ddot{\theta}_{qq}) \wedge \bar{B}$$

$$A_k(10) = -\, i_m \Lambda_{k,q+1} \bar{\Lambda}_{k+1,q} \dot{\theta}_{q,q+1} \wedge B \wedge \ddot{\eta}_{k+1,k} \wedge \bar{B}$$

$$A_k(11) = -\, i_m \Lambda_{k,q+1} \bar{\Lambda}_{k,q+1} \dot{\theta}_{q,q+1} \wedge B \wedge \ddot{\theta}_{q+1,q} \wedge \bar{B}$$

$$A_k(12) = i_m \bar{\Lambda}_{kq} \Lambda_{k,q+1} \dot{\theta}_{q,q+1} \wedge B \wedge (\ddot{\eta}_{kk} + \ddot{\theta}_{q+1,q+1}) \wedge \bar{B}$$

$$A_k(13) = i_m \bar{\Lambda}_{kq} \Lambda_{k-1,q+1} \dot{\theta}_{q,q+1} \wedge B \wedge \ddot{\eta}_{k,k-1} \wedge \bar{B}$$

$$A_k(14) = i_m\bar{\Lambda}_{kq}\Lambda_{kq}\dot{\theta}_{q,q+1} \wedge B \wedge \ddot{\theta}_{q+1,q} \wedge \bar{B}$$

$$A_k(15) = -\ i_m\bar{\Lambda}_{kq}\Lambda_{k,q+1}\dot{\theta}_{q,q+1} \wedge B \wedge \ddot{\theta}_{q+1,q+1} \wedge \bar{B}$$

$$A_k(16) = i_m\bar{\Lambda}_{kq}\Lambda_{k,q+1}\dot{\theta}_{q,q+1} \wedge B \wedge \ddot{\theta}_{qq} \wedge \bar{B}$$

Here we have

$$\sum_{j=0}^{q} (A_j(1) + A_j(4) + A_j(7) + A_j(8)) = 0$$

$$\sum_{k=0}^{p} (A_k(9) + A_k(12) + A_k(15) + A_k(16)) = 0$$

$$\sum_{j=0}^{q} A_j(2) = S \begin{matrix} p+1 & q \\ & | \\ p+1 & q \end{matrix} \mathbb{K}_p = (\Phi_{p+1,q} - \Phi_{pq})\mathbb{K}_p$$

$$\sum_{j=0}^{q} (A_j(3) + A_j(6)) = \Lambda_{p+1,q}\bar{\Lambda}_{p,q+1}\mathbb{K}_{pq}$$

$$\sum_{j=0}^{q} A_j(5) = -\ S \begin{matrix} p & q \\ & | \\ p & q \end{matrix} \mathbb{K}_p = -\ (\Phi_{pq} - \Phi_{p-1,q})\mathbb{K}_p$$

$$\sum_{k=0}^{p} (A_k(10) + A_k(13)) = \Lambda_{p,q+1}\bar{\Lambda}_{p+1,q}\mathbb{K}_{pq}$$

$$\sum_{k=0}^{p} A_k(11) = S \begin{matrix} p & q+1 \\ | & \\ p & q+1 \end{matrix} \mathbb{H}_q = (\Phi_{p,q+1} - \Phi_{p,q})\mathbb{H}_q$$

$$\sum_{k=0}^{p} A_k(14) = -\ S \begin{matrix} p & q \\ | & \\ p & q \end{matrix} \mathbb{H}_q = -\ (\Phi_{pq} - \Phi_{p,q-1})\mathbb{H}_q$$

Consequently we have

$$
(5.85)\qquad
\boxed{
\begin{aligned}
& m i_{m-1}\, dd^c\Phi_{pq} \wedge B \wedge \bar{B} \\
& = (\Phi_{p+1,q} - 2\Phi_{pq} + \Phi_{p-1,q})\mathbb{K}_p \\
& \quad + (\Phi_{p,q+1} - 2\Phi_{pq} + \Phi_{p,q-1})\mathbb{H}_q \\
& \quad + 2S^{p\ \ q+1}_{p+1\ \ q}\,\mathbb{K}_{pq}
\end{aligned}
}
$$

which proves Theorem 5.9.

From (5.97) and (5.98) we obtain:

$$
\begin{aligned}
& m i_{m-1} d\Phi_{pq} \wedge d^c\Phi_{pq} \wedge B \wedge \bar{B} \\
& = m i_{m-1}(-1)^{m-1}(i/2\pi)\partial\Phi_{pq} \wedge B\ \bar{\partial}\Phi_{pq} \wedge \bar{B} \\
& = -\left|\sum_{j=0}^{q} \Lambda_{pj}\bar{\Lambda}_{p+1,j}\right|^2 i_m \dot{\eta}_{p,p+1} \wedge B\ \ddot{\eta}_{p+1,p} \wedge \bar{B} \\
& \quad - \left|\sum_{k=0}^{p} \Lambda_{kq}\bar{\Lambda}_{k,q+1}\right|^2 i_m \dot{\theta}_{q,q+1} \wedge B \wedge \ddot{\theta}_{q+1,q} \wedge \bar{B} \\
& \quad - \sum_{j=0}^{q}\sum_{k=0}^{p} \bar{\Lambda}_{pj}\Lambda_{p+1,j}\Lambda_{kq}\bar{\Lambda}_{k,q+1} i_m \dot{\eta}_{p,p+1} \wedge B \wedge \ddot{\theta}_{q+1,q} \wedge \bar{B} \\
& \quad - \sum_{j=0}^{q}\sum_{k=0}^{p} \bar{\Lambda}_{kq}\Lambda_{k,q+1}\Lambda_{pj}\bar{\Lambda}_{p+1,j} i_m \dot{\theta}_{q,q+1} \wedge B \wedge \ddot{\eta}_{p+1,p} \wedge B \\
& = S^{p\ \ q\ \ p+1\ \ q}_{p+1\ \ q\ \ p\ \ q}\,\mathbb{K}_p + S^{p\ \ q\ \ p\ \ q+1}_{p\ \ q+1\ \ p\ \ q}\,\mathbb{H}_q + 2S^{p\ \ q\ \ p\ \ q+1}_{p+1\ \ q\ \ p\ \ q}\,\mathbb{K}_{pq}
\end{aligned}
$$

which proves Theorem 5.10. q.e.d.

If $p = 0$ and if g is constant, the formulas (5.85) and (5.86) are well known. See Cowen-Griffiths [123], Pit-Mann Wong [21] and Stoll [108]; but even if $p = 0$ and if g is not constant, the formulas are new and were first derived without the use of Frenet frames. The identity

$$S_{pq} = \Phi_{pq} - \Phi_{pq-1} - \Phi_{p-1,q} + \Phi_{p-1,q-1} \tag{5.101}$$

yields the Levi form of S^{pq}_{pq}

$$m i_{m-1}\, dd^c S^{pq}_{pq} \wedge B \wedge \bar{B} =$$

$$+ \left(S^{p+1,q}_{p+1,q} - S^{p,q}_{p,q}\right) \mathbb{K}_p - \left(S^{pq}_{pq} - S^{p-1,q}_{p-1,q}\right) \mathbb{K}_{p-1}$$

$$+ \left(S^{p,q+1}_{p,q+1} - S^{pq}_{pq}\right) \mathbb{H}_q - \left(S^{pq}_{pq} - S^{p,q-1}_{p,q-1}\right) \mathbb{H}_{q-1}$$

$$+ 2S^{p\ \ ,q+1}_{p+1,\ q} \mathbb{K}_{pq} - 2S^{p-1,q+1}_{p\ \ ,\ q} \mathbb{K}_{p-1,q}$$

$$- 2S^{p\ \ ,\ q}_{p+1,q-1} \mathbb{K}_{p,q-1} + 2S^{p-1,\ q}_{p\ \ ,q-1} \mathbb{K}_{p-1,q-1}\,.$$

§6 The Ahlfors Estimates

If g is constant, the well known Ahlfors estimates for Φ_q lead to the defect relation. We shall extend the Ahlfors estimates to the functions Φ_{pq} where g is not constant. As before, $f : M \longrightarrow \mathbb{P}(V)$ and $g : M \longrightarrow \mathbb{P}(V^*)$ are meromorphic maps and f is general for B. We use the curvature method of Cowen and Griffiths [21] as modified by Pit-Mann Wong [123] and Stoll [108]. We assume the notations and constructions of the last chapter.

LEMMA 6.1. Take $p \in \mathbb{Z}[0,\ell_g]$ and $q \in \mathbb{Z}[0,n]$. Then

$$dd^c \log \Phi_{pq} = (f_g \boxplus g_p)^*(\Omega_{q-1,p-1}) - f_q^*(\Omega_q) - g_p^*(\Omega_p) \tag{6.1}$$

$$m i_{m-1}\, dd^c \log \Phi_{pq} \wedge B \wedge \bar{B} = \mathbb{H}_{pq} - \mathbb{K}_p - \mathbb{H}_q\,. \tag{6.2}$$

PROOF. Take reduced representations $\mathfrak{w} : U \longrightarrow V$ of f and $\mathfrak{w} : U \longrightarrow V^*$ of g. Then we have

$$\begin{aligned} dd^c \log \Phi_{pq} &= dd^c \log \|\mathfrak{w}_{\underline{q}} \boxplus \mathfrak{w}_{\underline{p}}\|^2 - dd^c \log \|\mathfrak{w}_{\underline{q}}\|^2 - dd^c \log \|\mathfrak{w}_{\underline{p}}\|^2 \\ &= (f_q \boxplus g_p)^*(\Omega_{q-1,p-1}) - f_q^*(\Omega_q) - g_p^*(\Omega_p) \end{aligned}$$

which proves (6.1). Now (4.20) and (4.46) imply (6.2). q.e.d.

Observe that $\mathbb{H}_{pq} \geqslant 0$, $\mathbb{K}_p \geqslant 0$ and $\mathbb{H}_q \geqslant 0$, which makes estimates possible. We will assume that $\Phi_{pq} \not\equiv 0$, which is the case if and only if the pair (f,g) is free of order (p,q).

<u>**LEMMA**</u> **6.2.** Take $p \in \mathbb{Z}[0,\ell_g]$ and $q \in \mathbb{Z}[0,n]$ and $\beta \in \mathbb{R}(0,1]$. Then

$$\begin{aligned}
(6.3)\qquad & m i_{m-1}\, dd^c \Phi_{pq}^{\beta} \wedge B \wedge \bar{B} \\
&= \beta^2 \Phi_{pq}^{\beta-1} (\Phi_{p+1,q} + \Phi_{p-1,q}) \mathbb{K}_p \\
&\quad + \beta^2 \Phi_{pq}^{\beta-1} (\Phi_{p,q+1} + \Phi_{p,q-1}) \mathbb{H}_q \\
&\quad + 2\beta^2 \Phi_{pq}^{\beta-1}\, S^{p \quad q+1}_{p+1 \quad q}\, \mathbb{K}_{pq} \\
&\quad + \beta(1-\beta)\Phi_{pq}^{\beta}\mathbb{H}_{pq} - \beta(\beta+1)\Phi_{pq}^{\beta}\mathbb{K}_p - \beta(\beta+1)\Phi_{pq}^{\beta}\mathbb{H}_q \,.
\end{aligned}$$

<u>**PROOF**</u>. A trivial calculation shows

$$(6.4)\qquad d\Phi_{pq} \wedge d^c \Phi_{pq} = \Phi_{pq}\, dd^c \Phi_{pq} - \Phi_{pq}^2\, dd^c \log \Phi_{pq} \,.$$

Therefore (5.85) and (6.2) imply

$$\begin{aligned}
& m i_{m-1}\, dd^c \Phi_{pq}^{\beta} \wedge B \wedge \bar{B} \\
&= \beta \Phi_{pq}^{\beta-1}\, m i_{m-1}\, dd^c \Phi_{pq} \wedge B \wedge \bar{B} \\
&\quad + \beta(\beta-1)\Phi_{pq}^{\beta-2}\, m i_{m-1}\, d\Phi_{pq} \wedge d^c \Phi_{pq} \wedge B \wedge \bar{B} \\
&= \beta^2 \Phi_{pq}^{\beta-1}\, m i_{m-1}\, dd^c \Phi_{pq} \wedge B \wedge \bar{B}
\end{aligned}$$

$$+ \beta(1 - \beta)\Phi_{pq}^{\beta} \ mi_{m-1} \ dd^c \log \Phi_{pq} \wedge B \wedge \bar{B}$$

$$= \beta^2\Phi_{pq}^{\beta-1}(\Phi_{p+1,q} - 2\Phi_{pq} + \Phi_{p-1,q})\mathbb{K}_p$$

$$+ \beta^2\Phi_{pq}^{\beta-1}(\Phi_{p,q+1} - 2\Phi_{pq} + \Phi_{p,q-1})\mathbb{H}_q$$

$$+ \beta^2\Phi_{pq}^{\beta-1} \ 2S_{p+1, \ \ p}^{p \ \ , \ \ q+1} \ \mathbb{K}_{pq}$$

$$+ \beta(1 - \beta)\Phi_{pq}^{\beta}(\mathbb{H}_{pq} - \mathbb{K}_p - \mathbb{H}_q)$$

$$= \beta^2\Phi_{pq}^{\beta-1}(\Phi_{p+1,q} + \Phi_{p-1,q})\mathbb{K}_p$$

$$+ \beta^2\Phi_{pq}^{\beta-1}(\Phi_{p,q+1} + \Phi_{p,q-1})\mathbb{H}_q$$

$$+ 2\beta^2\Phi_{pq}^{\beta-1} \ S_{p+1, \ \ q}^{p \ \ , \ \ q+1} \ \mathbb{K}_{pq}$$

$$+ \beta(1 - \beta)\Phi_{pq}^{\beta} \ \mathbb{H}_{pq}$$

$$- \beta(\beta + 1)\Phi_{pq}^{\beta} \ \mathbb{K}_p - \beta(\beta + 1)\Phi_{pq}^{\beta} \ \mathbb{H}_q \ . \qquad \text{q.e.d.}$$

LEMMA 6.3. Take $p \in \mathbb{Z}[0,\ell_g]$, $q \in \mathbb{Z}[0,n]$ and $\beta \in \mathbb{R}(0,1]$. Then

$$(1 + \Phi_{pq}^{\beta})^2 mi_{m-1} \ dd^c \log (1 + \Phi_{pq}^{\beta}) \wedge B \wedge \bar{B} \tag{6.5}$$

$$= \beta^2\Phi_{pq}^{\beta-1}(\Phi_{p+1,q} + \Phi_{p-1,q})\mathbb{K}_p$$

$$+ \beta^2\Phi_{pq}^{\beta-1}(\Phi_{p,q+1} + \Phi_{p,q-1})\mathbb{H}_q$$

$$+ 2\beta^2 \Phi_{pq}^{\beta-1} S_{p+1,\ q}^{p\ ,q+1} \mathbb{K}_{pq} + \beta(1 - \beta + \Phi_{pq}^{\beta})\Phi_{pq}^{\beta}\mathbb{H}_{pq}$$

$$- \beta(\beta + 1 + \Phi_{pq}^{\beta})\Phi_{pq}^{\beta} \mathbb{K}_p - \beta(\beta + 1 + \Phi_{pq}^{\beta})\Phi_{pq}^{\beta}\mathbb{H}_q \cdot$$

$$- \beta(\beta + 1 + \Phi_{pq}^{\beta})\Phi_{pq}^{\beta}\mathbb{K}_p - \beta(\beta + 1 + \Phi_{pq}^{\beta})\Phi_{pq}^{\beta}\mathbb{H}_q$$

PROOF. We have

$$(1 + \Phi_{pq}^{\beta})^2 \, mi_{m-1} \, dd^c \log(1 + \Phi_{pq}^{\beta}) \wedge B \wedge \bar{B}$$

$$= (1 + \Phi_{pq}^{\beta}) mi_{m-1} \, dd^c \Phi_{pq}^{\beta} \wedge B \wedge \bar{B} - mi_{m-1} \, d\Phi_{pq}^{\beta} \wedge d^c \Phi_{pq}^{\beta} \wedge B \wedge \bar{B}$$

$$= mi_{m-1}(dd^c \Phi_{pq}^{\beta} \wedge B \wedge \bar{B} + \beta \Phi_{pq}^{2\beta} \, dd^c \log \Phi_{pq} \wedge B \wedge \bar{B})$$

Now (6.3) and (6.2) imply (6.5). q.e.d.

LEMMA 6.4. Take $p \in \mathbb{Z}[0,\ell_g]$, $q \in \mathbb{Z}[0,n]$, $\beta \in \mathbb{R}(0,1]$ and $0 \leqslant \epsilon \in \mathbb{R}$. Then we have

$$dd^c \log(1 + (\Phi_{pq} + \epsilon)^{\beta}) + \beta g^*(\Omega_p) + \beta f_g^*(\Omega_q) \geqslant 0. \tag{6.6}$$

PROOF. Abbreviate $v = (\Phi_{pq} + \epsilon)^{\beta}$. Then Lemma 6.1 implies

$$dd^c v = v dd^c \log v + (1/v) dv \wedge d^c v \geqslant v dd^c \log v$$

$$dd^c \log(1 + v) = (1 + v)^{-2}((1 + v) dd^c v - dv \wedge d^c v)$$

$$= (1 + v)^{-2}(dd^c v + v^2 dd^c \log v)$$

$$\geqslant v(1 + v)^{-1} \, dd^c \log v$$

$$= \beta v(1+v)^{-1}\, dd^c \log(\epsilon + \Phi_{pq})$$

$$= \beta v(1+v)^{-1}(\epsilon + \Phi_{pq})^{-2}(\epsilon dd^c \Phi_{pq} + \Phi_{pq}^2\, dd^c \log \Phi_{pq})$$

$$\geqslant \beta v(1+v)^{-1}(\epsilon + \Phi_{pq})^{-1}\, \Phi_{pq}\, dd^c \log \Phi_{pq}$$

$$\geqslant - \beta v(1+v)^{-1}(\epsilon + \Phi_{pq})^{-1}\, \Phi_{pq}(g_p^*(\Omega_p) + f_g^*(\Omega_q)$$

$$\geqslant - \beta(g_p^*(\Omega_p) + f_g^*(\Omega_q)) \,. \qquad \text{q.e.d.}$$

LEMMA 6.5. Take $p \in \mathbb{Z}[0,\ell_g]$ and $q \in \mathbb{Z}[0,n]$. Then we have

$$\text{(6.7)} \qquad 2|S^{p\;\;,q+1}_{p+1,\;\;q}|\mathbb{K}_{pq} \leqslant S^{p+1,q}_{p+1,q}\,\mathbb{K}_p + S^{p,q+1}_{p,q+1}\,\mathbb{H}_q \,.$$

PROOF. Let $\mathfrak{w} : U \longrightarrow V$ and $\mathfrak{v}\mathfrak{w} : U \longrightarrow V^*$ be reduced representations. Then

$$2|S^{p\;\;,q+1}_{p+1,\;\;q}|\mathbb{K}_{pq} = |\Lambda_{pq}\bar{\Lambda}_{p+1,q} + \bar{\Lambda}_{pq}\Lambda_{p+1,q}|K_p H_q \chi^m$$

$$\leqslant 2|\Lambda_{p,q+1}|\;|\Lambda_{p+1,q}|K_p H_q \chi^m$$

$$\leqslant |\Lambda_{p+1,q}|^2 K_p^2 \chi^m + |\Lambda_{p,q+1}|^2 H_q^2 \chi^m$$

$$= S^{p+1,q}_{p+1,q}\,\mathbb{K}_p + S^{p,q+1}_{p,q+1}\,\mathbb{H}_q \qquad \text{q.e.d.}$$

For $q \in \mathbb{Z}[0,\ell_g]$ and $q \in \mathbb{Z}[0,n]$, define

$$\text{(6.8)} \qquad \Xi_{pq} = \Phi_{p+1,q}\,\mathbb{K}_p + \Phi_{p,q+1}\,\mathbb{H}_q + 2S^{p+1,q}_{p\;\;\;,q+1}\,\mathbb{K}_{pq} \,.$$

LEMMA 6.6. If $p \in \mathbb{Z}[0,\ell_g]$ and $q \in \mathbb{Z}[0,n]$, then

$$(6.9) \qquad 0 \leqslant \Phi_{pq}(\mathbb{K}_p + \mathbb{H}_q) \leqslant \Xi_{pq} \leqslant (n+1)(\mathbb{K}_p + \mathbb{H}_q) + 2\mathbb{K}_{pq}$$

PROOF. (5.78) implies

$$(6.10) \qquad \Phi_{p+1,q} \geqslant \Phi_{pq} + S^{p+1,q}_{p+1,q} \qquad \Phi_{p,q+1} \geqslant \Phi_{pq} + S^{p,q+1}_{p,q+1}$$

Hence we have

$$\Xi_{pq} \geqslant \Phi_{pq}(\mathbb{K}_p + \mathbb{H}_q) + S^{p+1,q}_{p+1,q}\,\mathbb{K}_p + S^{p,q+1}_{p,q+1}\,\mathbb{H}_q + 2S^{p\;\;,q+1}_{p+1,\;\;q}\,\mathbb{K}_{pq}$$

$$\geqslant \Phi_{pq}(\mathbb{K}_p + \mathbb{H}_q) \geqslant 0.$$

Moreover, $\Phi_{p+1,q} \leqslant \mu \leqslant n+1$ and $\Phi_{p,q+1} \leqslant \mu \leqslant n+1$ and $|S^{p\;\;q+1}_{p+1\;\;q}| \leqslant 1$ imply

$$\Xi_{pq} \leqslant (n+1)(\mathbb{K}_p + \mathbb{H}_q) + 2\mathbb{K}_{pq}. \qquad \text{q.e.d.}$$

LEMMA 6.7. Take $p \in \mathbb{Z}[0,\ell_g]$, $q \in \mathbb{Z}[0,n]$ and $\beta \in \mathbb{R}(0,1]$. Define $\mu = \mathrm{Min}(p+1,q+1)$. Then we have

$$\beta^2\left(\frac{1}{\mu^2}\,\Phi_{pq}\right)^{\beta-1}\Xi_{pq}$$

$$\leqslant 4(n+1)^2(m i_{m-1}\, dd^c \log(1+\Phi^{\beta}_{pq}) \wedge B \wedge \bar{B} + \mathbb{K}_p + \mathbb{H}_q)$$

PROOF. Trivially $1 \leqslant \mu \leqslant n+1$ and $(1+\Phi^{\beta}_{pq})^2 \leqslant 4\mu^{2\beta}$. Now Lemma 6.3 and 6.4 imply

$$4(n + 1)^2(mi_{m-1}\, dd^c \log(1 + \Phi_{pq}^{\beta}) \wedge B \wedge \bar{B} + \mathbb{K}_q + \mathbb{H}_q)$$

$$\geqslant \mu^{2-2\beta}(1 + \Phi_{pq}^{\beta})^2(mi_{m-1}\, dd^c \log(1 + \Phi_{pq}^{\beta}) \wedge B \wedge \bar{B} + \mathbb{K}_p + \mathbb{H}_q)$$

$$= \mu^{2-2\beta}(\beta^2\Phi_{pq}^{\beta-1}((\Phi_{p+1,q} + \Phi_{p-1,q})\mathbb{K}_p + (\Phi_{p,q+1} + \Phi_{p,q-1})\mathbb{H}_q)$$

$$+ 2\beta^2\Phi_{pq}^{\beta-1}\, S_{p+1,\ q}^{p\ ,q+1}\ \mathbb{K}_{pq} + \beta(1 - \beta + \Phi_{pq}^{\beta})\Phi_{pq}^{\beta}\ \mathbb{H}_{pq}$$

$$- (\beta(\beta + 1) + \Phi_{pq}^{\beta})\Phi_{pq}^{\beta}(\mathbb{K}_p + \mathbb{H}_q) + (1 + \Phi_{pq}^{\beta})^2(\mathbb{K}_p + \mathbb{H}_q))$$

$$\geqslant \mu^{2-2\beta}(\beta^2\Phi_{pq}^{\beta-1}\ \Xi_{pq} + ((1 + \Phi_{pq}^{\beta})^2 - \beta(\beta + \Phi_{pq}^{\beta})\Phi_{pq}^{\beta})(\mathbb{K}_p$$
$$+ \mathbb{H}_q))$$

$$\geqslant \beta^2\left[\frac{1}{\mu^2}\ \Phi_{pq}\right]^{\beta-1}\ \Xi_{pq}\,. \qquad \text{q.e.d.}$$

Let (M,τ) be a parabolic manifold of dimension m. Let $B \not\equiv 0$ be a holomorphic form of degree $m - 1$ on M. The exhaustion τ is said to **majorize** the holomorphic form B, if for every $r > 0$ there exists a constant $c \geqslant 1$ such that

$$\textbf{(6.11)} \qquad 0 \leqslant mi_{m-1}\, B \wedge \bar{B} \leqslant c\upsilon^{m-1} \quad \text{on} \quad M[r].$$

The infinum of all these constants is called $Y_0(r)$. Then $Y_0(r) \geqslant 1$. The function Y_0 increases. Define

$$\textbf{(6.12)} \qquad Y(r) = \lim_{r<t\to r} Y_0(t).$$

Then $Y(r) \geqslant Y_0(r) \geqslant 1$. The increasing function Y is called the **majorant** associated to τ and B. if $r > 0$, then

$$\textbf{(6.13)} \qquad mi_{m-1}\, B \wedge \bar{B} \leqslant Y(r)\upsilon^{m-1} \quad \text{on} \quad M[r]$$

(6.14) $$m i_{m-1} B \wedge \bar{B} \leq (Y \circ \sqrt{\tau}) \upsilon^{m-1} \quad \text{on } M.$$

Since υ^{m-1} may not be positive definite, the existence of the constant c in (6.11) is not assured. If $\upsilon^{m-1} > 0$, then c exists, however $\upsilon^{m-1} > 0$ implies $\upsilon > 0$ and (M,τ) is biholomorphically isometric to $(\mathbb{C}^m, \tau_0)$ where $\tau_0(z) = \|z\|^2$. If $\phi : M \longrightarrow \mathbb{C}^m$ is proper, surjective and holomorphic and if $\tau = \tau_0 \circ \phi = \|\phi\|^2$, then (M,τ) is parabolic and a holomorphic form $B \not\equiv 0$ of bidegree $(m-1,0)$ exists on M such that τ majorizes B with

(6.15) $$Y(r) \leq 1 + r^{2n-2}$$

and such that f is general for B. If $m = 1$, that is, if M is an open parabolic Riemann surface, we take $B = 1$, then $m i_{m-1} B \wedge \bar{B} = 1 = \upsilon^{m-1}$ and τ majorizes B with $Y \equiv 1$. The idea of an estimate (6.14) was first introduced by Stoll [93] and later refined in [100].

Now the following **general assumptions** shall be made:

(A1) Let M be a connected, complex manifold of dimension m.
(A2) Let τ be a parabolic exhaustion of M.
(A3) Let V be a hermitian vector space of dimension $n + 1 > 1$.
(A4) Let $f : M \longrightarrow \mathbb{P}(V)$ and $g : M \longrightarrow \mathbb{P}(V^*)$ be meromorphic maps.
(A5) Let B be a holomorphic form of bidegree $(m-1,0)$ on M.
(A6) Assume that τ majorizes B with majorant Y.
(A7) Assume that f is general for B.

Here (A7) implies that $B \not\equiv 0$ and that f is linearly non-degenerate.

LEMMA 6.8. Assume that (A1) and (A2) hold. Let N be a complex manifold. Let $\phi : N \longrightarrow M$ be a meromorphic map with indeterminancy I_ϕ. Let w be a function of class C^2 on $M \times N$. On $M - I_\phi$ define u by $u(x) = w(x, \phi(x))$ for all $x \in M - I_\phi$. Take $0 < s < r \in \mathfrak{L}_\tau$ with $s \in \mathfrak{L}_\tau$. Then we have

(6.16) $$\int_s^r \int_{M[t]} dd^c u \wedge \upsilon^{m-1} \frac{dt}{t^{2m-1}} = \frac{1}{2} \int_{M\langle r\rangle} u\sigma - \frac{1}{2} \int_{M\langle s\rangle} u\sigma.$$

PROOF. Let Γ be the closed graph of ϕ. Let $\pi : \Gamma \longrightarrow M$ and $\psi : \Gamma \longrightarrow N$ be the projections. Then $\phi \circ \pi = \psi$ and $w = u \circ \pi$ on $\Gamma - \pi^{-1}(I_\phi)$. On Γ, a parabolic exhaustion $\hat{\tau} = \tau \circ \pi$ is given such that $\Gamma[t] = \pi^{-1}(M[t])$ and $\Gamma\langle r\rangle = \pi^{-1}(M\langle r\rangle)$. Because π is a modification we have

$$\int_s^r \int_{M[t]} dd^c u \wedge \upsilon^{m-1} \frac{dt}{t^{2m-1}} = \int_s^r \int_{\Gamma[t]} dd^c w \wedge \hat{\upsilon}^{m-1} \frac{dt}{t^{2m-1}}.$$

If $r \in \mathfrak{L}_\tau$, we obtain

$$\int_{M\langle r\rangle} u\sigma = \int_{\Gamma\langle r\rangle} w\hat{\sigma}.$$

If $0 < s < r \in \mathfrak{L}_\tau$ and $s \in \mathfrak{L}_\tau$, then $\Gamma(r)$ and $\Gamma(s)$ are Stokes domains (Tung [112]). Therefore we have

$$\int_s^r \int_{M[t]} dd^c u \wedge \upsilon^{m-1} \frac{dt}{t^{2m-1}} = \int_s^r \int_{\Gamma[t]} dd^c w \wedge \hat{\upsilon}^{m-1} \frac{dt}{t^{2m-1}}$$

$$\int_s^r \int_{\Gamma\langle t\rangle} d^c w \wedge \hat{\upsilon}^{m-1} \frac{dt}{t^{2m-1}} = \int_s^r \int_{\Gamma\langle t\rangle} d^c w \wedge \hat{\omega}^{m-1} \frac{dt}{t}$$

$$= \frac{1}{2} \int_{\Gamma[r]-\Gamma[s]} d^c \log \hat{\tau} \wedge d^c w \wedge \hat{\omega}^{m-1}$$

$$= \frac{1}{2} \int_{\Gamma(r)-\Gamma[s]} dw \wedge d^c \log \hat{\tau} \wedge \hat{\omega}^{m-1}$$

$$- \frac{1}{2} \int_{\Gamma(r)-\Gamma[s]} dw \wedge \hat{\sigma}$$

$$- \frac{1}{2} \int_{\Gamma\langle r\rangle} w\hat{\sigma} - \frac{1}{2} \int_{\Gamma\langle s\rangle} w\hat{\sigma}$$

$$- \frac{1}{2} \int_{M\langle r\rangle} u\sigma - \frac{1}{2} \int_{M\langle s\rangle} u\sigma. \qquad \text{q.e.d.}$$

LEMMA 6.9. Assume that (A1) – (A7) hold. Take $p \in \mathbb{Z}[0,\ell_g]$, $q \in \mathbb{Z}[0,n]$, $\beta \in \mathbb{R}(0,1]$ and $r \in \mathbb{R}$ and $s \in \mathbb{R}$ with $0 < s < r$. Then we have

$$\int_s^r \int_{M[t]} dd^c \log(1 + \Phi_{pq}^{\beta}) \wedge \upsilon^{m-1} \frac{dt}{t^{2m-1}} \leqslant (n+1)c_{\beta} . \tag{6.17}$$

PROOF. Define $\mu = \mathrm{Min}(p+1,q+1)$. Define $N = G_q(V) \times G_p(V^*)$. A meromorphic map $\phi : M \longrightarrow N$ is defined such that

$$\phi(x) = (f_q(x), g_p(x)) \qquad \text{for all} \quad x \in M - (I_{f_q} \cup I_{g_p}) \tag{6.18}$$

Take $\epsilon > 0$. A function w of class C^∞ is defined on $M \times N$ by

$$w(x,y,z) = \log(1 + (\mu^2 \, \square\, y \,\dot{\boxplus}\, z \,\square^2 + \epsilon)^{\beta})$$

for all $x \in M$ and $(y,z) \in N$. Define u by $u(x) = w(x, \phi(x))$ for $x \in M - (I_{f_q} \cup I_{g_p})$. Then we have

$$u(x) = w(x, f_q(x), g_p(x))$$

$$= \log(1 + (\mu^2 \, \square\, f_q(x) \,\dot{\boxplus}\, g_p(x) \,\square^2 + \epsilon)^{\beta})$$

$$= \log(1 + (\Phi_{pq} + \epsilon)^{\beta}).$$

Take $0 < s < r \in \mathcal{E}_T$ with $s \in \mathcal{E}_T$. Then

$$\int_s^r \int_{M[t]} dd^c \log(1 + (\Phi_{pq} + \epsilon)^\beta) \wedge \upsilon^{m-1} \frac{dt}{t^{2m-1}}$$

$$= \frac{1}{2} \int_{M\langle r\rangle} \log(1 + (\Phi_{pq} + \epsilon)^\beta)\sigma - \frac{1}{2} \int_{M\langle s\rangle} \log(1 + (\Phi_{pq} + \epsilon)^\beta)\sigma$$

$$\leqslant \frac{c_1}{2} \log(1 + (\mu + \epsilon)^\beta) \leqslant \frac{c_1}{2} \log(n + 2 + \epsilon).$$

The estimate (6.6) and Fatou's Lemma imply

$$\int_s^r \int_{M[t]} dd^c \log(1 + \Phi_{pq}^\beta) \wedge \upsilon^{m-1} \frac{dt}{t^{2m-1}}$$

$$= \int_s^r \int_{M[t]} \lim_{0<\epsilon\to 0} (dd^c \log(1 + \Phi_{pq} + \epsilon)^\beta) \wedge \upsilon^{m-1} \frac{dt}{t^{2m-1}}$$

$$\leqslant \lim_{\epsilon\to 0} \int_s^r \int_{M[t]} dd^c \log(1 + (\Phi_{pq} + \epsilon)^\beta) \wedge \upsilon^{m-1} \frac{dt}{t^{2m-1}}$$

$$\leqslant \frac{c_1}{2} \log(n + 2) \leqslant c_1(n + 1).$$

Since $\mathcal{E}_T$ is dense in $\mathbb{R}^+$ we obtain (6.17) for all $0 < s < r$. q.e.d.

<u>**THEOREM**</u> **6.10.** <u>**Ahlfors Estimates**</u>. Assume that (A1) – (A7) hold. Take $p \in \mathbb{Z}[0,\ell_g]$, $q \in \mathbb{Z}[0,n]$, $\beta \in \mathbb{R}(0,1]$ and $0 < s < r$. Assume that (f,g) is free of order (p,q). Then

(6.19)
$$\boxed{\begin{aligned}&\beta^2 \int_s^r \int_{M[t]} \left[\frac{1}{\mu^2}\,\Phi_{pq}\right]^{\beta-1} \Xi_{pq}\,\frac{dt}{t^{2m-1}}\\ &\leqslant 4(n+1)^3\,Y(r)(T_{f_q}(r,s) + T_{g_p}(r,s) + C_1)\end{aligned}}$$

<u>**PROOF.**</u> Lemma 6.7, (6.6) with $\epsilon = 0$ and (6.13) imply

$$\begin{aligned}
&\beta^2\left[\frac{1}{\mu^2}\,\Phi_{pq}\right]^{\beta-1}\Xi_{pq}\\
&\qquad\leqslant 4(n+1)^2(\mathrm{mi}_{m-1}\,dd^c\log(1+\Phi_{pq}^{\beta}) \wedge B \wedge \bar{B} + \mathbb{K}_p + \mathbb{H}_q)\\
&\qquad= 4(n+1)^2\,\mathrm{mi}_{m-1}(dd^c\log(1+\Phi_{pq}^{\beta}) + g_p^*(\Omega_p) + f_q(\Omega_q)) \wedge B \wedge \bar{B}\\
&\qquad\leqslant 4(n+1)^2\,Y(r)(dd^c\log(1+\Phi_{pq}^{\beta}) + g_p^*(\Omega_p) + f_q(\Omega_q)) \wedge \upsilon^{m-1}
\end{aligned}$$

on $M[r]$. Hence we have

$$\begin{aligned}
&\beta^2 \int_s^r \int_{M[t]} \left[\frac{1}{\mu^2}\,\Phi_{pq}\right]^{\beta-1} \Xi_{pq}\,\frac{dt}{t^{2m-1}}\\
&\qquad\leqslant 4(n+1)^2 Y(r)\left[\int_s^r \int_{M[t]} dd^c\log(1+\Phi_{pq}^{\beta}) \wedge \upsilon^{m-1}\,\frac{dt}{t^{2m-1}}\right.\\
&\qquad\qquad\left. + T_{g_p}(r,s) + T_{f_q}(r,s)\right]\\
&\qquad\leqslant 4(n+1)^2 Y(r)((n+1)C_1 + T_{g_p}(r,s) + T_{f_q}(r,s))\\
&\qquad\leqslant 4(n+1)^3 Y(r)(C_1 + T_{g_p}(r,s) + T_{f_q}(r,s)).
\end{aligned}$$

q.e.d.

We will establish an asymptotic Ahlfors Estimate which will involve the characteristic of f and g only. First we will recall some well known results.

LEMMA 6.11. Let A be a non-negative, increasing function on $\mathbb{R}_+$. For $0 < s < r$ define

$$T(r,s) = \int_s^r A(t)\,\frac{dt}{t} \tag{6.20}$$

and $A(\infty) = \lim_{t\to\infty} A(t) \leqslant \infty$. Then we have

$$\frac{T(r,s)}{\log r} \longrightarrow A(\infty) \qquad \text{for } r \longrightarrow \infty. \tag{6.21}$$

The proof is left to the reader.

LEMMA 6.12. Assume that (A1) – (A6) are satisfied. Take $p \in \mathbb{Z}[0,\ell_f)$. Then

$$0 < \frac{1}{r^{2m-2}} \int_{M(r)} \mathbb{H}_p \leqslant Y(r)A_{f_p}(r) \leqslant Y(r)A_{f_p}(\infty) \tag{6.22}$$

$$T_{f_p}(r,s) \longrightarrow \infty \qquad \text{for } r \longrightarrow \infty\,. \tag{6.23}$$

PROOF. Take $r > 0$. Then $\mathbb{H}_p > 0$ on $M(r)$ except on a thin analytic subset. Therefore we have

$$0 < \mathbb{H}_p = mi_{m-1}\, f_p^*(\Omega_p) \wedge B \wedge \bar{B} \leqslant Y(r) f_p^*(\Omega_p) \wedge \upsilon^{m-1} \tag{6.24}$$

which implies (6.22) and by Lemma 6.11 we also obtain (6.22). q.e.d.

For $0 < s < r$ we have

$$\int_0^r \int_{M[t]} \mathbb{H}_p \frac{dt}{t^{2m-1}} \leqslant Y(r)T_{f_p}(r,s)\,. \tag{6.25}$$

Take $0 \leqslant s \in \mathbb{R}$. Let ϕ and ψ be real valued functions on $\mathbb{R}(s,+\infty)$. We write $\phi \underset{\bullet}{\leqslant} \psi$ if there exists a subset E of finite measure in $\mathbb{R}_+$ such that $\phi(x) \leqslant \psi(x)$ for all $x \in \mathbb{R}(s,+\infty) - E$. The calculus of $\underset{\bullet}{\leqslant}$ is self evident. In particular $\varphi \underset{\bullet}{\leqslant} \psi$ implies

$$\liminf_{r\to\infty} \phi(r) \leqslant \limsup_{r\to\infty} \psi(r). \tag{6.26}$$

Nevanlinna [67] established the following well known result.

<u>LEMMA</u> 6.13. Take $\epsilon > 0$ and $s > 0$. Let $\phi \geqslant 0$ be a function which is integrable over $\mathbb{R}[s,r]$ for each $r > s$. Define $F : \mathbb{R}[s,+\infty] \longrightarrow \mathbb{R}_+$ by

$$F(x) = \int_s^x \phi(t)dt \tag{6.27}$$

for $x \geqslant s$. Then $\phi \underset{\bullet}{\leqslant} F^{1+\epsilon}$.

<u>PROPOSITION</u> 6.14. Assume that (A1) – (A6) are satisfied. Take $p \in \mathbb{Z}[0,\ell_f)$ and $\epsilon > 0$. Then $h_p^2\sigma$ is integerable over $M\langle r\rangle$ for almost all $r \in \ell_\tau$ and

$$\int_{M\langle r\rangle} h_p^2\sigma \underset{\bullet}{\leqslant} r^{\epsilon(2m-1)}(Y(r)T_{f_p}(r,s))^{(1+\epsilon)^2} \tag{6.28}$$

$$S_{f_p}(r) \underset{\bullet}{\leqslant} (c_1/2)(1+\epsilon)^2(\log T_{f_p}(r,s) + \log Y(r)) + (c_1/2)\epsilon \log r \tag{6.29}$$

where $S_{f_p}(r)$ is defined by (4.31) and (4.32).

PROOF. Fubini's Theorem implies

$$\int_{M[r]} \mathbb{H}_p = \int_{M[r]} h_p \upsilon^m = m \int_{M[r]} h_p^2 \tau^{m-1}\, d\tau \wedge \sigma$$

$$= 2m \int_0^r \left[\int_{M\langle t\rangle} h_p^2 \sigma \right] t^{2m-1}\, dt$$

which shows that $h_p^2 \sigma$ is integrable over $M\langle t\rangle$ for almost all $t \in \mathcal{L}_\tau$. (6.25) and Lemma 6.13 imply

$$\int_{M\langle r\rangle} h_p^2 \sigma \lessdot (1/(2m)) r^{1-2m} \left[\int_{M[r]} \mathbb{H}_p \right]^{1+\epsilon}$$

$$\int_{M[r]} \mathbb{H}_p \lessdot r^{2m-1} (Y(r) T_{f_p}(r,s))^{1+\epsilon} .$$

Hence

$$\int_{M\langle r\rangle} h_p^2 \sigma \lessdot (1/(2m)) r^{\epsilon(2m-1)} (Y(r) T_{f_p}(r,s))^{(1+\epsilon)^2}$$

which implies (6.28). Also we have

$$S_{f_p}(r) = \int_{M\langle r\rangle} \log h_p \sigma = (c/2) \log \left[\frac{1}{c} \int_{M\langle r\rangle} h_p^2 \sigma) \right]$$

$$\lessdot (c/2)(1+\epsilon)^2 (\log T_{f_p}(r,s) + \log Y(r)) + (2m-1)(c/2)\epsilon \log r$$

$$+ (c/2)\log(c/2)$$

$$\lessdot (c/2)(1+\epsilon)^2 (\log T_{f_p}(r,s) + \log Y(r)) + mc\epsilon \log r.$$

Replacing c by $c/(2m)$ implies (6.39). q.e.d.

The following proposition was proved in Stoll [108] Proposition 10.9. For completeness sake, the proof shall be repeated here.

PROPOSITION 6.15. Assume that (A1) – (A6) hold. Take $p \in \mathbb{Z}[0,\ell_f]$ and $\epsilon > 0$. Abbreviate

$$Q_\epsilon(r,s) = \log Y(r) + \mathrm{Ric}_\tau(r,s) + \epsilon c_\nu \log r \tag{6.30}$$

for $0 < s < r$. Then

$$T_{f_p}(r,s) \lessdot 3^p\, T_f(r,s) + \tfrac{1}{2}(3^p - 1)Q_\epsilon(r,s).$$

PROOF. The estimate (6.31) is trivial for $p = 0$. Assume that $0 \leqslant p < \ell_p$ and that (6.31) holds for p. Then (6.31) shall be proved for $p + 1$. W.l.o.g. we can assume that $0 < \epsilon < \sqrt{2} - 1$. The Plücker Difference Formula (4.33) and (6.29) imply

$$\begin{aligned} T_{f_{p+1}}(r,s) &\leqslant 2T_{f_p}(r,s) + S_{f_p}(r) - S_{f_p}(s) + \mathrm{Ric}_\tau(r,s) \\ &\lessdot 2T_{f_p}(r,s) + (c_\nu/2)(1+\epsilon)^2 \log T_{f_p}(r,s) + c_\nu \log Y(r) \\ &\quad + \mathrm{Ric}_\tau(r,s) + \epsilon c_\nu \log r \\ &\leqslant 3T_{f_p}(r,s) + Q_\epsilon(r,s) \\ &\leqslant 3^{p+1}\, T_f(r,s) + \tfrac{1}{2}(3^{p+1} - 1)Q_\epsilon(r,s)\,. \end{aligned}$$

q.e.d.

THEOREM 6.16. **Ahlfors Estimates**. Assume that (A1) – (A7) hold. Take $\beta \in \mathbb{R}(0,1)$, $\epsilon > 0$ and $s > 0$. Take $p \in \mathbb{Z}[0,\ell_g]$ and $q \in \mathbb{Z}[0,\ell_f]$. Assume that (f,g) is free of order (p,q). Then we have

(6.32)

$$\beta^2 \int_s^r \int_{M[t]} \left[\frac{1}{\mu^2}\Phi_{pq}\right]^{\beta-1} \Xi_{pq} \frac{dt}{t^{2m-1}}$$

$$\leqq 4(n+1)^3(3^q T_f(r,s) + 3^p T_g(r,s) + \tfrac{1}{2}(3^p + 3^q - 2)Q_\epsilon(r,s)) .$$

The proof is easily obtained from Theorem 6.10 and Proposition 6.15. We need a refined version.

THEOREM 6.17. **Ahlfors Estimates.** Assume that (A1) – (A7) hold. Take $p \in \mathbb{Z}[0,\ell_g]$ and $q \in \mathbb{Z}[0,\ell_f]$. Assume that (f,g) is free of order (p,q). Take $\epsilon > 0$ and $s > 0$. Define $\beta : \mathbb{R}_+ \longrightarrow \mathbb{R}(0,1)$ by

(6.33)

$$\frac{1}{\beta(r)} = \begin{cases} 1 + T_{f_q}(r,s) + T_{g_p}(r,s) + m_{f_q \dot{\boxplus} g_p}(s) & \text{if } r \geqq s \\ 1 + m_{f_q \dot{\boxplus} g_p}(s) & \text{if } 0 \leqq r < s . \end{cases}$$

Define M^+ by (4.26). On M^+ define ξ_{pq} bt $\Xi_{pq} = \xi_{pq}\upsilon^m$. On $M - M^+$ put $\xi_{pq} = 0$. Then $\xi_{pq} \geqq 0$ on M. For almost all $r > 0$ the integral

(6.34)
$$F(r) = \int_{M\langle r\rangle} \left[\frac{1}{\mu^2}\Phi_{pq}\right]^{\beta(r)-1} \xi_{pq}\sigma$$

exists. We have the estimate

$\log^+ F(r)$

(6.35) $$\leqq 2(1+\epsilon)(\log T_f(r,s) + \log^+ T_g(r,s) + \log Y(r) + \log^+ \mathrm{Ric}_\tau(r,s)) + C_\epsilon \log r.$$

PROOF. Define $\eta = \beta \circ \sqrt{\tau} : M \longrightarrow (0,1)$. For $r > 0$ the following integral exists (Theorem 6.10)

$$F_0(r) = \int_{M[r]} \left[\frac{1}{\mu^2}\Phi_{pq}\right]^{\beta(r)-1} \Xi_{pq} \geqq \int_{M[r]} \left[\frac{1}{\mu^2}\Phi_{pq}\right]^{\eta-1} \Xi_{pq}$$

$$\geqq \int_{M[r]} \left[\frac{1}{\mu^2}\Phi_{pq}\right]^{\eta-1} \xi_{pq} \upsilon^m$$

$$= m \int_{M[r]} \left[\frac{1}{\mu^2}\Phi_{pq}\right]^{\eta-1} \xi_{pq} \tau^{m-1} d\tau \wedge \sigma = F_1(r)$$

$$F_1(r) = 2m \int_s^r \int_{M\langle t\rangle} \left[\frac{1}{\mu^2}\Phi_{pq}\right]^{\beta(t)-1} \xi_{pq}\sigma) t^{2m-1} dt$$

$$= 2m \int_0^r F(t) t^{2m-1}\, dt.$$

Hence the integral $F(t)$ exists for almost all $t > 0$. Lemma 6.13 implies

$$F(r) \leqq 2mF(r) \leqq r^{1-2m} F_1(r)^{1+\epsilon}$$

Define

$$F_2(r) = \int_s^r F_1(t) \frac{dt}{t^{2m-1}} \qquad \text{for } r > s\,.$$

Then

$$F_1(r) \lessdot r^{2m-1}(F_2(r))^{1+\epsilon}$$

$$F(r) \lessdot r^{\epsilon(2m-1)}(F_1(r))^{(1+\epsilon)^2}$$

where

$$F_2(r) \leqslant \int_s^r F_0(t)\frac{dt}{t^{2m-1}} \leqslant \frac{4(n+1)^3}{\beta(r)^2} Y(r)(T_{f_q}(r,s) + T_{g_p}(r,s) + c_1) .$$

Take a constant $c_0 > 1 + c_1 + m_{f_q \dot{\boxplus} g_p}(s)$. Then

$$F_2(r) \leqslant 4(n+1)^3 Y(r)(T_{f_q}(r,s) + T_{g_p}(r,s) + c_0)^3 \qquad \text{if} \quad r > s .$$

consequently we obtain

$$\begin{aligned}
\log^+ F(r) &\lessdot (1+\epsilon)^2 \log^+ F_2(r) + c_1(2m-1)\log^+ r \\
&\lessdot (1+\epsilon)^2 \log Y(r) + 3(1+\epsilon)^2 \log^+(T_{f_q}(r,s) + T_{g_p}(r,s)) \\
&\quad + \epsilon(2m-1)\log^+ r + \log(4(n+1)^3) \\
&\quad + 3(1+\epsilon)^2(\log c_0 + \log 2) \\
&\lessdot (1+\epsilon)^2 \log Y(r) + 2m\epsilon \log^+ r \\
&\quad + 3(1+\epsilon)^2 \log^+(3^q T_f(r,s) + 3^p T_g(r,s) + \tfrac{1}{2}(3^p + 3^q - 2)Q_\epsilon(r,s)) \\
&\lessdot (1+\epsilon)^2 \log Y(r) + 2m\epsilon \log^+ r \\
&\quad + 3(1+\epsilon)^2(\log^+ T_f(r,s) + \log^+ T_g(r,s) + \log^+ Q_\epsilon(r,s))
\end{aligned}$$

Here we have

$$\log^+ Q_\epsilon(r,s) = \log^+(c_1 \log Y(r) + \mathrm{Ric}_\tau(r,s) + \epsilon c_1 \log r)$$

$$\leq \log^+ \log Y(r) + \log^+ \mathrm{Ric}_\tau(r,s) + \log^+ \log r$$

$$+ \log^+ c_1 + \log^+ \epsilon c_1 + \log 3$$

$$\underset{\cdot}{\leq} \frac{\epsilon}{3(1+\epsilon)^2} \log Y(r) + \log^+ T_g(r,s) + \frac{m\epsilon \log r}{3(1+\epsilon)^2} .$$

We obtain

$$\log^+ F(r) \underset{\cdot}{\leq} ((1+\epsilon)^2 + \epsilon) \log Y(r) + 4m\epsilon \log r$$

$$+ 3(1+\epsilon)^2(\log T_f(r,s) + \log^+ T_g(r,s) + \log^+ \mathrm{Ric}_\tau(r,s).$$

Without loss of generality we can assume that $0 < \epsilon < 1$. Then $(1+\epsilon)^2 + \epsilon \leq 1 + 4\epsilon$. Hence replacing ϵ by $\frac{\epsilon}{4m}$ we obtain

$$\log^+ F(r) \leq 3(1+\epsilon)(\log T_f(r,s) + \log^+ T_g(r,s) + \log Y(r)$$

$$+ \log^+ \mathrm{Ric}_\tau(r,s)) + \epsilon \log r . \qquad \text{q.e.d.}$$

On M^+ we have $\mathbb{K}_p = k_p^2 \upsilon^m$ and $\mathbb{H}_q = h_q^2 \upsilon^m$ and $\mathbb{K}_{pq} = k_p h_q \upsilon^m$. Therefore we obtain

(6.36) $$\xi_{pq} = \Phi_{p+1,q}\, k_p^2 + \Phi_{p,q+1}\, h_q^2 + 2S_{p\;\;,q+1}^{p+1,\;q}\, k_p h_q$$

on M^+. Define $\rho_{pq} = 1$ on $M - M^+$ and

(6.37) $$0 \leq \rho_{pq} = \frac{\Phi_{p,q+1} h_q^2}{\xi_{pq}} \quad \text{on } M^+$$

If $S^{p+1,\ q}_{p\ ,\ q+1} \geqslant 0$, then $\rho_{pq} \leqslant 1$ which in some sense measures attraction between f and g. If $S^{p+1,\ q}_{p\ ,q+1} h_q \leqslant -\Phi_{p+1,q}\, k_p$, then $\rho_{pq} \geqslant 1$, which in some sense measures distraction between f and g. In general, ρ_{pq} measures the twisting of f and g. Asymptotically we define the **obstruction terms**

(6.38)
$$P_{pq}(r) = \frac{1}{2} \int_{M\langle r\rangle} \log \rho_{pq}\, \sigma$$

(6.39)
$$R_{pq}(r) = \int_{M\langle r\rangle} \log\left[1 + \frac{k_p}{h_q}\right] \sigma$$

whenever the integrals exist. Here P_{pq} measures the twist between f and g, and R_{pq} the relative growth. Both terms will appear in the Second Main Theorem as unwelcome guests. If $k_p \equiv 0$, then $\rho_{pq} \equiv 1$ and $P_{pq} \equiv 0 \equiv R_{pq}$. This observation will lead to an important application (Chapter 9).

PROPOSITION 6.18. Assume that (A1) (A7) are satisfied. Take $p \in \mathbb{Z}[0,\ell_g]$ and $q \in \mathbb{Z}[0,n)$. Assume that (f,g) is free of order (p,q). Then $P_{pq}(r)$ and $R_{pq}(r)$ exist for almost all $r > 0$. We have the estimates

(6.40)
$$P_{pq}(r) \leqslant m_{f_q \dot{\boxplus} g_p}(r) - m_{f_{q+1} \dot{\boxplus} g_p}(r) + c_n$$

(6.41)
$$-c_n \log(n+1) \leqslant m_{f_{q+1} \dot{\boxplus} g_p}(r) + R_{pq}(r) + P_{pq}(r)$$
$$\leqslant m_{f_q \dot{\boxplus} g_p}(r) + 2c_n \log 2.$$

If $s > 0$ and $\epsilon > 0$, then we have

$$(6.42)\qquad R_{pq}(r) + S_{f_q}(r) \leqq (c_1/2)(1+\epsilon)^2(\log^+ T_{f_q}(r,s) + \log^+ T_{g_p}(r,s) + \log Y(r)) + (c_1/2)\epsilon \log r.$$

PROOF. The inequality (6.9) implies

$$0 \leqslant \Phi_{pq}(k_p^2 + h_q^2) \leqslant \xi_{pq} \leqslant (k_p + h_q)^2 \leqslant 2(n+1)(k_p^2 + h_q^2)$$

on M^+. Therefore

$$(6.43)\qquad \frac{\Phi_{p,q+1} h_q^2}{(n+1)(h_q + k_p)^2} \leqslant \rho_{pq} \leqslant \frac{\Phi_{p,q+1} h_q^2}{\Phi_{pq}(k_p^2 + h_q^2)} \leqslant \frac{\Phi_{p,q+1}}{\Phi_{pq}}$$

on M^+. Also we have

$$(6.44)\qquad 0 \leqslant \log\left(1 + \frac{k_p}{h_q}\right) \leqslant \log^+ h_q + \log^+ k_p - \log h_q + \log 2.$$

For almost all $r \in \mathcal{L}_\tau$, the forms $(\log h_q)\sigma$ and $(\log k_q)\sigma$ are integrable over $M\langle r\rangle$. The same is true for $(\log^+ h_q)\sigma$ and $(\log^+ k_q)\sigma$ consequently, $\log(1 + (k_p/h_q))\sigma$ is integrable over $M\langle r\rangle$ and $R_{pq}(r)$ exists for almost all $r \in \mathcal{L}_\tau$. We have

$$\log \Phi_{p,q+1} - \log\left(1 + \frac{k_p}{h_q}\right)^2 - \log(n+1) \leqslant \log \rho_{pq}$$

$$\log \rho_{pq} \leqslant \log \Phi_{p,q+1} - \log \Phi_{pq}.$$

Define $\mu(p,q) = \mathrm{Min}(p+1,q+1)$. Take $r \in \mathcal{L}_\tau$. Hence

$$\frac{1}{2}\int_{M\langle r\rangle} \log \Phi_{pq}\,\sigma = -\, m_{f_q \dot{\boxplus} g_p}(r) + c_1 \log \mu(p,q)$$

exists. Therefore $(\log \rho_{pq})\sigma$ is integrable over $M\langle r\rangle$ and $P_{pq}(r)$ exists for almost all $r \in \ell_\tau$. We have the estimates

$$- m_{f_{q+1}\dot{\boxplus} g_p}(r) + c_n \log \frac{\mu(p,q)}{n+1} - R_{pq}(r) \leqq P_{pq}(r)$$

$$P_{pq}(r) \leqq - m_{f_{q+1}\dot{\boxplus} g_p}(r) + m_{f_q\dot{\boxplus} g_p}(r) + c_n \log \frac{\mu(p,q+1)}{\mu(p,q)}$$

which implies (6.40) and the left hand side of (6.41). Also we have

$$\begin{aligned}\log \rho_{pq} &\leqq \log \Phi_{p,q+1} - \log \Phi_{pq} - \log\left(1 + \frac{k_p^2}{h_q^2}\right)\\ &\leqq \log \Phi_{p,q+1} - \log \Phi_{pq} - 2\log\left(1 + \frac{k_p}{h_q}\right) + \log 2\end{aligned}$$

and

$$\begin{aligned}P_{pq}(r) \leqq\ & m_{f_q\dot{\boxplus} g_p}(r) - m_{f_{q+1}\dot{\boxplus} g_p}(r) - R_{pq}(r) + (c_n/2)\log 2\\ &+ c_n \log \frac{\mu(p,q+1)}{\mu(p,q)}\end{aligned}$$

where $\mu(p,q+1)/\mu(p,q) \leqq 2$. Hence we obtain the right side estimate of (6.41). q.e.d.

PROPOSITION 6.19. Assume that (A1) – (A7) are satisfied. Take $p \in \mathbb{Z}[0,\ell_g]$ and $q \in \mathbb{Z}[0,n)$. Assume that (f,g) is free of order (p,q). Take $s > 0$ and $\epsilon \in \mathbb{R}(0,1)$. Then we have

(6.45)

$$\begin{aligned}&m_{f_q\dot{\boxplus} g_p}(r) - m_{f_{q+1}\dot{\boxplus} g_p}(r) + S_{f_q}(r)\\ &\quad \underset{\bullet}{\leqq} P_{pq}(r) + 2c_n(1+\epsilon)(\log T_f(r,s) + \log^+ T_g(r,s) + \log Y(r) + \log^+ \mathrm{Ric}_\tau(r,s))\\ &\qquad + \epsilon \log r\end{aligned}$$

PROOF. Define β by (6.33) and $F(r)$ by (6.34). Then we have

$$m_{f_q \dot{\boxplus} g_p}(r) - m_{f_{q+1} \dot{\boxplus} g_p}(r) + S_{f_q}(r)$$

$$= (1 - \beta(r)) m_{f_q \dot{\boxplus} g_p}(r) + S_{f_q}(r) + \beta(r) m_{f_q \dot{\boxplus} g_p}(r) - m_{f_{q+1} \dot{\boxplus} g_p}(r)$$

$$= \frac{1}{2} \int_{M\langle r\rangle} \log\left[\left(\frac{1}{\mu(p,q)^2}\Phi_{pq}\right)^{\beta(r)-1}\left(\frac{1}{\mu(p,q+1)^2}\Phi_{p,q+1}\right)h_q^2\right]\sigma$$

$$+ \beta(r) m_{f_q \dot{\boxplus} g_p}(r)$$

$$= \frac{1}{2} \int_{M\langle r\rangle} \log\left[\left(\frac{1}{\mu(p,q)^2}\Phi_{pq}\right)^{\beta(r)-1}\xi_{pq}\right]\sigma + P_{pq}(r)$$

$$+ c_1 \log \frac{1}{\mu(p,q+1)} + \beta(r) m_{f_q \dot{\boxplus} g_p}(r)$$

$$\leqslant (c_1/2)\log\left[\frac{1}{c_1}F(r)\right] + P_{pq}(r) + \beta(r) m_{f_q \dot{\boxplus} g_p}(r)$$

$$\underset{\cdot}{\leqslant} 2c_1(1+\epsilon)(\log T_f(r,s) + \log^+ T_g(r,s) + \log Y(r) + \log^+ \mathrm{Ric}_\tau(r,s))$$

$$+ \frac{c_1}{2}\epsilon \log r + P_{pq}(r) + \beta(r)(T_{f_q}(r,s) + T_{g_p}(r,s) + m_{f_q \dot{\boxplus} g_p}(s))$$

$$\underset{\cdot}{\leqslant} 2c_1(1+\epsilon)(\log T_f(r,s) + \log^+ T_g(r,s) + \log Y(r) + \log^+ \mathrm{Ric}_\tau(r,s))$$

$$+ \frac{c_1}{2}\epsilon \log r + P_{pq}(r) + 1$$

$$\lessdot\ 2c_1(2+\epsilon)(\log T_f(r,s) + \log^+ T_g(r,s) + \log Y(r) + \log^+ \mathrm{Ric}_\tau(r,s))$$

$$+ P_{pq}(r) + c_1\epsilon \log r.$$

If we replace c_1 by $(\epsilon/(1+c_1)) < \epsilon$, we obtain (6.44). q.e.d.

We will extend Proposition 6.19 to the Second Main Theorem by using a family of meromorphic associated maps g_p in general position.

§7 General Position

The properties of general position have been discussed in many places, for instance, Ahlfors [1], Weyl [119], Stoll [93] and Wu [126]. We hope to avoid the index acrobatics by which the reader is often confused. However the theory will remain difficult enough. Here the product to sum estimates contain constants which we have to know explicitly, since our target spaces move. Thus we are confronted with a difficult task, which however can be solved satisfactorily.

Let V be a complex vector space of dimension $n + 1 > 1$. Let $A \neq \emptyset$ be a subset of V. The <u>span</u> (or <u>linear</u> <u>hull</u>) of A is the intersection of all linear subspaces containing A and is denoted by $\mathrm{span}(A;V)$.

<u>LEMMA</u> 7.1. The following conditions are equivalent:

a) We have $\#(A \cap L) \leqslant \dim L$ for every proper linear subspace L of V.

b) We have $\#B = \dim \mathrm{span}(B,V)$ for every subset $B \neq \emptyset$ of A with $\#B \leqslant \dim V$.

c) Every subset $B \neq \emptyset$ of A with $\#B \leqslant \dim V$ is linear independent.

3) If $\emptyset \neq B \subseteq A$ with $\#B \leqslant \dim V$, then

$$\dim \bigcap_{\alpha \in B} E[\alpha] = n + 1 - \#B \tag{7.1}$$

where $E[\alpha] = \ker \alpha$ is a linear subspace of dimension n in V^*.

<u>PROOF</u>. a) $\Rightarrow$ b): Take $\emptyset \neq B \subseteq A$ with $\#B \leqslant \dim V$. Define $L = \mathrm{span}(B,V)$. If $\dim L < n + 1$, then L is proper and

$$\dim L \leqslant \#B \leqslant \#(A \cap L) \leqslant \dim L .$$

Hence $\dim \operatorname{span}(B,V) = \#B$. If $\dim L = n + 1$, then

$$\dim L \leqslant \#B \leqslant \dim V = \dim L$$

and $\dim \operatorname{span}(B,V) = \#B$ follows again.

b) $\Leftrightarrow$ c). Trivially, $\#B = \dim \operatorname{span}(B,V)$ if and only if B is linearly independent.

b) $\Rightarrow$ d). Define $E[B] = \bigcap_{\mathfrak{a} \in B} E[\mathfrak{a}]$. Let L be a linear subspace of V^*, then

$$L^\circ = \{\mathfrak{z} \in V \mid \langle \mathfrak{z},\mathfrak{w}\rangle = 0 \quad \forall \ \mathfrak{w} \in L\}$$

is a linear subspace of V with $\dim L^\circ + \dim L = \dim V$ and $L^{\circ\circ} = L$. Moreover

$$\begin{aligned} E[B] &= \{\mathfrak{z} \in V^* \mid \langle \mathfrak{a},\mathfrak{z}\rangle = 0 \quad \forall \ \mathfrak{a} \in B\} \\ &= \{\mathfrak{z} \in V^* \mid \langle \mathfrak{w},\mathfrak{z}\rangle = 0 \quad \forall \ \mathfrak{w} \in \operatorname{span}(B,V)\} \\ &= \operatorname{span}^\circ(B,V). \end{aligned}$$

Hence, if $\emptyset \neq B \subseteq A$ and $\#B \leqslant \dim V$, then $\#B = \dim \operatorname{span}(B,V)$ and

$$\dim E[B] = \dim V - \dim \operatorname{span}(B,V) = n + 1 - \#B.$$

d) $\Rightarrow$ a). Let L be a proper, linear subspace of V. A subset B_0 of $A \cap L$ exists such that $\operatorname{span}(B_0,V) = \operatorname{span}(A \cap L)$ and

$$\#B_0 = \dim \operatorname{span}(B_0,V) = \dim \operatorname{span}(A \cap L) \leqslant \dim L \leqslant n$$

Assume that $B_0 \neq A \cap L$. Take $\mathfrak{b} \in A \cap L - B_0$. Define $B = B_0 \cup \{\mathfrak{b}\}$. Then $\#B \leqslant n + 1$ and $B \subseteq A$. Therefore

$$\dim \bigcap_{\mathfrak{a} \in B} E[\mathfrak{a}] = n + 1 - \#B.$$

Since $\bigcap_{\mathfrak{a} \in B} E[\mathfrak{a}] = \mathrm{span}^{\circ}(B,V)$ we obtain that $\dim \mathrm{span}(B,V) = \#B$. Also we have

$$\mathrm{span}(A \cap L) = \mathrm{span}(B_0,V) \subseteq \mathrm{span}(B,V) \subseteq \mathrm{span}(A \cap L)$$

or

$$1 + \#B_0 = \#B = \dim \mathrm{span}(B,V) = \dim \mathrm{span}(A \cap L) = \#B_0$$

which is a contradiction. Hence $B_0 = A \cap L$. Hence

$$\#A \cap L = \dim \mathrm{span}(A \cap L) \leqslant \dim L \qquad \text{q.e.d.}$$

A subset $A \neq \emptyset$ of V is said to be in **general position** if and only if one and therefore all the conditions a), b), c) and d) of Lemma 7.1 are satisfied.

Assume that a hermitian metric is given on V. Abbreviate $k = \#A$. If $k \leqslant n + 1$ write $A = \{\mathfrak{a}_1, \ldots, \mathfrak{a}_k\}$. Then

$$\Gamma(A) = \frac{\|\mathfrak{a}_1 \wedge \cdots \wedge \mathfrak{a}_k\|}{\|\mathfrak{a}_1\| \cdots \|\mathfrak{a}_k\|} \tag{7.2}$$

does not depend on the enumeration. If $n + 1 < k = \#A \leqslant +\infty$, let $\mathfrak{T}[n,A]$ be the set of all injective maps from $\mathbb{Z}[0,n]$ into A. Define

$$\Gamma(A) = \inf \left\{ \frac{\|\tau(0) \wedge \cdots \wedge \tau(n)\|}{\|\tau(0)\| \cdots \|\tau(n)\|} \;\middle|\; \tau \in \mathfrak{T}[n,A] \right\} \tag{7.3}$$

$$\Gamma(A) = \inf\{\Gamma(\tau(\mathbb{Z}[0,n])) \mid \tau \in \mathfrak{T}[n,A]\}. \tag{7.4}$$

Then $0 \leqslant \Gamma(A) \leqslant 1$. The number $\Gamma(A)$ is called the **gauge** of A.

LEMMA 7.2. Let $A \neq \emptyset$ be a subset of V. If $\Gamma(A) > 0$ then A is in general position. If A is finite and in general position, then $\Gamma(A) > 0$.

PROOF. Define $k = \#A$. If $k \leqslant n + 1$, then A is in general position if and only if A is linearly independent (see Lemma 7.1 a)) which is the case if and only if $\Gamma(A) > 0$.

Assume that $k > n + 1$ and $\Gamma(A) > 0$. Take $\emptyset \neq B \subseteq A$ with $\#B \leqslant n + 1$. Then $\hat{B} \subseteq A$ exists with $B \subseteq \hat{B}$ and $\#\hat{B} = n + 1$. A bijective map $\tau : \mathbb{Z}[0,n] \longrightarrow \hat{B}$ exists. By (7.4) we have $0 < \Gamma(A) \leqslant \Gamma(\hat{B})$. Hence $\hat{B}$ is linearly indendent. Therefore B is linearly independent. By Lemma 7.1 c), A is in general position.

Assume that A is in general position and $n + 1 < k < \infty$. Then $\tau \in \mathfrak{T}[n,A]$ exists such that $B = \tau(\mathbb{Z}[0,n]) \subseteq A$ and $\Gamma(A) = \Gamma(B)$. (See (7.4).) Since $\#B = n + 1$, the set B is linearly independent by Lemma 7.1 c). Hence $0 < \Gamma(B) = \Gamma(A)$; q.e.d.

LEMMA 7.3. If $\emptyset \neq A \subset V$ with $\#A \leqslant n + 1$, then $\hat{A}$ exists such that $A \subseteq \hat{A} \subset V$ with $\#\hat{A} = n + 1$ and such that $\Gamma(\hat{A}) = \Gamma(A)$.

PROOF. If $\#A = n + 1$, take $\hat{A} = A$. If $k + 1 = \#A \leqslant n$, let W be the linear subspace of V perpendicular to span(A,V). Then $A = \{\mathfrak{a}_0, \ldots ,\mathfrak{a}_k\}$ and $\tilde{\mathfrak{a}}_p, \ldots ,\tilde{\mathfrak{a}}_{n+1}$ is an orthonormal base of W with $p \leqslant k + 1$. Then $(\mathfrak{a}_j | \tilde{\mathfrak{a}}_q) = 0$ if $0 \leqslant j \leqslant k$ and $p \leqslant q \leqslant n + 1$. Define $\hat{A} = \{\mathfrak{a}_0, \ldots ,\mathfrak{a}_k,\tilde{\mathfrak{a}}_{k+1}, \ldots ,\tilde{\mathfrak{a}}_n\}$. Then

$$\Gamma(\hat{A}) = \frac{\|\mathfrak{a}_0 \wedge \cdots \wedge \mathfrak{a}_k \wedge \tilde{\mathfrak{a}}_{k+1} \wedge \cdots \wedge \tilde{\mathfrak{a}}_n\|}{\|\mathfrak{a}_0\| \cdots \|\mathfrak{a}_k\| \, \|\tilde{\mathfrak{a}}_{k+1}\| \cdots \|\tilde{\mathfrak{a}}_n\|}$$

$$= \frac{\|\mathfrak{a}_0 \wedge \cdots \wedge \mathfrak{a}_k\|}{\|\mathfrak{a}_0\| \cdots \|\mathfrak{a}_k\|} = \Gamma(A). \quad \text{q.e.d.}$$

A subset $A \neq \emptyset$ of $\mathbb{P}(V)$ is said to be in **general position** if there exists a subset $\hat{A} \neq \emptyset$ of V_* in general position such that

$\mathbb{P} : \hat{A} \longrightarrow A$ is bijective. If $A \neq \emptyset$ is a subset of $\mathbb{P}(V)$ and if $\hat{A} \subseteq V_*$ such that $\mathbb{P} : \hat{A} \longrightarrow A$ is bijective, $\Gamma(A) = \Gamma(\hat{A})$ is independent of the choice of $\hat{A}$. If $\Gamma(A) > 0$, then A is in general position. If A is finite and in general position, then $\Gamma(A) > 0$. If $\hat{A} \subseteq V_*$ and $\mathbb{P} : \hat{A} \longrightarrow A$ is bijective, then A is in general position if and only if $\hat{A}$ is in general position. Let $\mathrm{span}(A, \mathbb{P}(V))$ be the intersection of all projective planes in $\mathbb{P}(V)$ containing A. Then $\mathrm{span}(A, \mathbb{P}(V))$ is a projective plane. If $\hat{A} \subseteq V_*$ such that $\mathbb{P}(\hat{A}) = A$, then

$$\mathrm{span}(A, \mathbb{P}(V)) = \mathbb{P}(\mathrm{span}(\hat{A}, V)). \tag{7.5}$$

Hence Lemma 7.1 translates into:

LEMMA 7.4. The following conditions are equivalent if $A \subseteq \mathbb{P}(V)$:

a) We have $\#(A \cap L) \leqslant 1 + \dim L$ for every proper plane L in $\mathbb{P}(V)$.
b) If $\emptyset \neq B \subseteq A$ with $\#B \leqslant \dim V$, then $\#B = 1 + \dim \mathrm{span}(B, \mathbb{P}(V))$.
c) Every subset $B \neq \emptyset$ of A with $\#B \leqslant \dim V$ is linearly independent.
d) If $\emptyset \neq B \subseteq A$ with $\#B \leqslant \dim V$, then

$$\dim \bigcap_{a \in B} \ddot{E}[a] = n - \#B. \tag{7.4}$$

e) A is in general position.

Take $p \in \mathbb{Z}[0, n]$ and $\emptyset \neq \mathfrak{g} \subseteq G_p(V^*)$. Then $\mathfrak{g} \subseteq \mathbb{P}(\bigwedge_{p+1} V^*)$ and as such it is defined when $\mathfrak{g}$ is in general position. Then Lemmata 7.1 and 7.4 remain true if V is replaced by $\bigwedge_{p+1} V$. Here

$\ddot{E}[a] = \ddot{E}[a, \underset{p+1}{\wedge} V]$ is a hyperplane in $\mathbb{P}(\underset{p+1}{\wedge} V)$ and is not to be mistaken for $\ddot{E}[a,V]$ which is a plane of dimension $n - p - 1$ in $\mathbb{P}(V)$.

REMARK. As the following example shows, the statement d) of Wu [126] page 181 is wrong: Take $V = \mathbb{C}^4$, $n = 3$, $h = 1$, $i = 3$. Let $\mathfrak{n}_0, \mathfrak{n}_1, \mathfrak{n}_2, \mathfrak{n}_3$ be the standard base of $\mathbb{C}^4$. Define $a_j = \mathbb{P}(\mathfrak{n}_j \wedge \mathfrak{n}_3)$ for $j = 0, 1, 2$. Then $\{a_0, a_1, a_2\} = \{A^h\}$ is in general position. We have

$$E(a_1) \cap E(a_2) \cap E(a_3) = \{\mathbb{P}(\mathfrak{n}_3)\} \neq \emptyset$$

while statement d) of Wu [126] would imply that the intersection is empty since $h - i + 1 = -1 < 0$ and $i = 3 = h + 2$.

We assume that a hermitian metric on V is given. Take $q \in \mathbb{Z}[0,n]$ and $b \in G_q(V)$. Take $m \in \mathbb{Z}$ with $0 \leqslant m \leqslant \text{Min}(p,q)$. If $a \in G_p(V^*)$, then $E(a) \subseteq V^*$ and $G_m(E(a)) \subseteq G_m(V^*)$. Hence we can define

$$\mathfrak{A}(b,m) = \{a \in \mathfrak{A} \mid \Box\, b \mathbin{\dot{\llcorner}} x\, \Box = 0 \quad \forall\ x \in G_m(E(a))\}. \tag{7.7}$$

If $r \geqslant 0$, we define

$$\mathfrak{A}(b,m,r) = \{a \in \mathfrak{A} \mid \Box\, b \;\dot{\boxplus}^{m+1}\; a\, \Box \leqslant r\}. \tag{7.8}$$

We shall estimate the number of elements of $\mathfrak{A}(b,m,r)$ for sufficiently small r and we shall give an estimate for the bound of r.

LEMMA 7.5. Take $q \in \mathbb{Z}[0,n]$ and $p \in \mathbb{Z}[0,n]$. Define $\mu = \text{Min}(p+1,q+1)$. Take $m \in \mathbb{Z}[0,\mu)$. Take $a \in G_p(V^*)$ and $b \in G_q(V)$. Then $\Box\, b \;\dot{\boxplus}^{m+1}\; a\, \Box = 0$ if and only if $\Box\, b \mathbin{\dot{\llcorner}} x\, \Box = 0$ for all $x \in G_m(E(a))$; i.e.

(7.9) $$\mathfrak{G}(b,m,0) = \mathfrak{G}(b,m).$$

<u>PROOF</u>. Take $\mathfrak{b} \in \tilde{G}_q(V)$ and $\mathfrak{a} \in G_p(V^*)$ with $\|\mathfrak{b}\| = 1 = \|\mathfrak{a}\|$ such that $b = \mathbb{P}(\mathfrak{b})$ and $a = \mathbb{P}(\mathfrak{a})$. Let $\mathfrak{a}_0, \ldots, \mathfrak{a}_p$ and $\mathfrak{b}_0, \ldots, \mathfrak{b}_q$ be orthonormal systems such that $\mathfrak{a} = \mathfrak{a}_0 \wedge \ldots \wedge \mathfrak{a}_p$ and $\mathfrak{b} = \mathfrak{b}_0 \wedge \ldots \wedge \mathfrak{b}_q$.

<u>CASE</u> 1: $p \leqslant q$. Then $\mu = p + 1$. By Lemma 1.7 we have

$$\begin{bmatrix} p+1 \\ m+1 \end{bmatrix}^2 \Box\, b \,\dot{\boxplus}^{m+1}\, a \,\Box^2 = \begin{bmatrix} p+1 \\ m+1 \end{bmatrix}^2 \|\mathfrak{b} \cdot \boxplus^{m+1}\, \mathfrak{a}\|^2 = \sum_{\lambda \in \mathfrak{S}[m,p]} \|\mathfrak{b} \llcorner \mathfrak{a}_\lambda\|^2.$$

Hence $\Box\, b \,\dot{\boxplus}^{m+1}\, a \,\Box = 0$ if and only if $\mathfrak{b} \llcorner \mathfrak{a}_\lambda = 0$ for all $\lambda \in \mathfrak{S}[m,p]$. Since $(\mathfrak{a}_\lambda)_{\lambda \in \mathfrak{S}[m,p]}$ is a base of $\bigwedge_{m+1} E(a)$, we see that $\mathfrak{b} \llcorner \mathfrak{a}_\lambda = 0$ for all $\lambda \in \mathfrak{S}[m,p]$ if and only if $\mathfrak{b} \llcorner \mathfrak{x} = 0$ for all $\mathfrak{x} \in \tilde{G}_m(E(a))$ which is the case if and only if $\Box\, b \dot{\llcorner} x \,\Box = 0$ for all $x \in G_m(E(a))$. This proves the lemma in the first case.

<u>CASE</u> 2: $p > q$. Then $\mu = q + 1$. By Lemma 1.5 we have

$$\begin{bmatrix} q+1 \\ m+1 \end{bmatrix}^2 \Box\, b \,\dot{\boxplus}^{m+1}\, a \,\Box^2 = \begin{bmatrix} q+1 \\ m+1 \end{bmatrix}^2 \|\mathfrak{b}\, \boxplus^{m+1}\, \mathfrak{a}\|^2 = \sum_{\lambda \in \mathfrak{S}[m,q]} \|\mathfrak{a} \llcorner \mathfrak{b}_\lambda\|^2.$$

Hence $\Box\, b \,\dot{\boxplus}^{m+1}\, a \,\Box = 0$ if and only if $\mathfrak{a} \llcorner \mathfrak{b}_\lambda = 0$ for all $\lambda \in \mathfrak{S}[m,q]$. We have

(7.10) $$\mathfrak{a} \llcorner \mathfrak{b}_\lambda = \sum_{\psi \in \mathfrak{S}[m,p]} \operatorname{sign}(\psi, \psi^\perp) \langle \mathfrak{a}_\psi, \mathfrak{b}_\lambda \rangle \mathfrak{a}_{\psi^\perp}.$$

Hence $\mathfrak{a} \llcorner \mathfrak{b}_\lambda = 0$ if and only if $\langle \mathfrak{a}_\psi, \mathfrak{b}_\lambda \rangle = 0$ for all $\psi \in \mathfrak{S}[m,p]$.

Hence $\Box\, b \,\dot{\boxplus}^{m+1}\, a \,\Box^2 = 0$ if and only if $\langle \mathfrak{a}_\psi, \mathfrak{b}_\lambda \rangle = 0$ for all $\psi \in \mathfrak{S}[m,p]$ and all $\lambda \in \mathfrak{S}[m,q]$. We have

(7.11) $$\mathfrak{b} \llcorner \mathfrak{a}_\psi = \sum_{\lambda \in \mathfrak{P}[m,q]} \operatorname{sign}(\lambda,\lambda^\perp) \langle \mathfrak{a}_\psi, \mathfrak{b}_\lambda \rangle \mathfrak{b}_{\lambda^\perp} .$$

Hence $\mathfrak{b} \llcorner \mathfrak{a}_\psi = 0$ if and only if $\langle \mathfrak{a}_\psi, \mathfrak{b}_\lambda \rangle = 0$ for all $\lambda \in \mathfrak{P}[m,q]$. Consequently $\square\, b \,\dot{\boxplus}^{m+1}\, a \,\square^2 = 0$ if and only if $\mathfrak{b} \llcorner \mathfrak{a}_\psi = 0$ for all $\psi \in \mathfrak{P}[m,p]$ which is the case if and only if $\square\, b \dot{\llcorner} x \,\square = 0$ for all $x \in G_m(E(a))$ (see Case 1). q.e.d.

Take p and q in $\mathbb{Z}[0,n]$ and $m \in \mathbb{Z}[0,p] \cap \mathbb{Z}[0,q]$. Define

(7.12) $$D(p,q,m) = \begin{bmatrix} n+1 \\ p+1 \end{bmatrix} - \sum_{j=0}^{p-m} \begin{bmatrix} q+1 \\ m+j+1 \end{bmatrix} \begin{bmatrix} n-q \\ p-m-j \end{bmatrix} .$$

Let $\iota : \mathbb{Z}[0,m] \longrightarrow \mathbb{Z}[0,p]$ be the inclusion map. Define

(7.13) $$\mathfrak{T} = \{\tau \in \mathfrak{P}[p,n] \mid \tau \circ \iota \in \mathfrak{P}[m,q]\}.$$

Take $b \in G_q(V)$ and define

(7.14) $S[b,p,m]$

$$= \{\mathfrak{z} \in \textstyle\bigwedge_{p+1} V^* \mid \langle \mathfrak{y} \wedge \mathfrak{z}', \mathfrak{z} \rangle = 0 \quad \forall \begin{Bmatrix} \mathfrak{y} \in \tilde{G}_m(E(b)) \\ \mathfrak{z}' \in \tilde{G}_{p-m-1}(V) \end{Bmatrix} \}$$

LEMMA 7.6. $S[b,p,m]$ is a linear subspace of $\bigwedge_{p+1} V^*$ with

(7.15) $$\dim S[b,p,m] = D(p,q,m) < \begin{bmatrix} n+1 \\ p+1 \end{bmatrix}$$

Moreover, if $\mathfrak{b}_0, \ldots, \mathfrak{b}_n$ is a base of V such that $\mathfrak{b} = \mathfrak{b}_0 \wedge \ldots \wedge \mathfrak{b}_q$ and $\mathbb{P}(\mathfrak{b}) = b$, then

(7.16) $$S[b,p,m] = \{\mathfrak{z} \in \textstyle\bigwedge_{p+1} V^* \mid \langle \mathfrak{b}_\tau, \mathfrak{z} \rangle = 0 \quad \forall\ \tau \in \mathfrak{T}\}.$$

PROOF. Take a base $\mathfrak{b}_0, \dots, \mathfrak{b}_n$ of V with $\mathfrak{b} = \mathfrak{b}_0 \wedge \dots \wedge \mathfrak{b}_q$ such that $\mathbb{P}(\mathfrak{b}) = b$. Define

$$\text{(7.17)} \qquad \tilde{S}[b,p,m] = \{\mathfrak{t} \in \bigwedge_{p+1} V^* \mid \langle \mathfrak{b}_\tau, \mathfrak{t} \rangle = 0 \quad \forall \ \tau \in \mathfrak{T}\}.$$

Obviously $S[b,p,m]$ and $\tilde{S}[b,p,m]$ are linear subspaces of $\bigwedge_{p+1} V^*$. Take $\mathfrak{t} \in S[b,p,m]$. Take $\tau \in \mathfrak{T}$. Then $\nu = \tau \circ \iota \in \mathfrak{S}[m,q]$. Hence $\mathfrak{b}_\nu \in \tilde{G}_m(E(b))$ and $\mathfrak{z} = \mathfrak{b}_{\tau(m+1)} \wedge \dots \wedge \mathfrak{b}_{\tau(p)} \in \tilde{G}_{p-m-1}(V)$. Thus $\langle \mathfrak{b}_\tau ; \mathfrak{t} \rangle = \langle \mathfrak{b}_\nu \wedge \mathfrak{z} ; \mathfrak{t} \rangle = 0$ by the definition of $S[b,p,m]$. Therefore $\mathfrak{t} \in \tilde{S}[b,p,m]$.

Take $\mathfrak{t} \in \tilde{S}[b,p,m]$. Take $\mathfrak{w} \in \tilde{G}_m(E(b))$ and $\mathfrak{z} \in \tilde{G}_{p-m-1}(V)$. Then

$$\mathfrak{w} = \sum_{\nu \in \mathfrak{S}[m,q]} y_\nu \mathfrak{b}_\nu \qquad \mathfrak{z} = \sum_{\lambda \in \mathfrak{S}[p-m-1,n]} z_\lambda \mathfrak{b}_\lambda .$$

We obtain

$$\text{(7.18)} \qquad \mathfrak{w} \wedge \mathfrak{z} = \sum_{\nu \in \mathfrak{S}[m,q]} \sum_{\lambda \in \mathfrak{S}[p-m-1,n]} y_\nu z_\lambda \mathfrak{b}_\nu \wedge \mathfrak{b}_\lambda .$$

CLAIM 1: If $\nu \in \mathfrak{S}[m,q]$ and $\lambda \in \mathfrak{S}[p-m-1,n]$, then

$$\text{(7.19)} \qquad \langle \mathfrak{b}_\nu \wedge \mathfrak{b}_\lambda , \mathfrak{t} \rangle = 0$$

PROOF OF CLAIM 1. If $\mathfrak{b}_\nu \wedge \mathfrak{b}_\lambda = 0$, then (7.19) is trivial. Hence we can assume that $\mathfrak{b}_\nu \wedge \mathfrak{b}_\lambda \neq 0$. An injective map $\Upsilon : \mathbb{Z}[0,p] \longrightarrow \mathbb{Z}[0,n]$ is defined by

$$\Upsilon(x) = \begin{cases} \nu(x) & \text{if} \quad x \in \mathbb{Z}[0,m] \\ \lambda(x - m - 1) & \text{if} \quad x \in \mathbb{Z}[m+1,p]. \end{cases}$$

There is a bijective map $\pi : \mathbb{Z}[0,p] \longrightarrow \mathbb{Z}[0,p]$ such that $\tau = \Upsilon \circ \pi$ is increasing. Naturally, τ is injective. Hence $\tau \in \mathfrak{P}[p,n]$. Define $\psi = \pi^{-1}$. Then $\tau \circ \psi = \Upsilon$. If $0 \leqslant z \leqslant y \leqslant m$, then

$$\tau(\psi(z)) = \Upsilon(z) = \nu(z) \leqslant \nu(y) = \Upsilon(y) = \tau(\psi(y)).$$

Since τ is injective and increasing we have $\psi(z) \leqslant \psi(y)$. Thus $\psi \mid \mathbb{Z}[0,m]$ is increasing. In particular, $m \leqslant \psi(m)$. Take any $x \in \mathbb{Z}[0,m]$ and let $\iota : \mathbb{Z}[0,m] \longrightarrow \mathbb{Z}[0,p]$ be the inclusion. Then

$$0 \leqslant \tau(\iota(x)) = \tau(x) \leqslant \tau(m) \leqslant \tau(\psi(m)) = \Upsilon(m) = \nu(m) \leqslant q.$$

Hence $\tau \circ \iota \in \mathfrak{P}[m,q]$. By (7.13) we have $\tau \in \mathfrak{T}$. Therefore

$$<\mathfrak{G}_\nu \wedge \mathfrak{G}_\lambda, \mathfrak{t}> = <\mathfrak{G}_\Upsilon, \mathfrak{t}> = \text{sign } \pi <\mathfrak{G}_\tau, \mathfrak{t}> = 0$$

by (7.17). Claim 1 is proved.

Claim 1 and (7.17) show that $<\varphi \wedge \mathfrak{z}, \mathfrak{t}> = 0$. Therefore $\mathfrak{t} \in S[b,p,m]$. We have shown that $\tilde{S}[b,p,m] = S[b,p,m]$ which proves (7.16).

Since $(\mathfrak{G}_\tau)_{\tau \in \mathfrak{P}[p,n]}$ is a base of $\underset{p+1}{\wedge} V$, the family $(\mathfrak{G}_\tau)_{\tau \in \mathfrak{T}}$ is linearly independent and spans a linear subspace L of dimension $\#\mathfrak{T}$. Then $S[b,p,m] = L^\circ$ and $\dim S[b,p,m] = \dim \underset{p+1}{\wedge} V - \dim L$. Hence $\dim S[b,p,m] = \begin{pmatrix} n+1 \\ p+1 \end{pmatrix} - \#\mathfrak{T}$. In order to prove (7.15) it remains to calculate the number of elements of $\mathfrak{T}$.

For each $s \in \mathbb{Z}[m,p]$ define

$$T_s = \left\{\tau \in \mathfrak{P}[p,n] \,\middle|\, \begin{cases} \tau(x) \leqslant q & \text{if } x \in \mathbb{Z}[0,s] \\ \tau(x) > q & \text{if } x \in \mathbb{Z}[s+1,p] \end{cases}\right\}.$$

Then $\mathfrak{T} = T_m \cup T_{m+1} \cup \ldots \cup T_p$ is a disjoint union. Therefore

$$\#\mathfrak{T} = \sum_{s=m}^{p} \#T_s .$$

Take $s \in \mathbb{Z}[m,p]$ and define a map

$$\rho_s : \mathfrak{q}[s,q] \times \mathfrak{q}[p - s - 1, n - q - 1] \longrightarrow T_s$$

If $\nu \in \mathfrak{q}[s,q]$ and $\mu \in \mathfrak{q}[p - s - 1, n - q - 1]$ define $\rho_s(\nu,\mu) = \tau$ by

$$\tau(x) = \begin{cases} \nu(x) & \text{if } x \in \mathbb{Z}[0,s] \\ \mu(x - s - 1) + q + 1 & \text{if } x \in \mathbb{Z}[s + 1, p] \end{cases}$$

Then $\tau \in T_s$. Obviously ρ_s is bijective. Therefore

$$\#T_s = (\#\mathfrak{q}[s,q])(\#\mathfrak{q}[p - s - 1, n - q - 1]) = \binom{q + 1}{s + 1}\binom{n - q}{p - s} .$$

We obtain

$$0 < \#\mathfrak{T} = \sum_{s=m}^{p} \binom{q + 1}{s + 1}\binom{n - q}{p - s} = \sum_{j=0}^{p-m} \binom{q + 1}{m + j + 1}\binom{n - q}{p - m - j}$$

$$\dim S[b,p,m] = \binom{n + 1}{p + 1} - \#\mathfrak{T}$$

$$= \binom{n + 1}{p + 1} - \sum_{j=0}^{p-m} \binom{q + 1}{m + j + 1}\binom{n - q}{p - m - j} < \binom{n + 1}{p + 1}.$$

q.e.d.

Define

(7.20) $$\ddot{S}[b,p,m] = \mathbb{P}(S[b,p,m])$$

LEMMA 7.7. Take p and q in $\mathbb{Z}[0,n]$. Take $m \in \mathbb{Z}[0,p] \cap \mathbb{Z}[0,q]$. Take $b \in G_q(V)$. Then we have

$$\ddot{S}[b,p,m] \cap G_p(V^*) = \{a \in G_p(V^*) \mid \square\, b \,\dot{\boxplus}^{m+1}\, a\, \square = 0\}. \tag{7.21}$$

PROOF. By Lemma 7.5 and symmetry $\square\, b \,\dot{\boxplus}^{m+1}\, a \,\square = 0$ if and only if $\square\, a \dot{\llcorner} y\, \square = 0$ for all $y \in G_m(E(b))$. Take $0 \neq \mathfrak{a} \in \tilde{G}_p(V^*)$ such that $\mathbb{P}(\mathfrak{a}) = a$. Then $\square\, a \dot{\llcorner} y\, \square = 0$ for all $y \in G_m(E(b))$ if and only if $\mathfrak{a} \llcorner \mathfrak{y} = 0$ for all $\mathfrak{y} \in \tilde{G}_m(E(b))$. Since $\tilde{G}_{p-m-1}(V)$ generates $\bigwedge_{p-m} V$, we have $\mathfrak{a} \llcorner \mathfrak{y} = 0$ if and only if $\langle \mathfrak{a} \llcorner \mathfrak{y}, \mathfrak{z}\rangle = 0$ for all $\mathfrak{z} \in \tilde{G}_{p-m-1}(V)$. Here $\langle \mathfrak{a} \llcorner \mathfrak{y}, \mathfrak{z}\rangle = \langle \mathfrak{a}; \mathfrak{y} \wedge \mathfrak{z}\rangle$. Hence $\mathfrak{a} \llcorner \mathfrak{y} = 0$ for all $\mathfrak{y} \in \tilde{G}_{p-m-1}(V)$, if and only if $\langle \mathfrak{y} \wedge \mathfrak{z}, \mathfrak{a}\rangle = 0$ for all $\mathfrak{y} \in \tilde{G}(E(b))$ and $\mathfrak{z} \in \tilde{G}_{p-m-1}(V)$, which is the case if and only if $\mathfrak{a} \in S[b,p,m]$. Hence $\square\, b \,\dot{\boxplus}^{m+1}\, a\, \square = 0$ if and only if $a \in \ddot{S}[b,p,m]$; q.e.d.

THEOREM 7.8. Take $p \in \mathbb{Z}[0,n]$ and $q \in \mathbb{Z}[0,n]$. Take $m \in \mathbb{Z}$ with $0 \leqslant m \leqslant \operatorname{Min}(p,q)$. Take $b \in G_q(V)$ and $\emptyset \neq \mathfrak{G} \subseteq G_p(V^*)$. Assume that $\mathfrak{G}$ is in general position. Then

$$\#\mathfrak{G}(b,m) \leqslant D(p,q,m). \tag{7.22}$$

PROOF. Since $\mathfrak{G}$ is in general position and since $\ddot{S}[b,p,m]$ is a proper projective plane in $\mathbb{P}(\bigwedge_{p+1} V)$, we have

$$\#\mathfrak{G}(b,m) \leqslant 1 + \dim \ddot{S}[b,p,m] = \dim S[b,p,m] = D(p,q,m) \qquad \text{q.e.d.}$$

Basically, our proof is the same as Wu's [126]. However, Wu does not use the product $\boxplus^{m+1}$ and the norm $\square\, b \,\dot{\boxplus}^{m+1}\, a\, \square$. These new concepts simplify the argument. The number $D(p,q,m)$ shall be calculated in certain cases.

LEMMA 7.9. Take p and q in $\mathbb{Z}[0,n]$. Take $m \in \mathbb{Z}[0,p] \cap \mathbb{Z}[0,q]$. Define $D(p,q,m)$ by (7.12). Abbreviate $D(p,q) = D(p,q,0)$. Then we have

$$(7.23)\quad 0 \leqslant D(p,q,m) = \sum_{j=0}^{m} \binom{q+1}{j}\binom{n-q}{p+1-j} < \binom{n+1}{p+1} < 2^{n+1}$$

$$(7.24)\quad D(p,q) = \binom{n-q}{p+1}$$

If $0 \leqslant m \leqslant q \leqslant p$, then we have

$$(7.25)\quad \sum_{j=0}^{p-m} \binom{q+1}{j+m+1}\binom{n-q}{p-m-j} \geqslant \binom{n-q}{p-j} > 0$$

$$(7.26)\quad D(p,q,m) \leqslant \binom{n+1}{p+1} - \binom{n-q}{p-q}$$

$$(7.27)\quad D(p,q,q) = \binom{n+1}{p+1} - \binom{n-q}{p-q}$$

$$(7.28)\quad D(p,p,p) = \binom{n+1}{p+1} - 1$$

$$(7.29)\quad D(p,0,0) = \binom{n}{p+1} = D(p,0).$$

If $0 \leqslant m \leqslant p \leqslant q$, then we have

$$(7.30)\quad \sum_{j=0}^{p-m} \binom{q+1}{j+m+1}\binom{n-q}{p-m-j} \geqslant \binom{q+1}{p+1} > 0$$

$$(7.31)\quad D(p,q,m) \leqslant \binom{n+1}{p+1} - \binom{q+1}{p+1}$$

$$(7.32)\quad D(p,q,p) = \binom{n+1}{p+1} - \binom{q+1}{p+1}$$

(7.33) $$D(0,q,0) = n - q = D(0,q)$$

PROOF. We have

$$\begin{bmatrix} n+1 \\ p+1 \end{bmatrix} = \sum_{j=0}^{p+1} \begin{bmatrix} q+1 \\ j \end{bmatrix} \begin{bmatrix} n-q \\ p+1-j \end{bmatrix}$$

$$= \sum_{j=0}^{m} \begin{bmatrix} q+1 \\ j \end{bmatrix} \begin{bmatrix} n-q \\ p+1-j \end{bmatrix} + \sum_{j=0}^{p-m} \begin{bmatrix} q+1 \\ j+1+m \end{bmatrix} \begin{bmatrix} n-q \\ p-m-j \end{bmatrix} .$$

Hence $D(p,q,m) = \sum_{j=0}^{m} \begin{bmatrix} q+1 \\ j \end{bmatrix} \begin{bmatrix} n-q \\ p+1-j \end{bmatrix}$ which proves (7.23) and in the case $m = 0$ also (7.24). If $0 \leqslant m \leqslant q \leqslant p$, take $j = q - m$ in the sum of (7.25) and the estimate (7.25) follows. If $m = q$, this is the only non-vanishing term in the sum and we have equality (7.27) which implies (7.28) and (7.29). If $0 \leqslant m \leqslant p \leqslant q$, take $j = p - m$ in the sum of (7.30) and we have the estimate (7.30). If $m = p$, this is the only non-vanishing term in the sum and we have equality (7.32) which implies (7.33) and (7.28). q.e.d.

LEMMA 7.10. Let W be a hermitian vector space of dimension h. Let $\mathfrak{a}_1, \ldots, \mathfrak{a}_h$ be vectors in W. Take $\varphi \in W^*$. Then we have

(7.34) $$\|\mathfrak{a}_1 \wedge \ldots \wedge \mathfrak{a}_h \llcorner \varphi\| = \|\mathfrak{a}_1 \wedge \ldots \wedge \mathfrak{a}_h\| \, \|\varphi\|$$

PROOF. Let $\mathfrak{n}_1, \ldots, \mathfrak{n}_h$ be an orthonormal base of W. Let $\mathfrak{n}_1^*, \ldots, \mathfrak{n}_h^*$ be the dual base. Define $y_\mu = \langle \mathfrak{n}_\mu, \varphi \rangle$ for $\mu \in \mathbb{N}[1,h]$. Then

$$\varphi = \sum_{\mu=1}^{m} y_\mu \mathfrak{n}_\mu^*$$

Define

$$\hat{\mathfrak{n}}_\mu = (-1)^{\mu-1} \mathfrak{n}_1 \wedge \ldots \wedge \mathfrak{n}_{\mu-1} \wedge \mathfrak{n}_{\mu+1} \wedge \ldots \wedge \mathfrak{n}_h$$

Then $\hat{\mathfrak{n}}_1, \ldots, \hat{\mathfrak{n}}_h$ is an orthonormal base of $\bigwedge_{h-1} V$. A number $A \in \mathbb{C}$ exists such that

$$\mathfrak{a}_1 \wedge \ldots \wedge \mathfrak{a}_h = A\mathfrak{n}_1 \wedge \ldots \wedge \mathfrak{n}_h$$

Therefore we have

$$\|\mathfrak{a}_1 \wedge \ldots \wedge \mathfrak{a}_h \llcorner \mathfrak{y}\| = |A| \, \|\mathfrak{n}_1 \wedge \ldots \wedge \mathfrak{n}_h \llcorner \mathfrak{y}\|$$

$$= |A| \, \Big\| \sum_{\mu=1}^{h} y_\mu \hat{\mathfrak{n}}_\mu \Big\|$$

$$= |A| \left[\sum_{\mu=1}^{h} |y_\mu|^2 \right]^{1/2}$$

$$= \|\mathfrak{a}_1 \wedge \ldots \wedge \mathfrak{a}_h\| \, \|\mathfrak{y}\| \, . \qquad \text{q.e.d.}$$

If $\mathfrak{a}_1, \ldots, \mathfrak{a}_h$ are linearly independent, then $\infty = \mathbb{P}(\mathfrak{a}_1 \wedge \ldots \wedge \mathfrak{a}_h)$ is the one and only element of $G_{h-1}(W) = \mathbb{P}(\bigwedge_h W)$. Lemma 7.10 implies

(7.35) $$\Box\, \infty \dot{\llcorner} y \,\Box = 1 \qquad \text{for all} \quad y \in \mathbb{P}(W^*) \, .$$

We are ready to prove the required estimate. Recall the definition of the gauge (7.2) – (7.4).

<u>THEOREM</u> **7.11.** Take p and q in $\mathbb{Z}[0,n]$. Take $m \in \mathbb{Z}[0,p] \cap \mathbb{Z}[0,q]$. Let $\mathfrak{g} \neq \emptyset$ be a subset of $G_p(V^*)$. Assume that $\mathfrak{g}$ is in general position. Take $b \in G_q(V)$. Take $r \in \mathbb{R}$ with

(7.36) $$0 < r \leqslant 2^{-3n-3} \Gamma(\mathfrak{g}) \, .$$

Then we have

(7.37) $$\#\mathfrak{g}(b,m,r) \leqslant D(p,q,m) \, .$$

PROOF. If $\mathfrak{q}$ is enlarged, $\#\mathfrak{q}(b,m,r)$ increases. By Lemma 7.3 we can assume that $\#\mathfrak{q} \geq \begin{bmatrix} n+1 \\ p+1 \end{bmatrix}$. We abbreviate

(7.38) $$\begin{bmatrix} n+1 \\ p+1 \end{bmatrix} = h \qquad d = D(p,q,m) \qquad \mu = \operatorname{Min}(p+1,q+1)$$

Define $\mathfrak{T}$ by (7.13). Then $0 < h - d = \#\mathfrak{T} \leq h$. Let $\mathfrak{b}_0, \ldots, \mathfrak{b}_n$ be an orthonormal base of V such that $\mathfrak{b} = \mathfrak{b}_0 \wedge \ldots \wedge \mathfrak{b}_q$ and $b = \mathbb{P}(\mathfrak{b})$.

CLAIM 1. Take $\alpha \in \tilde{G}_p(V^*)$ with $\|\alpha\| = 1$. Define $a = \mathbb{P}(\alpha) \in G_p(V^*)$. Take $\tau \in \mathfrak{T}$. Then

(7.39) $$|\langle \mathfrak{b}_\tau, \alpha \rangle| \leq \begin{bmatrix} \mu \\ m+1 \end{bmatrix} \Box\, b \mathbin{\dot{\boxplus}}^{m+1} a \,\Box\,.$$

PROOF OF CLAIM 1. Let $\iota : \mathbb{Z}[0,m] \longrightarrow \mathbb{Z}[0,p]$ be the inclusion map. Then $\tau \circ \iota = \nu \in \mathfrak{F}[m,q]$. Define $\rho \in \mathfrak{F}[p-m-1,n]$ by

$$\rho(x) = \tau(x+m-1) \qquad \text{for } x \in \mathbb{Z}[0,m-1]$$

Then $\mathfrak{b}_\tau = \mathfrak{b}_\nu \wedge \mathfrak{b}_\rho$. We have

$$|\langle \alpha, \mathfrak{b}_\tau \rangle| = |\langle \alpha \llcorner \mathfrak{b}_\nu, \mathfrak{b}_\rho \rangle| \leq \|\alpha \llcorner \mathfrak{b}_\nu\| \, \|\mathfrak{b}_\rho\| = \|\alpha \llcorner \mathfrak{b}_\nu\|$$

$$\leq \left[\sum_{\lambda \in \mathfrak{F}[m,q]} \|\alpha \llcorner \mathfrak{b}_\lambda\|^2 \right]^{1/2} = \begin{bmatrix} \mu \\ m+1 \end{bmatrix} \|\mathfrak{b} \boxplus^{m+1} \alpha\|$$

$$= \begin{bmatrix} \mu \\ m+1 \end{bmatrix} \Box\, b \mathbin{\dot{\boxplus}}^{m+1} a \,\Box$$

which proves Claim 1.

<u>Assume</u> <u>that</u> $\#\mathfrak{g}(b,m,r) > d$ where r is a real number satisfying (7.36).

Since $\#\mathfrak{g} \geq h$, there exists $a_j \in \mathfrak{g}$ for $j = 1, \ldots, h$ such that $a_j \in \mathfrak{g}(b,m,r)$ for $j = 1, \ldots, d+1$. Take $\alpha_j \in \tilde{G}_p(V^*)$ with $\|\alpha_j\| = 1$ such that $a_j = \mathbb{P}(\alpha_j)$. Since $\mathfrak{g}$ is in general position $\alpha_1, \ldots, \alpha_h$ are linearly independent and constitute a base of $\underset{p+1}{\wedge} V^*$. Therefore

$$L = \bigcap_{j=d+2}^{h} E[a_j, \underset{p+1}{\wedge} V] = \{\xi \in \underset{p+1}{\wedge} V \mid \langle \xi, \alpha_j \rangle = 0 \quad \forall \; j = d+2, \ldots, h\}$$

is a linear subspace of dimension $d+1$ of $\underset{p+1}{\wedge} V$. If we put $A = \{\alpha_j \mid j = d+2, \ldots, h\}$, then $L = \mathrm{span}^{\circ}(A, \underset{p+1}{\wedge} V^*)$. Define $B = \{\mathfrak{b}_\tau \mid \tau \in \mathfrak{T}\}$. Then $E = \mathrm{span}(B, \underset{p+1}{\wedge} V) = S[b,p,m]^{\circ}$ is a linear subspace of dimension $h - d$ of $\underset{p+1}{\wedge} V$. Therefore

$$\dim E \cap L \geq \dim E + \dim L - \dim \underset{p+1}{\wedge} V$$

$$= h - d + d + 1 - h = 1$$

There is a vector $\psi \in E \cap L$ with $\|\psi\| = 1$. Then

$$\psi = \sum_{\tau \in \mathfrak{T}} y_\tau \mathfrak{b}_\tau \qquad 1 = \|\psi\|^2 = \sum_{\tau \in \mathfrak{T}} |y_\tau|^2$$

$$\langle \psi, \alpha_j \rangle = 0 \qquad \text{for } j = d+2, \ldots, h$$

The exterior product and the interior product over $\underset{p+1}{\wedge} V$ are denoted by $\boldsymbol{\wedge}$ and $\boldsymbol{\llcorner}$ respectively to distinguish it from the operation $\wedge$ and $\llcorner$ over V. Abbreviate

$$\hat{\alpha}_j = (-1)^{j-1} \alpha_1 \wedge \dots \wedge \alpha_{j-1} \wedge \alpha_{j+1} \wedge \dots \wedge \alpha_h$$

for $j \in \mathbb{Z}[1,h]$. Then $\|\hat{\alpha}_j\| \leqslant 1$. We obtain

$$\Gamma(\mathfrak{q}) \leqslant \|\alpha_1 \wedge \dots \wedge \alpha_h\| = \|(\alpha_1 \wedge \dots \wedge \alpha_h) \llcorner \varphi\|$$

$$= \|\sum_{j=1}^{h} <\varphi,\alpha_j>\hat{\alpha}_j\| = \|\sum_{j=1}^{d+1} <\varphi,\alpha_j>\hat{\alpha}_j\|$$

$$\leqslant \sum_{j=1}^{d+1} |<\varphi,\alpha_j>| = \sum_{\tau\in\mathfrak{T}} | \sum_{\tau\in\mathfrak{T}} y_\tau <\beta_\tau,\alpha_j>|$$

$$\leqslant \sum_{j=1}^{d+1} \left[\sum_{\tau\in\mathfrak{T}} |y_\tau|^2\right]^{1/2} \left[\sum_{\tau\in\mathfrak{T}} |<\beta_\tau,\alpha_j>|^2\right]^{1/2}$$

$$= \sum_{j=1}^{d+1} \left[\sum_{\tau\in\mathfrak{T}} |<\beta_\tau,\alpha_j>|^2\right]^{1/2}$$

$$\leqslant \sum_{j=1}^{d+1} \sqrt{h-d} \begin{bmatrix} \mu \\ m+1 \end{bmatrix} \square\, b\, \dot{\boxplus}^{m+1} a_j \,\square\, .$$

Here $d+1 \leqslant h < 2^{n+1}$ and $\sqrt{h-d} \leqslant \sqrt{h} < 2^n + 1$ and $\begin{bmatrix} \mu \\ m+1 \end{bmatrix} < 2^\mu \leqslant 2^{n+1}$. Since $a_j \in \mathfrak{q}(b,m,r)$ for $j = 1, \dots, d+1$, we have

$$\square\, b\, \dot{\boxplus}^{m+1} a_j \,\square \leqslant r \qquad \text{for } j = 1, \dots, d+1.$$

Consequently we obtain

$$0 < \Gamma(\mathfrak{q}) < 2^{3n+3} r \leqslant \Gamma(\mathfrak{q})$$

which is impossible. Therefore $\#\mathfrak{q}(b,m,r) \leqslant d = D(p,q,m)$ if $2^{3n+3} r \leqslant \Gamma(\mathfrak{q})$. q.e.d.

The interval (7.36) is much too small, but it is amazing that such a simple estimate holds for the difficult inequality (7.37) to be true. If $p = 0 = m$, then 2^{-3n-3} can be replaced by $1/(n+1)$.

Now, the <u>product to sum estimates</u> can be established.

<u>THEOREM</u> 7.12. Take p and q in $\mathbb{Z}[0,n]$. Take $m \in \mathbb{Z}[0,p] \cap \mathbb{Z}[0,q]$. Let $\mathfrak{g} \neq \emptyset$ be a finite subset of $G_p(V^*)$. Assume that $\mathfrak{g}$ is in general position. Define $k = \#\mathfrak{g}$. Let $\beta : \mathfrak{g} \longrightarrow \mathbb{R}(0,1)$ and $P : \mathfrak{g} \longrightarrow \mathbb{R}[0,1]$ be functions. For $x \in \mathbb{R}$ define $x^+ = \mathrm{Max}(0,x)$. Also define

(7.40)
$$c_{pq}(\mathfrak{g}) = \left[\left(\frac{1}{2}\right)^{3(n+1)}\Gamma(\mathfrak{g})\right]^{\frac{2(k-D(p,q,m))^+}{D(p,q,m)}}$$

Take $b \in G_q(V)$. Then we have

(7.41)
$$D(p,q,m)c_{pq}(\mathfrak{g})\left[\prod_{a\in\mathfrak{g}} \frac{P(a)}{\square\, b\, \dot{\boxplus}^{m+1}\, a\, \square^{2-2\beta(a)}}\right]^{\frac{1}{D(p,q,m)}}$$
$$\leqslant \sum_{a\in\mathfrak{g}} \frac{P(a)}{\square\, b\, \dot{\boxplus}^{m+1}\, a\, \square^{2-2\beta(a)}} + (D(p,q,m) - k)^+ .$$

<u>PROOF</u>. <u>Case</u> 1: $0 < k \leqslant D(p,q,m)$. Then $c_{pq}(\mathfrak{g}) = 1$. We can consider the product in (7.41) extended over $D(p,q,m)$ factors by counting $D(p,q,m) - k$ additional factors 1. Then (7.41) is true, since the arithmetic mean majorizes the geometric mean.

<u>Case</u> 2: $k > D(p,q,m)$. Then $(D(p,q,m) - k)^+ = 0$. Abbreviate $\theta = \left(\frac{1}{2}\right)^{3n+3}\Gamma(\mathfrak{g})$ and $\lambda = \frac{2(k - D(p,q,m))}{D(p,q,m)}$. Then $c_{pq}(\mathfrak{g}) = \theta^\lambda$. Because $\#\mathfrak{g}(b,m,\theta) \leqslant D(p,q,m) \leqslant \#\mathfrak{g}$, we can find a subset $\mathfrak{h}$ of $\mathfrak{g}$ such that $\mathfrak{h} \geqslant \mathfrak{g}(b,m,\theta)$ and such that $\#\mathfrak{h} = D(p,q,m)$. If $a \in \mathfrak{g} - \mathfrak{h}$, then

$$\square\, b\, \dot{\boxplus}^{m+1}\, a\, \square^{2-2\beta(a)} \geqslant \square\, b\, \dot{\boxplus}^{m+1}\, a\, \square^{2} \geqslant \theta^2 .$$

Since $0 \leqslant P \leqslant 1$, we obtain

$$D(p,q,m)c_{pq}(\mathfrak{g})\left[\prod_{a\in\mathfrak{g}} \frac{P(a)}{\Box\; b\; \dot{\boxplus}^{m+1}\; a\; \Box^{2-2\beta(a)}}\right]^{\frac{1}{D(p,q,m)}}$$

$$\leqslant D(p,q,m)c_{pq}(\mathfrak{g})\theta^{-\lambda}\left[\prod_{a\in\mathfrak{h}} \frac{P(a)}{\Box\; b\; \dot{\boxplus}^{m+1}\; a\; \Box^{2-2\beta(a)}}\right]^{\frac{1}{D(p,q,m)}}$$

$$\leqslant \sum_{a\in\mathfrak{h}} \frac{P(a)}{\Box\; b\; \dot{\boxplus}^{m+1}\; a\; \Box^{2-2\beta(a)}} \leqslant \sum_{a\in\mathfrak{g}} \frac{P(a)}{\Box\; b\; \dot{\boxplus}^{m+1}\; a\; \Box^{2-2\beta(a)}} .$$

q.e.d.

Let V be a hermitian vector space of dimension $n + 1 > 1$. Let M be a connected, complex manifold of dimension m. Let $\mathfrak{g} \neq \emptyset$ be a finite set of meromorphic maps $g : M \longrightarrow \mathbb{P}(V^*)$. Define $k = \#\mathfrak{g}$. The <u>indeterminacy</u> of $\mathfrak{g}$ is defined by

(7.42) $$I_{\mathfrak{g}} = \bigcup_{g\in\mathfrak{g}} I_g .$$

The set $I_{\mathfrak{g}}$ is analytic with $\dim I_{\mathfrak{g}} \leqslant m - 2$. If $\emptyset \neq \mathfrak{h} \subseteq \mathfrak{g}$, then $I_{\mathfrak{h}} \subseteq I_{\mathfrak{g}}$. For each $z \in M - I_{\mathfrak{g}}$, define the <u>orbit</u>

(7.43) $$\mathfrak{g}(z) = \{g(z) \mid g \in \mathfrak{g}\} .$$

If $\emptyset \neq \mathfrak{h} \subseteq \mathfrak{g}$, then $\mathfrak{h}(z) \subseteq \mathfrak{g}(z) \subseteq \mathbb{P}(V^*)$. The <u>fusion</u> of $\mathfrak{g}$ is defined by

(7.44) $$\mathfrak{J}_{\mathfrak{g}} \equiv I_{\mathfrak{g}} \cup \{z \in M - I_{\mathfrak{g}} \mid \#\mathfrak{g}(z) < k\} .$$

Then $\mathfrak{J}_{\mathfrak{g}} \neq M$. If $\emptyset \neq \mathfrak{h} \subseteq \mathfrak{g}$, then $\mathfrak{J}_{\mathfrak{h}} \subseteq \mathfrak{J}_{\mathfrak{g}}$. If $k = 1$, then $\mathfrak{J}_{\mathfrak{g}} = I_{\mathfrak{g}}$.

The <u>gauge</u> of $\mathfrak{g}$ is the function $\Gamma(\mathfrak{g}) : M - \mathfrak{J}_{\mathfrak{g}} \longrightarrow \mathbb{R}[0,1]$ defined by

(7.45) $$\Gamma(\mathfrak{g})(z) = \Gamma(\mathfrak{g}(z)) \quad \text{for} \quad z \in M - \mathfrak{J}_{\mathfrak{g}} .$$

If $\emptyset \neq \mathfrak{h} \subseteq \mathfrak{g}$, then $\Gamma(\mathfrak{g}) \leqslant \Gamma(\mathfrak{h})$ on $M - \mathfrak{J}_{\mathfrak{g}}$. The __degeneracy__ of $\mathfrak{g}$ is defined by

(7.46) $$\Delta(\mathfrak{g}) = \mathfrak{J}_{\mathfrak{g}} \cup \{z \in M - \mathfrak{J}_{\mathfrak{g}} \mid \Gamma(\mathfrak{g})(z) = 0\} .$$

If $\emptyset \neq \mathfrak{h} \subseteq \mathfrak{g}$, then $\Delta(\mathfrak{h}) \subseteq \Delta(\mathfrak{g})$. If $z \in M - \mathfrak{J}_{\mathfrak{g}}$, then $k = \#\mathfrak{g}(z)$ and $z \in \Delta(\mathfrak{g})$ if and only if $\mathfrak{g}(z)$ is not in general position.

The set $\mathfrak{g}$ is said to be in __general position__ if $\Delta(\mathfrak{g}) \neq M$ which is the case if and only if $\mathfrak{g}(z_0)$ is in general position for some point $z_0 \in M - \mathfrak{J}_{\mathfrak{g}}$.

For each $g \in \mathfrak{g}$, let L_g be the hyperplane section bundle of g and let F_g be the representation section of g. Take $\emptyset \neq \mathfrak{h} \subseteq \mathfrak{g}$ with $\#\mathfrak{h} = p \leqslant n + 1$. Enumerate $\mathfrak{h} = \{g_1, \dots, g_p\}$ and abbreviete $G_j = F_{g_j}$ and $L_j = L_{g_j}$ for $j = 1, \dots, p$. An exterior product

(7.47)
$$\wedge = \wedge \otimes \mathrm{Id} : (V_M^* \otimes L_1) \oplus \dots \oplus (V_M^* \otimes L_p) \longrightarrow (\textstyle\bigwedge_p V_M^*) \otimes L_1 \otimes \dots \otimes L_p$$

is defined, which gives a holomorphic section

(7.48) $$G_1 \wedge \dots \wedge G_p \in \Gamma(M, (\textstyle\bigwedge_p V_M^*) \otimes L_1 \otimes \dots \otimes L_p) .$$

The zeroset $Z(\mathfrak{h}) = Z(G_1 \wedge \dots \wedge G_p)$ of the section $G_1 \wedge \dots \wedge G_p$ is an analytic subset of M which does not depend on the enumeration of $\mathfrak{h}$.

For each set A let $\mathfrak{P}(A)$ be the set of subsets of A. For $0 \leqslant s \in \mathbb{Z}$ define

(7.49) $$\mathfrak{P}_s(A) = \{B \in \mathfrak{P}(A) \mid \#B = s\} .$$

LEMMA 7.13. If $k = 1$, then $\mathfrak{J}_{\mathfrak{g}} = I_{\mathfrak{g}}$. If $k > 1$, then

$$(7.50) \qquad \mathfrak{J}_{\mathfrak{g}} = \bigcup_{\mathfrak{h}\in\mathfrak{P}_2(\mathfrak{g})} Z(\mathfrak{h}) .$$

In both cases, $\mathfrak{J}_{\mathfrak{g}}$ is a thin analytic subset of M.

PROOF. The case $k = 1$ is trivial. Assume that $k \geq 2$. Define

$$W = \bigcup_{\mathfrak{h}\in\mathfrak{P}_2(\mathfrak{g})} Z(\mathfrak{h}) .$$

Take $z \in \mathfrak{J}_{\mathfrak{g}}$. If $z \in I_{\mathfrak{g}}$, then $z \in I_g$ for some $g \in \mathfrak{g}$. Hence $F_g(z) = 0$. Take $h \in \mathfrak{g}$ with $h \neq g$. Then $\mathfrak{h} = \{g,h\} \in \mathfrak{P}_2(\mathfrak{g})$ and $F_g(z) \wedge F_h(z) = 0$. Therefore $z \in Z(\mathfrak{h}) \subseteq W$. If $z \in \mathfrak{J}_{\mathfrak{g}} - I_{\mathfrak{g}}$, then $\#\mathfrak{g}(z) < k$. Therefore $g \in \mathfrak{g}$ and $h \in \mathfrak{g}$ with $g \neq h$ exist such that $g(z) = h(z)$. Take reduced representations $\mathfrak{w}_g : U \longrightarrow V^*$ and $\mathfrak{w}_h : U \longrightarrow V^*$ of g and h respectively such that $z \in U$. Since $z \in \mathfrak{J}_{\mathfrak{g}} - I_{\mathfrak{g}}$ we have $\mathfrak{w}_g(z) \neq 0 \neq \mathfrak{w}_h(z)$ and

$$(7.51) \qquad \mathbb{P}(\mathfrak{w}_g(z)) = g(z) = h(z) = \mathbb{P}(\mathfrak{w}_h(z))$$

which implies $\mathfrak{w}_g(z) \wedge \mathfrak{w}_h(z) = 0$. Therefore

$$(7.52) \qquad F_g(z) \wedge F_h(z) = (\mathfrak{w}_g(z) \wedge \mathfrak{w}_h(z)) \otimes \mathfrak{w}_g^{\Delta}(z) \otimes \mathfrak{w}_h^{\Delta}(z) = 0 .$$

We have $\mathfrak{h} = \{g,h\} \in \mathfrak{P}_2(\mathfrak{g})$ and $z \in Z(\mathfrak{h}) \subseteq W$.

We have shown that $\mathfrak{J}_{\mathfrak{g}} \subseteq W$. Take $z \in W$. Then $z \in Z(\mathfrak{h})$ for some $\mathfrak{h} = \{g,h\} \in \mathfrak{P}_2(\mathfrak{g})$ where $g \neq h$. Take reduced representations $\mathfrak{w}_g : U \longrightarrow V^*$ of g and $\mathfrak{w}_h : U \longrightarrow V^*$ of h with $z \in U$. Then (7.52) holds. Hence $\mathfrak{w}_g(z) \wedge \mathfrak{w}_h(z) = 0$. If $z \in I_{\mathfrak{g}}$, then $z \in \mathfrak{J}_{\mathfrak{g}}$. Assume that $z \in Z(\mathfrak{h}) - I_{\mathfrak{g}}$. Then $\mathfrak{w}_g(z) \neq 0 \neq \mathfrak{w}_h(z)$ and

$$g(z) = \mathbb{P}(\mathfrak{v}_g(z)) = \mathbb{P}(\mathfrak{v}_h(z)) = h(z)$$

which implies $\#\mathfrak{g}(z) < k$. Hence $z \in \mathfrak{J}_{\mathfrak{g}}$. We have shown that $\mathfrak{J}_{\mathfrak{g}} \supseteq W$. Together we obtain $\mathfrak{J}_{\mathfrak{g}} = W$. Hence (7.50) is proved and the set $\mathfrak{J}_{\mathfrak{g}}$ is analytic. Since $\mathfrak{J}_{\mathfrak{g}} \neq M$, the set $\mathfrak{J}_{\mathfrak{g}}$ is thin analytic. q.e.d.

LEMMA 7.14. If $k \leqslant n+1$, then $\Delta(\mathfrak{g}) = Z(\mathfrak{g})$. If $k > n+1$, then

$$\Delta(\mathfrak{g}) = \bigcup_{\mathfrak{h} \in \mathfrak{P}_{n+1}(\mathfrak{g})} Z(\mathfrak{h}) . \tag{7.53}$$

In both cases, $\Delta(\mathfrak{g})$ is analytic.

PROOF. Case 1: Assume that $k \leqslant n+1$. Let $\mathfrak{g} = \{g_1, \dots, g_k\}$ be an enumeration of $\mathfrak{g}$. Abbreviate $G_j = F_{g_j}$ for $j = 1, \dots, k$. Then $Z(\mathfrak{g}) = Z(G_1 \wedge \dots \wedge G_k)$. Take $z \in M$. Let U be an open, connected neighborhood of z and $\mathfrak{v}_j : U \longrightarrow V^*$ be a reduced representation of g_j for $j = 1, \dots, k$. Then

$$Z(\mathfrak{g})\,|\,U = \{x \in U \mid \mathfrak{v}_1(x) \wedge \dots \wedge \mathfrak{v}_k(x) = 0\} \tag{7.54}$$

Subcase a: Assume that $z \in I_{\mathfrak{g}}$. Then $j \in \mathbb{Z}[1,k]$ exists such that $z \in I_{g_j}$. Hence $\mathfrak{v}_j(z) = 0$. By (7.54) $z \in Z(\mathfrak{g})$.

Subcase b: Assume that $z \in \mathfrak{J}_{\mathfrak{g}} - I_{\mathfrak{g}}$. Then $j \in \mathbb{Z}[1,k]$ and $p \in \mathbb{Z}[1,k]$ with $j < p$ exist such that $g_j(z) = g_p(z)$. Here $\mathfrak{v}_j(z) \neq 0 \neq \mathfrak{v}_k(z)$ and $\mathbb{P}(\mathfrak{v}_j(z)) = g_j(z) = g_p(z) = \mathbb{P}(\mathfrak{v}_p(z))$. Hence $\mathfrak{v}_j(z) \wedge \mathfrak{v}_p(z) = 0$ with $j < p$. By (7.54) $z \in Z(\mathfrak{g})$.

<u>Subcase</u> c: <u>Assume</u> <u>that</u> $z \in \Delta(\mathfrak{g}) - \mathfrak{J}_{\mathfrak{g}}$. Then $\mathfrak{w}_j(z) \neq 0$ for all $j \in \mathbb{Z}[1,k]$ and $g_j(z) = \mathfrak{P}(\mathfrak{w}_j(z)) \neq \mathfrak{P}(\mathfrak{w}_p(z)) = g_p(z)$ for $1 \leqq j < p \leqq k$. Also $k = \#\mathfrak{g}(z)$ and $\Gamma(\mathfrak{g}(z)) = 0$. Define $\tilde{\mathfrak{g}}(z) = \{\mathfrak{w}_1(z), \ldots, \mathfrak{w}_k(z)\}$. The map $\mathbb{P} : \tilde{\mathfrak{g}}(z) \longrightarrow \mathfrak{g}(z)$ is bijective. Hence $\Gamma(\tilde{\mathfrak{g}}(z)) = \Gamma(\mathfrak{g}(z)) = 0$ and $\tilde{\mathfrak{g}}(z)$ is not in general position. Therefore $\mathfrak{w}_1(z) \wedge \ldots \wedge \mathfrak{w}_k(z) = 0$, which implies $z \in Z(\mathfrak{g})$.

<u>Subcase</u> d: <u>Assume</u> <u>that</u> $z \in Z(\mathfrak{g}) - \mathfrak{J}_{\mathfrak{g}}$. Define $\tilde{\mathfrak{g}}(z)$ as in subcase c. Then $\tilde{\mathfrak{g}}(z) \subset V_*^*$ and $\#\tilde{\mathfrak{g}}(z) = k$. The map $\mathbb{P} : \tilde{\mathfrak{g}}(z) \longrightarrow \mathfrak{g}(z)$ is bijective. Since $z \in Z(\mathfrak{g})$, we see that $\Gamma(\mathfrak{g}(z)) = \Gamma(\tilde{\mathfrak{g}}(z)) = 0$. Hence $z \in \Delta(\mathfrak{g}) - \mathfrak{J}_{\mathfrak{g}}$.

<u>Subcase</u> e: <u>Assume</u> <u>that</u> $z \in Z(\mathfrak{g}) \cap \mathfrak{J}_{\mathfrak{g}}$. Then $z \in \mathfrak{J}_{\mathfrak{g}} \subseteq \Delta(\mathfrak{g})$.

These subcases show, that $\Delta(\mathfrak{g}) = Z(\mathfrak{g})$ which proves Case 1.

<u>Case</u> 2: <u>Assume</u> <u>that</u> $k > n + 1$. Define

$$W = \bigcup_{\mathfrak{h} \in \mathfrak{P}_{n+1}(\mathfrak{g})} Z(\mathfrak{h}) .$$

If $\mathfrak{h} \in \mathfrak{P}_{n+1}(\mathfrak{g})$, then $Z(\mathfrak{h}) = \Delta(\mathfrak{h}) \subseteq \Delta(\mathfrak{g})$. Hence $W \subseteq \Delta(\mathfrak{g})$. Take $z \in \Delta(\mathfrak{g})$.

<u>Subcase</u> a: <u>Assume</u> <u>that</u> $z \in I_{\mathfrak{g}}$. then $g \in \mathfrak{g}$ exists such that $z \in I_g$. Then $\mathfrak{h} \in \mathfrak{P}_{n+1}(\mathfrak{g})$ exists such that $g \in \mathfrak{h}$. Then $z \in I_g \subseteq I_{\mathfrak{h}} \subseteq Z(\mathfrak{h}) \subseteq W$.

<u>Subcase</u> b: <u>Assume</u> <u>that</u> $z \in \mathfrak{J}_{\mathfrak{g}} - I_{\mathfrak{g}}$. Then $k \geqq 2$ and $g \in \mathfrak{g}$ and $h \in \mathfrak{g}$ exist such that $g \neq h$ but $g(z) = h(z)$. Also $\mathfrak{h} \in \mathfrak{P}_{n+1}(\mathfrak{g})$ exists such that $\{g,h\} \subseteq (\mathfrak{h})$. Then $z \in \mathfrak{J}_{\mathfrak{h}} - I_{\mathfrak{h}} \subseteq Z(\mathfrak{h}) \subseteq W$.

<u>Subcase</u> c: <u>Assume</u> <u>that</u> $z \in \Delta(\mathfrak{g}) - I_{\mathfrak{g}}$. Then $\#\mathfrak{g}(z) = k$, but $\mathfrak{g}(z)$ is not in general position. Hence $\mathfrak{h} \in \mathfrak{P}_{n+1}(\mathfrak{g})$ exists such that $\mathfrak{h}(z)$ is not in general position. Here $\#\mathfrak{h}(z) = n + 1$ and $\mathfrak{h}(z) \subseteq \mathfrak{g}(z)$. Hence $z \in \Delta(\mathfrak{h}) - \mathfrak{J}_{\mathfrak{h}} \subseteq Z(\mathfrak{h}) \subseteq W$.

Together we obtain $\Delta(\mathfrak{g}) = W$. In particular, $\Delta(\mathfrak{g})$ is analytic. q.e.d.

<u>Assume</u> <u>that</u> $k \leqslant n + 1$. Enumerate $\mathfrak{g} = \{g_1, \ldots, g_k\}$. Then

(7.53) $$\Box\, \mathfrak{g}\, \Box = \Box\, g_1 \mathbin{\dot\wedge} \ldots \mathbin{\dot\wedge} g_k\, \Box$$

is defined and of class C^∞ on $M - I_{\mathfrak{g}}$ and does not depend on the enumeration. If $z \in \mathfrak{J}_{\mathfrak{g}} - I_{\mathfrak{g}}$, then $\Box\, \mathfrak{g}\, \Box\, (z) = \Box\, g_1(z) \mathbin{\dot\wedge} \ldots \mathbin{\dot\wedge} g_k(z)\, \Box = 0$. If $z \in M - \mathfrak{J}_{\mathfrak{g}}$ then

(7.54) $$\Gamma(\mathfrak{g}(z)) = \Gamma(\mathfrak{g})(z) = \Box\, \mathfrak{g}\, \Box\, (z) = \Box\, g_1(z) \mathbin{\dot\wedge} \ldots \mathbin{\dot\wedge} g_k(z)\, \Box \leqslant 1.$$

Hence $\Gamma(\mathfrak{g})$ extends to a function of class C^∞ on $M - I_{\mathfrak{g}}$ by setting $\Gamma(\mathfrak{g})(z) = 0$ for $z \in \mathfrak{J}_{\mathfrak{g}} - I_{\mathfrak{g}}$.

Assume that $\mathfrak{g}$ is in general position. Then abbreviate $G_j = F_{g_j}$ for $j = 1, \ldots, k$ and $G_1 \wedge \ldots \wedge G_k \not\equiv 0$. The zero divisor

(7.55) $$\mu_{\mathfrak{g}} = \mu_{G_1 \wedge \ldots \wedge G_k} = \mu_{g \dot\wedge \ldots \dot\wedge g_k}$$

does not depend on the enumeration of $\mathfrak{g}$. Its support is the union of the $(m-1)$-dimensional branches of $Z(\mathfrak{g}) = \Delta(\mathfrak{g})$. Hence $\operatorname{supp} \mu_{\mathfrak{g}} \subseteq Z(\mathfrak{g}) = \Delta(\mathfrak{g})$. If $k = n + 1$, then $G_1 \wedge \ldots \wedge G_k$ is a section in a line bundle and we have $\operatorname{supp} \mu_{\mathfrak{g}} = Z(\mathfrak{g}) = \Delta(\mathfrak{g})$.

Also the meromorphic map

$$\wedge \mathfrak{g} = g_1 \wedge \ldots \wedge g_k : M \longrightarrow G_{k-1}(V^*) \tag{7.56}$$

does not depend on the enumeration $\mathfrak{g}$. If $k = n + 1$, then $\wedge \mathfrak{g}$ is constant.

Now, assume that (M,τ) is a parabolic manifold. Still under the assumptions that $k = \#\mathfrak{g} \leq n + 1$ and that $\mathfrak{g}$ is in general position, we define the <u>counting</u> and <u>valence</u> <u>function</u> by

$$n_{\mathfrak{g}}(t) = n_{\mu_{\mathfrak{g}}}(t) \qquad N_{\mathfrak{g}}(r,s) = N_{\mu_{\mathfrak{g}}}(r,s) \tag{7.57}$$

for $t \geq 0$ and $0 < s < r$. If $r \in \mathcal{L}_\tau$ we define the <u>compensation</u> <u>function</u> and <u>gauge</u> <u>measure</u> <u>function</u> by

$$m_{\mathfrak{g}}(r) = \int_{M\langle r\rangle} \log \frac{1}{\square\, \mathfrak{g}\, \square}\, \sigma \qquad \Gamma_{\mathfrak{g}}(r) = \int_{M\langle r\rangle} \log \frac{1}{\Gamma(\mathfrak{g})}\, \sigma \tag{7.58}$$

provided the integrals exist, which will be shown below. Trivially $m_{\mathfrak{g}}(r) = \Gamma_{\mathfrak{g}}(r)$, which however fails if $k > n + 1$.

The First Main Theorem for the exterior product proved in Section 3, 1. Special case gives us immediately:

<u>THEOREM</u> 7.15. <u>First Main Theorem for general position</u> ($k \leq n + 1$). Let (M,τ) be a parabolic manifold of dimension m. Let V be a hermitian vector space of dimension $n + 1 > 1$. Let $\mathfrak{g} \neq \emptyset$ be a finite set of meromorphic maps $g : M \longrightarrow \mathbb{P}(V^*)$ with $\#\mathfrak{g} = k \leq n + 1$. Assume that $\mathfrak{g}$ is in general position. Then $m_{\mathfrak{g}}(r) = \Gamma_{\mathfrak{g}}(r)$ exists for all $r \in \mathcal{L}_\tau$. If $0 < s < r \in \mathcal{L}_\tau$ with $s \in \mathcal{L}_\tau$, then

$$\sum_{g\in\mathfrak{g}} T_g(r,s) = N_{\mathfrak{g}}(r,s) + m_{\mathfrak{g}}(r) - m_{\mathfrak{g}}(s) + T_{\wedge\mathfrak{g}}(r,s) \tag{7.59}$$

If $k = n + 1$, then $\wedge \mathfrak{g}$ is constant and $T_{\wedge\mathfrak{g}} = 0$. Hence

(7.60) $$\sum_{g \in \mathfrak{g}} T_g(r,s) = N_{\mathfrak{g}}(r,s) + M_{\mathfrak{g}}(r) - m_{\mathfrak{g}}(s) .$$

Now, assume that $k = \#\mathfrak{g} > n + 1$. If $z \in M - \mathfrak{Z}_{\mathfrak{g}}$, then

(7.61) $$\Gamma(\mathfrak{g})(z) = \Gamma(\mathfrak{g}(z)) = \inf\{\Gamma(\mathfrak{h}(z)) \mid \mathfrak{h} \in \mathfrak{P}_{n+1}(\mathfrak{g})\} \leq 1$$

by the definition of the gauge. If $z \in \mathfrak{Z}_{\mathfrak{g}} - I_{\mathfrak{g}}$, then $\Gamma(\mathfrak{h})(z)$ is also defined and the infinimum is zero. Hence we define $\Gamma(\mathfrak{g})(z) = 0$ for $z \in \mathfrak{Z}_{\mathfrak{g}} - I_{\mathfrak{g}}$. For $z \in M - I_{\mathfrak{g}}$ we have

(7.62) $$\Gamma(\mathfrak{g})(z) = \inf\{\Gamma(\mathfrak{h})(z) \mid \mathfrak{h} \in \mathfrak{P}_{n+1}(\mathfrak{g})\} \leq 1$$

<u>Assume that $\mathfrak{g}$ is in general position</u>. Then $\Gamma(\mathfrak{g}) > 0$ on $M - \Delta(\mathfrak{g})$. Also (7.53) implies

(7.63) $$\Delta(\mathfrak{g}) = \bigcup_{\mathfrak{h} \in \mathfrak{P}_{n+1}(\mathfrak{g})} \operatorname{supp} \mu_{\mathfrak{h}}$$

Also we define the divisor

(7.64) $$\mu_{\mathfrak{g}} = \sum_{\mathfrak{h} \in \mathfrak{P}_{n+1}(\mathfrak{g})} \mu_{\mathfrak{h}}$$

with $\operatorname{supp} \mu_{\mathfrak{g}} = \Delta(\mathfrak{g})$. We define the distance

(7.65) $$\Box\, \mathfrak{g}\, \Box = \prod_{\mathfrak{h} \in \mathfrak{P}_{n+1}(\mathfrak{g})} \Box\, \mathfrak{h}\, \Box \leq 1 .$$

Since $\Box\, \mathfrak{h}\, \Box = \Gamma(\mathfrak{h})$ for $\mathfrak{h} \in \mathfrak{P}_{n+1}(\mathfrak{g})$, we obtain

(7.66) $$\Box\, \mathfrak{g}\, \Box \leq \Gamma(\mathfrak{g})$$

from (7.62) and (7.65).

Now, assume that (M,τ) is a parabolic manifold. Still under the assumption that $k = \#\mathfrak{g} > n+1$ and that $\mathfrak{g}$ is in general position, we define the __counting function__

(7.67) $$n_{\mathfrak{g}}(t) = n_{\mu_{\mathfrak{g}}}(t) = \sum_{\mathfrak{b}\in\mathfrak{P}_{n+1}(\mathfrak{g})} n_{\mathfrak{b}}(t) \geq 0 \qquad \text{for} \quad t > 0$$

and the __valence function__

(7.68) $$N_{\mathfrak{g}}(r,s) = N_{\mu_{\mathfrak{g}}}(r,s) = \sum_{\mathfrak{b}\in\mathfrak{P}_{n+1}(\mathfrak{g})} N_{\mathfrak{b}}(r,s) \geq 0 \qquad \text{for} \quad 0 < s < r$$

and the __compensation function__

(7.69) $$m_{\mathfrak{g}}(r) = \sum_{\mathfrak{b}\in\mathfrak{P}_{n+1}(\mathfrak{g})} m_{\mathfrak{b}}(r) = \int_{M\langle r\rangle} \log \frac{1}{\Box\,\mathfrak{g}\,\Box}\,\sigma$$

for $r \in \mathcal{L}_{\tau}$. Hence the __gauge measure function__ $\Gamma_{\mathfrak{g}}$ is defined for all $r \in \mathcal{L}_{\tau}$ by

(7.70) $$0 \leq \Gamma_{\mathfrak{g}}(r) = \int_{M\langle r\rangle} \log \frac{1}{\Gamma(\mathfrak{g})}\,\sigma \leq m_{\mathfrak{g}}(r)\,.$$

The estimate (7.70) is crude but helpful.

Observe that

(7.71) $$\#\mathfrak{P}_{n+1}(\mathfrak{g}) = \binom{k}{n+1}\,.$$

If $g \in \mathfrak{g}$ is given, then

(7.72) $$\#(\mathfrak{b} \in \mathbb{P}_{n+1}(\mathfrak{g}) \mid g \in \mathfrak{b}) = \binom{k-1}{n}\,.$$

Therefore (7.60), (7.68) and (7.69) imply:

THEOREM 7.16. **First Main Theorem of General Position** $(k > n + 1)$.

Let (M,τ) be a parabolic manifold of dimension m. Let V be a hermitian vector space of dimension $n + 1 > 1$. Let $\mathfrak{g} \neq \emptyset$ be a finite set of meromorphic maps $g : M \longrightarrow \mathbb{P}(V^*)$ with $\#\mathfrak{g} = k > n + 1$. Assume that $\mathfrak{g}$ is in general position. Take $0 < s < r \in \mathfrak{L}_\tau$ with $s \in \mathfrak{L}_\tau$. Then

(7.73) $$\binom{k-1}{n} \sum_{g \in \mathfrak{g}} T_g(r,s) = N_{\mathfrak{g}}(r,s) + m_{\mathfrak{g}}(r) - m_{\mathfrak{g}}(s)$$

(7.74) $$0 \leq \Gamma_{\mathfrak{g}}(r) \leq m_{\mathfrak{g}}(r) \leq \binom{k-1}{n} \sum_{g \in \mathfrak{g}} T_g(r,s) + m_{\mathfrak{g}}(s) < \infty$$

The identity (7.73) extends $m_{\mathfrak{g}}$ to a continuous function on $\mathbb{R}^+$ such that (7.73) holds for all $0 < s < r$. Let $f : M \longrightarrow \mathbb{P}(V)$ be a meromorphic map such that $T_g(r,s)/T_f(r,s) \longrightarrow 0$ for $r \longrightarrow \infty$ for all $g \in \mathfrak{g}$. Then $\Gamma_{\mathfrak{g}}(r)/T_f(r,s) \longrightarrow 0$ for $r \longrightarrow \infty$. Although the estimate (7.74) is crude, it is surprising that such an explicit and still useful estimate exists for the function $\Gamma_{\mathfrak{g}}$ which measures the decline of general position for $r \longrightarrow \infty$.

A family $\mathfrak{g} = (g_\lambda)_{\lambda \in \Lambda}$ of meromorphic maps $g_\lambda : M \longrightarrow \mathbb{P}(V^*)$ with finite index set Λ is said to be in **general position** if $g_\lambda \neq g_\mu$ for all $\lambda \in \Lambda, \mu \in \Lambda$ with $\lambda \neq \mu$ and if $\{g_\lambda \mid \lambda \in \Lambda\}$ is in general position.

§8. The Second Main Theroem

Our theory was guided by the symmetry between f and g. For the Second Main Theorem we have to leave this convincing arrangement and assume that there is a family of meromorphic maps $g : M \longrightarrow \mathbb{P}(V^*)$. This assumption is due to the very nature of the Second Main Theorem as exemplified in the case where the maps g are constant. We will make the following **general assumptions**:

(B1) Let M be a connected, complex manifold of dimension m.

(B2) Let τ be a parabolic exhaustion of M.

(B3) Let V be a hermitian vector space of dimension $n + 1 > 1$.

(B4) Let $f : M \longrightarrow \mathbb{P}(V)$ be a meromorphic map.

(B5) Let $\mathfrak{g}$ be a finite set of meromorphic maps $g : M \longrightarrow \mathbb{P}(V^*)$.

(B6) Let B be a holomorphic form of bidegree $(m - 1,0)$ on M.

(B7) Assume that τ majorizes B with majorant Y.

(B8) Assume that f is general for B.

By (B8) we have $\ell_f = n$. Define $k = \#\mathfrak{g} > 0$. Define $\ell_{\mathfrak{g}} = \min\{\ell_g \mid g \in \mathfrak{g}\}$. For $p \in \mathbb{Z}[0,\ell_{\mathfrak{g}}]$ define

(8.1) $$\mathfrak{g}_p = \{g_p \mid g \in \mathfrak{g}\}$$

as the family of associated maps of degree p of $\mathfrak{g}$. Again put $x^+ = \mathrm{Max}(0,x)$ for all $x \in \mathbb{R}$. The dependence on $g \in \mathfrak{g}$ is indicated by affixing g. For instance, $\Xi_{pq}(g)$, $\xi_{pq}(g)$, $\rho_{pq}(g)$, $\mathbb{K}_p(g)$, $k_p(g)$, $\Phi_{pq}(g)$, $\Phi_q(g)$, $\Psi_p(g)$, $P_{pq}(r,g)$, $R_{pq}(r,g)$ etc.

PROPOSITION 8.1. Assume that (B1) – (B8) hold. Take $p \in \mathbb{Z}[0,\ell_{\mathfrak{g}}]$ and $q \in \mathbb{Z}[0,n]$. Assume that $\mathfrak{g}_p$ is in general position. Assume that (f,g) is free of order (p,q) for all $g \in \mathfrak{g}$. For $\epsilon > 0$ we have the estimate

(8.2)
$$D(p,q) + S_{f_q}(r) + \sum_{g\in\mathfrak{g}} (m_{f_q \dot{\boxplus} g_p}(r) - m_{f_{q+1}\dot{\boxplus} g_p}(r))$$

$$\underset{\bullet}{\leqq} (k - D(p,q))^+(\Gamma_{\mathfrak{g}_p}(r) + \sum_{g\in\mathfrak{g}} R_{pq}(r,g)) + \sum_{g\in\mathfrak{g}} P_{pq}(r,g)$$

$$+ 3D(p,q)k\varsigma(1 + \epsilon)(\log T_f(r,s) + \log Y(r) + \log^+ \mathrm{Ric}_\tau(r,s))$$
$$+ 2D(p,q)\varsigma(1 + \epsilon) \sum_{g\in\mathfrak{g}} \log^+ T_g(r,s) + \epsilon \log r \,.$$

PROOF. Define $\mu = \mathrm{Min}(p+1,q+1)$. Define

(8.3)
$$k_p(\mathfrak{g}) = \sum_{g\in\mathfrak{g}} k_p(g) \,.$$

By (6.9) we have

(8.4)
$$0 \leqq \frac{\xi_{pq}(g)}{(n + 1)(k_p(\mathfrak{g}) + h_q)^2} \leqq \frac{\xi_{pq}(g)}{(n + 1)(k_p(g) + h_q)^2} \leqq 1 \,.$$

CASE 1. Assume that $\#\mathfrak{g} = k \geqq D(p,q)$. Abbreviate $d = k - D(p,q) \geqq 0$ and define $\mathcal{B}(r,g)$ by (6.33). Put

(8.5)
$$c_0 = D(p,q)^{-D(p,q)}\, 4^{(3n+3)d}(n + 1)^d \,.$$

Then Theorem 7.12 gives us the estimate

$$\Gamma(\mathfrak{G}_p)^{2d} \prod_{g\in\mathfrak{G}} \left[\frac{1}{\mu^2}\,\Phi_{pq}(g)\right]^{\beta(r,g)-1} \frac{\xi_{pq}(g)}{(k_p(\mathfrak{G}) + h_q)^2}$$

$$\leqslant c_0\left[\sum_{g\in\mathfrak{G}} \left[\frac{1}{\mu^2}\,\Phi_{pq}(g)\right]^{\beta(r,g)-1} \frac{\xi_{pq}(g)}{(k_p(\mathfrak{G}) + h_q)^2}\right]^{D(p,q)}$$

which implies

$$\prod_{g\in\mathfrak{G}} \Box\, f_q \dot{\boxplus} g_p \,\Box^{2\beta(r,g)-2}\xi_{pq}(g)$$

$$\leqslant c_0\Gamma(\mathfrak{G}_p)^{-2d}(k_g(\mathfrak{G}) + h_q)^{2d}\left[\sum_{g\in\mathfrak{G}} \Box\, f_q \dot{\boxplus} g_p \,\Box^{2\beta(r,g)-2}\xi_{pq}(g)\right]^{D(p,q)}.$$

The definition (6.37) converts this estimate to

$$h_q^{2D(p,q)} \prod_{g\in\mathfrak{G}} \frac{\Box\, f_{q+1} \dot{\boxplus} g_p \,\Box^2}{\Box\, t_q \dot{\boxplus} g_p \,\Box^2}$$

$$\leqslant h_q^{2D(p,q)} \prod_{g\in\mathfrak{G}} \frac{\Phi_{p,q+1}(g)}{\Box\, f_p \dot{\boxplus} g_q \,\Box^2}$$

$$= h_q^{-2d} \left[\prod_{g\in\mathfrak{G}} \frac{\Phi_{p,q+1}(g)h_q^2}{\xi_{pq}(g)}\right]\left[\prod_{g\in\mathfrak{G}} \Box\, f_q \dot{\boxplus} g_p \,\Box^{2\beta(r,g)-2}\xi_{pq}(g)\right]$$

$$\cdot \left[\prod_{g\in\mathfrak{G}} \Box\, f_q \dot{\boxplus} g_p \,\Box^{-2\beta(r,g)}\right]$$

$$\leqslant c_0\Gamma(\mathfrak{G}_p)^{-2d}\left[1 + \frac{k_p(\mathfrak{G})}{h_q}\right]^{2d}\left[\prod_{g\in\mathfrak{G}} \rho_{pq}(g)\right]\left[\prod_{g\in\mathfrak{G}} \Box\, f_q \dot{\boxplus} g_p \,\Box^{-2\beta(r,g)}\right]$$

$$\cdot \left[\sum_{g\in\mathfrak{G}} \Box\, f_q \dot{\boxplus} g_p \,\Box^{2\beta(r,g)-2}\xi_{pq}(g)\right]^{D(p,q)}.$$

Take the logarithm, integrate over $M\langle r\rangle$ and divide by two. This gives us

(8.6)

$$D(p,q)S_{f_q}(r) + \sum_{g\in\mathfrak{G}} \left[m_{f_g \dot{\boxplus} g_p}(r) - m_{f_{q+1}\dot{\boxplus} g_p}(r) \right]$$

$$\leqslant \log c_0 + d\left[\Gamma_{\mathfrak{G}_p}(r) + \int_{M\langle r\rangle} \log\left[1 + \frac{k_p(\mathfrak{G})}{h_q}\right]\sigma \right]$$

$$+ \sum_{g\in\mathfrak{G}} P_{pq}(r,g) + \sum_{g\in\mathfrak{G}} \beta(r,g) m_{f_q \dot{\boxplus} g_p}(r)$$

$$+ \frac{1}{2} D_{pq} \int_{M\langle r\rangle} \log\left[\sum_{g\in\mathfrak{G}} \left[\frac{1}{\mu^2}\, \Phi_{pq}(g)\right]^{\beta(r,g)-1} \xi_{pq}(g) \right]\sigma$$

where

(8.7)

$$\int_{M\langle r\rangle} \log\left[1 + \frac{k_p(\mathfrak{G})}{h_q}\right]\sigma \leqslant \sum_{g\in\mathfrak{G}} \int_{M\langle r\rangle} \log^+ \frac{k_p(g)}{h_q}\,\sigma + c_1 \log(k+1)$$

$$\leqslant \sum_{g\in\mathfrak{G}} \int_{M\langle r\rangle} \log\left[1 + \frac{k_p(g)}{h_q}\right]\sigma + c_1 \log(k+1)$$

$$= \sum_{g\in\mathfrak{G}} R_{pq}(r,g) + c_1 \log(k+1)\,.$$

Also we have

(8.8) $$\beta(r,g) m_{f_q \dot{\boxplus} g_p}(r) \leqslant \beta(r,g)(T_{f_q}(r,s) + T_{g_p}(r,s) + m_{f_q \dot{\boxplus} g_p}(s)) \leqslant 1$$

The Ahlfors estimate (6.35) implies

(8.9)

$$\int_{M\langle r\rangle} \log \sum_{g\in\mathfrak{g}} \left[\frac{1}{\mu^2}\Phi_{pq}(g)\right]^{\beta(r,g)-1}\xi_{pq}(g)\sigma$$

$$\leqslant c_1 \log \frac{1}{c_1}\int_{M\langle r\rangle} \sum_{g\in\mathfrak{g}}\left[\frac{1}{\mu^2}\Phi_{pq}(g)\right]^{\beta(r,g)-1}\xi_{pq}\sigma$$

$$\leqslant c_1 \sum_{g\in\mathfrak{g}} \log^+\left[\int_{M\langle r\rangle}\left[\frac{1}{\mu^2}\Phi_{pq}(g)\right]^{\beta(r,g)-1}\xi_{pq}\sigma\right] + c_1 \log\frac{k}{c_1}$$

$$\underset{\bullet}{\leqslant} 3c_1(1+\epsilon)k\,(\log T_f(r,s) + \log Y(r) + \log^+ \mathrm{Ric}_\tau(r,s))$$

$$+\, 3c_1(1+\epsilon)\sum_{g\in\mathfrak{g}}\log^+ T_g(r,s) + c_1 k\epsilon \log r + c_1 \log\frac{k}{c_1}$$

Now (8.6) – (8.9) yield

(8.10)

$$D(p,q)S_{f_q}(r) + \sum_{g\in\mathfrak{g}}(m_{f_q\dot{\boxplus} g_p}(r) - m_{f_{q+1}\dot{\boxplus} g_p}(r))$$

$$\underset{\bullet}{\leqslant} d\left[\Gamma_{\mathfrak{g}_p}(r) + \sum_{g\in\mathfrak{g}} R_{pq}(r,g)\right] + \sum_{g\in\mathfrak{g}} P_{pq}(r,g)$$

$$+\, 2c_1(1+\epsilon)k(\log T_f(r,s) + \log Y(r) + \log^+ \mathrm{Ric}_\tau(r,s))$$

$$+\, 2c_1(1+\epsilon)\sum_{g\in\mathfrak{g}}\log^+ T_g(r,s) + c_1(k+1)\epsilon \log r$$

where we absorpt the constants into the $\log r$ term. Take a constant $c_1 > 1 + c_1(k+1) > 1$ and replace ϵ by $\epsilon/c_1 < \epsilon$. Then (8.10) implies (8.2) in the case $k \geqslant D(p,q)$.

CASE 2. Assume that $0 < k < D(p,q)$. Abbreviate $d = D(p,q) - k$. Define $\beta(t,g)$ by (6.33). Put

$$c_2 = D(p,q)^{-D(p,q)}(n+1)^k . \tag{8.11}$$

Then Theorem 7.12 gives us the estimate

$$\prod_{g\in\mathfrak{g}}\left[\frac{1}{\mu^2}\Phi_{pq}(g)\right]^{\beta(r,g)-1}\frac{\xi_{pq}(g)}{(k_p(\mathfrak{g})+h_q)^2}$$

$$\leqslant c_1\left[d+\sum_{g\in\mathfrak{g}}\left[\frac{1}{\mu^2}\Phi_{pq}(g)\right]^{\beta(r,g)-1}\frac{\xi_{pq}(g)}{(k_p(\mathfrak{g})+h_q)^2}\right]^{D(p,q)}$$

which implies

$$\prod_{g\in\mathfrak{g}}\square\, f_q \boxplus g_p\, \square^{2\beta(r,g)-2}\xi_{pq}(g)$$

$$\leqslant \frac{c_1}{(k_p(\mathfrak{g})+h_q)^{2d}}\left[d(k_p(\mathfrak{g})+h_q)^2\right.$$

$$\left.+\sum_{g\in\mathfrak{g}}\left[\frac{1}{\mu^2}\Phi_{pq}(g)\right]^{\beta(r,g)-1}\xi_{pq}(g)\right]^{D(p,q)}$$

Since $d > 0$ and $h_q/(k_p(\mathfrak{g})+h_q) < 1$, we obtain

$$h_q^{2D(p,q)}\prod_{g\in\mathfrak{g}}\frac{\square\, f_{q+1}\,\dot{\boxplus}\, g_p\,\square^2}{\square\, f_q\,\dot{\boxplus}\, g_p\,\square^2}$$

$$\leqslant c_1\left[\prod_{g\in\mathfrak{g}}\rho_{pq}(g)\right]\left[\prod_{g\in\mathfrak{g}}\left[\frac{1}{\mu^2}\Phi_{pq}(g)\right]^{-\beta(r,g)}\right]$$

$$\cdot\left[d(k_p(\mathfrak{g})+h_q)^2+\sum_{g\in\mathfrak{g}}\left[\frac{\Phi_{pq}(g)}{\mu^2}\right]^{\beta(r,g)-1}\xi_{pq}(g)\right]^{D(p,q)} .$$

The operator $\frac{1}{2} \int\limits_{M\langle r\rangle} \log (\ )\sigma$ is applied:

(8.12)

$$D(p,q) + S_{f_q}(r) + \sum_{g \in \mathfrak{G}} (m_{f_q \dot{\boxplus} g_p}(r) - m_{f_{q+1} \dot{\boxplus} g_p}(r))$$

$$\leqslant \log c_1 + \sum_{g \in \mathfrak{G}} P_{pq}(r,g) + \sum_{g \in \mathfrak{G}} \beta(r,g) m_{f_q \dot{\boxplus} g_p}(r)$$

$$+ \frac{1}{2} D_{pq} \int\limits_{M\langle r\rangle} \log \Big[d(k_p(\mathfrak{G}) + h_q)^2$$

$$+ \sum_{g \in \mathfrak{G}} \left[\frac{1}{\mu^2} \Phi_{pq}(g)\right]^{\beta(r,g)-1} \xi_{pq}(g)\Big] \sigma$$

$$\leqslant \sum_{g \in \mathfrak{G}} P_{pq}(r,g) + k + \log c_1$$

$$+ \frac{\varsigma}{2} D(p,q) \log \Big[\frac{1}{\varsigma} \int\limits_{M\langle r\rangle} \Big[d(k_p(\mathfrak{G}) + h_q)^2$$

$$+ \sum_{g \in \mathfrak{G}} \left[\frac{1}{\mu^2} \Phi_{pq}(g)\right]^{\beta(r,g)-1} \xi_{pq}(g)\Big] \sigma$$

$$\leqslant c_2 + \sum_{g \in \mathfrak{G}} P_{pq}(r,g) + \frac{\varsigma}{2} D(p,q) \log^+ \int\limits_{M\langle r\rangle} (k_p(\mathfrak{G}) + h_q)^2 \sigma$$

$$+ \frac{\varsigma}{2} D(p,q) \sum_{g \in \mathfrak{G}} \log^+ \int\limits_{M\langle r\rangle} \left[\frac{1}{\mu^2} \Phi_{pq}(g)\right]^{\beta(r,g)-1} \xi_{pq} \sigma \ .$$

We have

$$(k_p(\mathfrak{G}) + h_q)^2 \leqslant (k+1)(h_q^2 + \sum_{g \in \mathfrak{G}} k_p(g)^2) \ .$$

From (6.28) we obtain

$$\int_{M\langle r\rangle} (k_p(\mathfrak{g}) + h_q)^2\sigma$$

$$\leqq (k+1) \int_{M\langle r\rangle} (h_q^2 + \sum_{g\in\mathfrak{g}} k_p(g)^2)\sigma$$

$$\underset{\cdot}{\leqq} (k+1) r^{(2m-1)\epsilon} Y(r)^{(1+\epsilon)^2} (T_{f_q}(r,s)^{(1+\epsilon)^2} + \sum_{g\in\mathfrak{g}} T_{g_p}(r,s)^{(1+\epsilon)^2}) .$$

Consequently we have

$$\log^+ \int_{M\langle r\rangle} (k_p(\mathfrak{g}) + h_q)^2\sigma$$

$$\underset{\cdot}{\leqq} (1+\epsilon)^2 (\log Y(r) + \log T_{f_q}(r,s) + \sum_{g\in\mathfrak{g}} \log^+ T_{g_p}(r,s))$$

$$+ \epsilon(2m-1)\log r + 2\log(k+1)$$

Now (6.31) implies

$$\log T_{f_q}(r,s) \underset{\cdot}{\leqq} \log^+ T_f(r,s) + \log^+ \log Y(r) + \log^+ \mathrm{Ric}_\tau(r,s)$$

$$+ \log^+ \log r + c_3$$

$$\underset{\cdot}{\leqq} \log^+ T_f(r,s) + \log Y(r) + \log^+ \mathrm{Ric}_\tau(r,s) + \epsilon \log r$$

$$\log T_{g_p}(r,s) \underset{\cdot}{\leqq} \log^+ T_g(r,s) + \log Y(r) + \log^+ \mathrm{Ric}_\tau(r,s) + \epsilon \log r .$$

Therefore we obtain

$$\log^+ \int_{M\langle r\rangle} (k_p(\mathfrak{g}) + h_q)^2 \sigma$$

$$\leq (1+\epsilon)^2(\log T_f(r,s) + (k+2)\log Y(r) + (k+1)\log^+ \mathrm{Ric}_\tau(r,s))$$

$$+ (1+\epsilon)^2 \sum_{g\in\mathfrak{g}} \log^+ T_g(r,s) + ((2m-1) + (k+1)(1+\epsilon)^2)\epsilon \log r$$

which implies

(8.13)

$$\log^+ \int_{M\langle r\rangle} (k_p(\mathfrak{g}) + h_q)^2 \sigma$$

$$\leq (1+\epsilon)(2+k)(\log T_f(r,s) + \log Y(r) + \log^+ \mathrm{Ric}_\tau(r,s))$$

$$+ (1+\epsilon) \sum_{g\in\mathfrak{g}} \log^+ T_g(r,s) + c \log r .$$

Now (6.35) gives us

(8.14)

$$\log^+ \int_{M\langle r\rangle} \left[\frac{1}{\mu^2} \Phi_{pq}(g)\right]^{\beta(r,g)-1} \xi_{pq}(g)\sigma$$

$$\leq 3(1+\epsilon)(\log T_f(r,s) + \log^+ T_g(r,s) + \log Y(r) + \log^+ \mathrm{Ric}_\tau(r,s))$$

$$+ \epsilon \log r .$$

Now (8.12), (8.13) and (8.14) imply

$$D(p,q) + S_{f_q}(r) + \sum_{g\in\mathfrak{g}} (m_{f_q \dot{\boxplus} g_p}(r) - m_{f_{q+1} \dot{\boxplus} g_p}(r))$$

$$\leq 3D(p,q)C(1+\epsilon)k(\log T_f(r,s) + \log Y(r) + \log^+ \mathrm{Ric}_\tau(r,s))$$

$$+ 2D(p,q)\varsigma(1 + \epsilon) \sum_{g \in \mathfrak{q}} \log^+ T_g(r,s) + \sum_{g \in \mathfrak{q}} P_{pq}(r,g)$$

$$+ D(p,q)(\varsigma/2)(1 + k)\epsilon \log r .$$

By a change in the value of ϵ we obtain (8.2). q.e.d.

Let F_q be the qth representation section of the map $f_q : M \longrightarrow G_q(V)$. Let $N_{F_q}(r,s)$ be the valence function of the zero divisor of F_q.

<u>LEMMA</u> **8.2.** Assume that (B1) – (B4), (B6) and (B8) are satisfied. Assume that $0 < s < r \in \mathcal{L}^0_\tau$ and $s \in \mathcal{L}^0_\tau$ and that $p \in \mathbb{Z}[0,\ell_{\mathfrak{q}}]$ and $q \in \mathbb{Z}[0,n]$. Then we have

(8.15)

$$\sum_{q=0}^{n-1} D(p,q)(S_{f_q}(r) - S_{f_q}(s)) + \begin{bmatrix} n + 1 \\ p + 2 \end{bmatrix} \mathrm{Ric}_\tau(r,s)$$

$$= \sum_{q=1}^{n-1} \begin{bmatrix} n - q - 1 \\ p - 1 \end{bmatrix} (N_{F_q}(r,s) + T_{f_q}(r,s))$$

$$+ N_{F_n}(r,s) - \left(\begin{bmatrix} n \\ p + 1 \end{bmatrix} + \begin{bmatrix} n - 1 \\ p \end{bmatrix} \right) T_f(r,s)$$

<u>PROOF</u>. Recall (4.17). Hence the Plücker Difference Formula (4.33) can be written as

$$(N_{F_{q-1}}(r,s) + T_{f_{q-1}}(r,s)) - 2(N_{F_q}(r,s) + T_{f_q}(r,s)) + (N_{F_{q+1}}(r,s) + T_{f_{q+1}}(r,s))$$

$$= S_{f_q}(r) - S_{f_q}(s) + \mathrm{Ric}_\tau(r,s)$$

Here $N_{F_{-1}}(r,s) = N_{F_0}(r,s) = T_{f_{-1}}(r,s) = T_{f_n}(r,s) = 0$ and $T_{f_0}(r,s) = T_f(r,s)$. Observe that

(8.16)
$$\sum_{q=0}^{n-1} D(p,q) = \sum_{q=0}^{n-1} \begin{bmatrix} n & - & q \\ p & + & 1 \end{bmatrix} = \begin{bmatrix} n & + & 1 \\ p & + & 2 \end{bmatrix}$$

If $x_0, x_1, \ldots, x_n$ are indeterminates over the ring $\mathbb{Z}$ and if $x_q = 0$ for $q \in \mathbb{Z} - \mathbb{Z}[0,n]$, then we have the algebraic identity

(8.17)
$$\sum_{q=0}^{n-1} \begin{bmatrix} n & - & q \\ p & + & 1 \end{bmatrix} (x_{q-1} - 2x_q + x_{q+1})$$

$$= \sum_{q=1}^{n-1} \begin{bmatrix} n - q - 1 \\ p - 1 \end{bmatrix} x_q + \begin{bmatrix} 1 \\ p + 1 \end{bmatrix} x_n - \left(\begin{bmatrix} n \\ p + 1 \end{bmatrix} + \begin{bmatrix} n - 1 \\ p \end{bmatrix} \right) x_0$$

Therefore we obtain

$$\sum_{q=1}^{n-1} \begin{bmatrix} n - q - 1 \\ p - 1 \end{bmatrix} (N_{F_q}(r,s) + T_{f_q}(r,s)) + \begin{bmatrix} 1 \\ p + 1 \end{bmatrix} N_{f_n}(r,s)$$

$$= \left(\begin{bmatrix} n \\ p + 1 \end{bmatrix} + \begin{bmatrix} n - 1 \\ p \end{bmatrix} \right) T_f(r,s)$$

$$+ \sum_{q=0}^{n-1} D(p,q)(S_{f_q}(r) - S_{f_q}(s)) + \begin{bmatrix} n + 1 \\ p + 2 \end{bmatrix} \mathrm{Ric}_{\tau}(r,s) . \qquad \text{q.e.d.}$$

By Lemma 1.11, $m_{f_n \dot{\boxplus} g_p}(r) - (c_n/2)\mathrm{l}\log(p + 1)$ is a constant. This constant and the constants $S_{f_q}(s)$ can be absorpt by the remainder term $\epsilon \log r$. Recall that (f,g) is free of order (q,p) for all $q \in \mathbb{Z}[0,n]$ if and only if (f,g) is free of order $(0,p)$ (Lemma 4.2). Therefore if we sum (8.2) over $q = 0,1, \ldots ,n - 1$, we obtain:

THEOREM 8.3. **Second Main Theroem.** Assume that (B1) – (B8) are satisfied. Take $p \in \mathbb{Z}[0, \ell_{\mathfrak{G}}]$. Assume that $\mathfrak{G}_p$ is in general position. Assume that (f,g) is free of order $(0,p)$ for all $g \in \mathfrak{G}$. Take $\epsilon > 0$ and $s > 0$. Then we have the estimate

(8.18)

$$\sum_{q=1}^{n-1}\begin{bmatrix} n & - & q & - & 1 \\ & p & - & 1 & \end{bmatrix}(N_{F_q}(r,s) + T_{f_q}(r,s)) + \begin{bmatrix} & 1 & \\ p & + & 1\end{bmatrix}N_{F_n}(r,s)$$

$$+ \sum_{g\in\mathfrak{G}} m_{g_p} \dot{L}_f(r)$$

$$\leqslant \left\{\begin{bmatrix} n \\ p + 1\end{bmatrix} + \begin{bmatrix} n - 1 \\ p \end{bmatrix}\right\} T_f(r,s) + \sum_{q=0}^{n-1} \sum_{g\in\mathfrak{G}} P_{pq}(r,g)$$

$$+ \sum_{q=0}^{n-1} (k - D(p,q))^+ (\Gamma_{\mathfrak{G}_p}(r) + \sum_{g\in\mathfrak{G}} R_{pq}(r,g))$$

$$+ \begin{bmatrix} n + 1 \\ p + 2\end{bmatrix} \mathrm{Ric}_\tau(r,s) + 2\begin{bmatrix} n + 1 \\ p + 2\end{bmatrix} c(1 + \epsilon) \sum_{g\in\mathfrak{G}} \log^+ T_g(r,s)$$

$$+ 3\begin{bmatrix} n + 1 \\ p + 2\end{bmatrix} k c(1 + \epsilon)(\log T_f(r,s) + \log Y(r) + \log^+ \mathrm{Ric}_\tau(r,s))$$

$$+ \epsilon \log r .$$

Formula (8.18) is extremely long and needs interpretation. Without doubt P_{pq} and R_{pq} are the worst terms. However if

(8.19) $$0 \leqslant p = \ell_{\mathfrak{G}} = \ell_g \quad \text{for all} \quad g \in \mathfrak{G} .$$

Then $k_p(g) = 0$. Hence $P_{pq}(r,g) = R_{pq}(r,g) = 0$ for all $r > 0$, all $q \in \mathbb{Z}[0,n)$ and $g \in \mathfrak{G}$. We obtain:

<u>THEOREM</u> 8.4. <u>Second Main Theorem</u> (Maximal version). Assume that (B1) – (B8) and (8.19) are satisfied. Assume that $\mathfrak{G}_p$ is in general position. Assume that (f,g) is free of order $(0,p)$ for all $g \in \mathfrak{G}$. Take $\epsilon > 0$ and $s > 0$. Then we have the estimate

(8.20)

$$\sum_{q=1}^{n-1}\binom{n-q-1}{p-1}(N_{F_q}(r,s)+T_{f_q}(r,s))+\binom{1}{p+1}N_{F_n}(r,s)$$

$$+\sum_{g\in\mathfrak{g}} m_{g_p}\dot{L}_f(r)$$

$$\leqq\left(\binom{n}{p+1}+\binom{n-1}{p}\right)T_f(r,s)+\sum_{q=0}^{n-1}(k-D(p,q))^+\Gamma_{\mathfrak{g}_p}(r)$$

$$+\binom{n+1}{p+2}\mathrm{Ric}_\tau(r,s)+2\binom{n+1}{p+2}c_1(1+\epsilon)\sum_{g\in\mathfrak{g}}\log^+T_g(r,s)$$

$$+3\binom{n+1}{p+2}kc_1(1+\epsilon)(\log T_f(r,s)+\log Y(r)+\log^+\mathrm{Ric}_\tau(r,s))$$

$$+\epsilon\log r\ .$$

There is a geometric condition which implies (8.19) if B is chosen properly. First we need the following result.

PROPOSITION 8.5. Let M be a connected, complex manifold of dimension m. Assume that there is a finite dimensional vector space W and a holomorphic map $\varphi : M \longrightarrow W$ and a point $x_0 \in M$ such that the differential $d\varphi(x_0) : \mathfrak{T}_{x_0}(M) \longrightarrow W$ of φ at x_0 is injective. Let $\mathfrak{h} \neq \emptyset$ be a finite set of linearly non-degenerate meromorphic maps $h : M \longrightarrow \mathbb{P}(V_h)$ where V_h is a complex vector space of dimension $n_h + 1$. Define $n = \mathrm{Max}(n_h \mid h \in \mathfrak{h})$. Then there exists a holomorphic differential form $\hat{B}$ of bidegree $(m-1,0)$ whose coefficients are polynomials of at most degree $n-1$ such that each $h \in \mathfrak{h}$ is general for $B = \varphi^*(\hat{B})$ that is $\ell_h = n_h$ for all $h \in \mathfrak{h}$.

PROOF. By Theorem 7.11 in [27] there exists a holomorphic form $\hat{B}_h$ if bidegree $(m-1,0)$ on W such that $\hat{B}_h$ has coefficients which are polynomials of degree $n_h - 1$ at most and such that $h : M \longrightarrow \mathbb{P}(V_h)$

is general for $B_h = \varphi(\hat{B}_h)$ for each $h \in \mathfrak{H}$. Define $k = \#\mathfrak{H}$. Let Y be a complex vector space of dimension k and let $\{\epsilon_h \mid h \in \mathfrak{H}\}$ be a base of Y. For $\eta \in Y$ define

$$\hat{B}(\eta) = \sum_{h\in\mathfrak{H}} \eta_h \hat{B}_h \qquad B(\eta) = \varphi^*(\hat{B}(\eta)) = \sum_{h\in\mathfrak{H}} \eta_h B_h$$

where $\eta = \sum_{h\in\mathfrak{H}} \eta_h \epsilon_h$. Then $\hat{B}(\eta)$ is a holomorphic form of bidegree $m-1$ on W whose coefficients are polynomials of at most degree $n-1$.

For each $h \in \mathfrak{H}$ take a reduced representation $\mathfrak{v}_h : U \longrightarrow V_h$ where U is an open, connected subset of M. Let $\mathfrak{n}_0^h, \ldots, \mathfrak{n}_{n_h}^h$ be a base of V_h and define $\mathfrak{n}^h = \mathfrak{n}_0^h \wedge \ldots \wedge \mathfrak{n}_{n_h}^h$. A holomorphic function $D_h : U \times Y \longrightarrow \mathbb{C}$ exists uniquely such that $\mathfrak{v}_{hn_h} = D_h(\square, \eta)\mathfrak{n}^h$ in respect to $B(\eta)$. Since h is general for $B(\epsilon_h) = B_h$, there is a point $z_h \in U$ such that $D_h(z_h, \epsilon_h) \neq 0$. Becouse $U \times Y$ is connected, there is a point $x_0 \in U$ and a vector $\beta \in Y$ such that $D_h(x_0, \beta) \neq 0$ for each $h \in H$. Then $\hat{B} = \hat{B}(\beta)$ is a holomorphic form of bidegree $(m-1,0)$ with polynomial coefficients of at most degree $n-1$. Define $B = \varphi^*(\hat{B}) = B(\beta)$. Since $\mathfrak{v}_{hn_h}(x_0) = D_h(x_0, \beta) \neq 0$ in respect to B, the meromorphic map $h : M \longrightarrow \mathbb{P}(V_h)$ is general for B for each $h \in \mathfrak{H}$. q.e.d.

<u>**COROLLARY**</u> **8.6.** Assume that (B1) – (B5) are satisfied. Assume there is a finite dimensional complex vector space W and and holomorphic map $\varphi : M \longrightarrow W$ and a point $x_0 \in M$ such that the differential $d\varphi(x_0) : \mathfrak{T}_{x_0}(M) \longrightarrow W$ of φ at x_0 is injective. Assume that f is linearly non-degenerate. Take $p \in \mathbb{Z}[0,n)$. For each $g \in \mathfrak{G}$ assume

that $g(M)$ is contained in a p-dimensional, projective plane in $\mathbb{P}(V^*)$ but that $g(M)$ is not contained in any $(p-1)$-dimensional projective plane in $\mathbb{P}(V^*)$. Then there is a holomorphic differential form $\hat{B}$ of bidegree $(m - 1,0)$ on W whose coefficients are polynomials of at most degree $n - 1$, such that $f : M \longrightarrow \mathbb{P}(V)$ is general for $B = \varphi^*(\hat{B})$ and such that $\ell_f = n$ and $\ell_g = p$ for all $g \in \mathfrak{G}$ in respect to B. Hence (B8) and (8.19) are satisfied.

PROOF. Let V_g^* be the smallest linear subspace of V^* such that $g(M) \subseteq \mathbb{P}(V_g^*)$. Then $\dim V_g^* = p + 1$ where $p < n$. The meromorphic map $g : M \longrightarrow \mathbb{P}(V_g^*)$ is linearly non-degenerate. By Proposition 8.5 there is a holomorphic form $\hat{B}$ of bidegree $(m - 1,0)$ on W whose coefficients are polynomials of at most degree $n - 1$, such that $f : M \longrightarrow \mathbb{P}(V)$ and $g : M \longrightarrow \mathbb{P}(V_g^*)$ are general for B. Hence $\ell_f = n$. If we consider g as a map into $\mathbb{P}(V^*)$, then $p \leqslant \ell_g$. Since $g(M) \subseteq \mathbb{P}(V_h^*)$ we have $\ell_g \leqslant p$. Hence $\ell_g = p$ for all $g \in \mathfrak{G}$. q.e.d.

We want to study the case $p = 0$. Then $\binom{1}{p+1} = 1$ and the term $N_{F_n}(r,s)$ appears in the Second Main Theorem. This term can be used to modify the Second Main Theorem.

Let $A \neq \emptyset$ be an analytic subset of pure dimension $m - 1$ of M. Then there exists one and only one divisor ν_A such that $\nu_A(x) = 1$ for each simple point $x \in \mathfrak{R}(A)$ of A and such that $\operatorname{supp} \nu_A = A$. Let ν be an divisor on M. Assume that $S = \operatorname{supp} \nu \neq \emptyset$. Let $\mathfrak{A}$ be the set of branches of S. For each $A \in \mathfrak{A}$ there is an integer P_A such that $\nu(x) = P_A$ for all $x \in A \cap \mathfrak{R}(S)$. The family $\mathfrak{A}$ is locally finite. We have

$$\nu = \sum_{A \in \mathfrak{A}} P_A \nu_A . \tag{8.21}$$

Then $\nu \geqslant 0$ if and only if $P_A > 0$ for all $A \in \mathfrak{A}$. Assume that $\nu \geqslant 0$. Take $n \in \mathbb{Z}$ with $n \geqslant 0$. The **truncated divisor** $\nu^{(n)}$ is defined by

(8.22) $$\nu^{(n)} = \sum_{A \in \mathfrak{A}} \mathrm{Min}(P_A, n)\nu_A .$$

Obviously, $0 \leqslant \nu^{(n)} \leqslant \nu$ and $\nu^{(0)} = 0$. If $\nu \equiv 0$, put $\nu^{(n)} \equiv 0$. If (M,τ) is a parabolic manifold, write

(8.23) $$n_\nu^{(n)}(t) = n_{\nu^{(n)}}(t) \qquad N_\nu^{(n)}(r,s) = N_{\nu^{(n)}}(r,s) .$$

By a combination of the methods of L. Smiley [25] Lemma 3.1 and B. Shiffman [22], [23] (1.14), we obtain:

THEOREM 8.7. Assume that (B1), (B3), (B4), (B5), (B6) and (B8) hold. Let $\mathfrak{g}$ be in general position with $\#\mathfrak{g} = k \geqslant n+1$. Assume that $\ell_g = 0$ for all $g \in \mathfrak{g}$. Let F_n be the nth representation section of f and let μ_{F_n} be its divisor. Let $\mu_{\mathfrak{g}}$ be the general position divisor by (7.56). Then

(8.24) $$\sum_{g \in \mathfrak{g}} \mu_{f,g} \leqslant \mu_{F_n} + \mu_{\mathfrak{g}} + \sum_{g \in \mathfrak{g}} \mu_{(f,g)}^{(n)} .$$

REMARK 1. In (8.24) only μ_{F_n} depends on B.

REMARK 2. If (M,τ) is a parabolic manifold, then (8.24) implies

(8.25) $$\sum_{g \in \mathfrak{g}} N_{f,g}(r,s) \leqslant N_{F_n}(r,s) + N_{\mathfrak{g}}(r,s) + \sum_{g \in \mathfrak{g}} N_{f,g}^{(n)}(r,s) .$$

PROOF. We have $\mathfrak{P}_{n+1}(\mathfrak{g}) = \{\mathfrak{h} \subseteq \mathfrak{g} \mid \#\mathfrak{h} = n+1\}$. Then $\Delta(\mathfrak{g})$ is defined by (7.44) and $I_{\mathfrak{g}}$ by (7.42). See also Lemma 7.13. We have $\mathrm{supp}\, \mu_{\mathfrak{g}} = \Delta(\mathfrak{g})$ since $k \geqslant n+1$. Hence either $\Delta(\mathfrak{g})$ is empty or $\Delta(\mathfrak{g})$ has pure dimension $m-1$. Let I_f be the indeterminacy of f. Define

(8.26) $$D = \mathrm{supp} \sum_{g \in \mathfrak{g}} \mu_{f,g} = \bigcup_{g \in \mathfrak{g}} \mathrm{supp}\, \mu_{f,g} .$$

If $D = \emptyset$, nothing is to be proved. Assume that $D \neq \emptyset$. Then D has pure dimension $m - 1$. The set

(8.27) $$S = \Sigma\,(D) \cup \Sigma(\Delta(\mathfrak{g})) \cup I_{\mathfrak{g}} \cup I_f \,.$$

is analytic with $\dim S \leqslant m - 2$. It suffices to verify (8.24) at every point of $D - S$.

Take $x_0 \in D - S$. Take $\mathfrak{h} \in \mathfrak{P}_{n+1}(\mathfrak{g})$. We claim that

(8.28) $$\sum_{g \in \mathfrak{h}} \mu_{f,g}(x_0) \leqslant \mu_{\mathfrak{h}}(x_0) + \mu_{F_n}(x_0) + \sum_{g \in \mathfrak{h}} \mu^{(n)}_{f,g}(x_0)$$

(8.29) $$\mathrm{Min}\{\mu_{f,g}(x_0) \mid g \in \mathfrak{h}\} \leqslant \mu_{\mathfrak{h}}(x_0) \,.$$

Enumerate $\mathfrak{h} = \{g_0, \dots, g_n\}$ such that $\nu_j = \mu_{f,g_j}(x_0)$ and

(8.30) $$\nu_0 \geqslant \nu_1 \geqslant \dots \geqslant \nu_n \,.$$

The maps $g_0, \dots, g_n$ and f are holomorphic at x_0. Take

(8.30) $$a \in \mathbb{P}(V) - \bigcup_{j=0}^{n} \ddot{E}[g_j(x_0)]$$

There is an open, connected neighborhood U of x_0 such that the following conditions are satisfied.

(1) There is a chart $\mathfrak{z} = (z^1, \dots, z^m) : U \longrightarrow U'$ with $\mathfrak{z}(x_0) = 0$, where U' is a ball centered at 0 and where $x_0 \in U \subseteq M - S$.

(2) The set $U \cap D = Y \cap D - S = \{x \in U \mid z^1(x) = 0\}$ is connected.

(3) $g_0, \dots, g_n$ and f are holomorphic on U.

(4) $a \notin \ddot{E}[g_j(x)]$ for all $x \in U$ and $j = 0, \dots, n$.

Take $\mathfrak{a} \in V_*$ with $\mathbb{P}(\mathfrak{a}) = a$. By (3) and (4) there are reduced representations $\mathfrak{v} : U \longrightarrow V_*$ of f and $\tilde{\mathfrak{w}}_j : U \longrightarrow V_*^*$ of g_j such that $\langle \mathfrak{a}, \tilde{\mathfrak{w}}_j(x) \rangle \neq 0$ for all $x \in U$. Then

$$(8.32) \qquad \mathfrak{w}_j = \frac{\tilde{\mathfrak{w}}_j}{\langle \mathfrak{a}, \tilde{\mathfrak{w}}_j \rangle} : U \longrightarrow V_*^*$$

is a reduced representation of g_j with

$$(8.33) \qquad \mathfrak{w}_j' = \frac{\langle \mathfrak{a}, \tilde{\mathfrak{w}}_j \rangle \tilde{\mathfrak{w}}_j' - \langle \mathfrak{a}, \tilde{\mathfrak{w}}_j' \rangle \tilde{\mathfrak{w}}_j}{\langle \mathfrak{a}, \tilde{\mathfrak{w}}_j \rangle^2} = \frac{\tilde{\mathfrak{w}}_j \wedge \tilde{\mathfrak{w}}_j' \llcorner \mathfrak{a}}{\langle \mathfrak{a}, \tilde{\mathfrak{w}}_j \rangle^2}$$

Since $\ell_g = 0$, we have $\tilde{\mathfrak{w}}_j \wedge \tilde{\mathfrak{w}}_j' \equiv 0$. Hence $\mathfrak{w}_j' = 0$ on U for $j = 0, \ldots ,n$.

Define $Y_j = \langle \mathfrak{v}, \mathfrak{w}_j \rangle \not\equiv 0$ on U. Then $Y_j^{(\mu)} = \langle \mathfrak{v}^{(\mu)}, \mathfrak{w}_j \rangle$ for $0 \leqslant \mu \in \mathbb{Z}$. The holomorphic function Y_j vanishes of order ν_j on $U \cap D$. Define $\rho_j = \mathrm{Max}(\nu_j - n, 0)$. If $\rho_j > 0$, then $\nu_j > n$. Hence

$$(8.34) \qquad \mu_{Y_j^{(\lambda)}}(x) \geqslant \nu_j - \lambda \geqslant \rho_j \qquad \text{for } \lambda = 0, \ldots ,n \text{ and } x \in U \cap D$$

There are holomorphic functions X_j^λ on U such that

$$(8.35) \qquad Y_j^{(\lambda)} = X_j^\lambda z_1^{\rho_j} \qquad \text{for } \lambda = 0, \ldots ,n .$$

If $\rho_j = 0$, define $X_j^\lambda = Y_k^{(\lambda)}$ and (8.35) remains correct. Define

$$(8.36) \qquad Y = \det(Y_j^{(\lambda)}) \qquad X = \det(X_j^\lambda) \qquad \rho = \sum_{j=0}^{n} \rho_j .$$

Then

$$Y = z_1^{\rho} X \; . \tag{8.37}$$

Let $\mathfrak{n}_0, \ldots, \mathfrak{n}_n$ be a base of V. Let $\mathfrak{n}_0^*, \ldots, \mathfrak{n}_n^*$ be the dual base. Define $v_j = \langle \mathfrak{v}, \mathfrak{n}_j^* \rangle$ and $w_{\mu\nu} = \langle \mathfrak{n}_\mu, \mathfrak{w}_\nu \rangle$. Define

$$v = \det(v_j^{(\mu)}) \qquad w = \det(w_{\mu\nu}) \; . \tag{8.38}$$

Then we have

$$0 \not\equiv \mathfrak{v}_n = \mathfrak{v} \wedge \mathfrak{v}' \wedge \ldots \wedge \mathfrak{v}_n = v\mathfrak{n}_0 \wedge \ldots \wedge \mathfrak{n}_n \tag{8.39}$$

$$0 \not\equiv \mathfrak{w}_0 \wedge \mathfrak{w}_1 \wedge \ldots \wedge \mathfrak{w}_n = w\mathfrak{n}_0^* \wedge \ldots \wedge \mathfrak{n}_n^* \tag{8.40}$$

$$\langle \mathfrak{v}_n, \mathfrak{w}_0 \wedge \ldots \wedge \mathfrak{w}^{(n)} \rangle = v \cdot w \tag{8.39}$$

$$\mu_{F_n} | U = \mu_v^0 \qquad \mu_{\mathfrak{g}} | U = \mu_w^0 \tag{8.42}$$

and we have

$$v \cdot w = \langle \mathfrak{v} \wedge \mathfrak{v}' \wedge \ldots \wedge \mathfrak{v}^{(n)}, \mathfrak{w}_0 \wedge \ldots \wedge \mathfrak{w}_n \rangle \tag{8.43}$$

$$= \det \langle \mathfrak{v}^{(\mu)}, \mathfrak{w}_\nu \rangle = \det(Y_\nu^{(\mu)}) = Y = z_1^{\rho} X$$

which implies

$$\mu_{F_n} | U + \mu_{\mathfrak{g}} | U = \rho + \mu_X^0 \geq \rho = \sum_{j=0}^{n} \left(\mu_{f,g_j}(x_0) - \mu_{f,g_j}^{(n)}(x_0)\right) . \tag{8.44}$$

Hence we obtain (8.28).

Abbreviate

$$(8.45)\qquad \hat{\mathfrak{n}}_j^* = (-1)^j \mathfrak{n}_0^* \wedge \dots \wedge \mathfrak{n}_{j-1}^* \wedge \mathfrak{n}_{j+1}^* \wedge \dots \wedge \mathfrak{n}_n^*$$

$$(8.46)\qquad \hat{\mathfrak{w}} = \sum_{j=0}^{n} v_j \hat{\mathfrak{n}}_j^* : U \longrightarrow (\Lambda V^*)_*{}_n$$

$$(8.47)\qquad \hat{\mathfrak{w}}_j = (-1)^j \mathfrak{w}_0 \wedge \dots \wedge \mathfrak{w}_{j-1} \wedge \mathfrak{w}_{j+1} \wedge \dots \wedge \mathfrak{w}_n .$$

Obviously, (8.29) is correct if $\nu_n = 0$. Hence we assume that $\nu_n > 0$. Holomorphic functions $H_j \not\equiv 0$ exist on U such that $Y_j = z_1^{\nu_n} H_j$ on U. We have

$$(8.48)\qquad w \cdot \hat{\mathfrak{w}} = w \mathfrak{n}_0^* \wedge \dots \wedge \mathfrak{n}_n^* \llcorner \mathfrak{w} = \mathfrak{w}_0 \wedge \dots \wedge \mathfrak{w}_n \llcorner \mathfrak{w}$$

$$= \sum_{j=0}^{n} \langle \mathfrak{w}, \mathfrak{w}_j \rangle \hat{\mathfrak{w}}_j = \sum_{j=0}^{n} Y_j \hat{\mathfrak{w}}_j = z_1^{\nu_n} \sum_{j=0}^{n} H_j \hat{\mathfrak{w}}_j .$$

Since $\mathfrak{w}$ is a reduced representation, $\mu_{\hat{\mathfrak{w}}} = 0$. We obtain

$$(8.49)\qquad \mu_{\mathfrak{h}}(x_0) = \mu_w(x_0) \geqslant \nu_n = \operatorname{Min}(\mu_{f,g}(x_0) \mid g \in \mathfrak{h})$$

which proves (8.29).

Take an enumeration $\mathfrak{g} = \{g_1, \dots, g_k\}$ such that $\nu_j = \mu_{f,g_j}(x_0)$ and $\nu_1 \geqslant \nu_2 \geqslant \dots \geqslant \nu_k \geqslant 0$. For $j = n+1, \dots, k$ define

$$(8.50)\qquad \mathfrak{h}_j = \{g_1, \dots, g_n, g_j\} \in \mathfrak{P}_{n+1}(\mathfrak{g})$$

where $\mathfrak{h}_j \neq \mathfrak{h}_i$ if $j \neq i$. By (8.28) we obtain

$$(8.51)\qquad \sum_{j=1}^{n+1} \mu_{f,g_j}(x_0) \leqslant \mu_{\mathfrak{h}_{n+1}}(x_0) + \nu_{F_n}(x_0) + \sum_{j=1}^{n+1} \mu_{f,g_j}^{(n)}(x_0) .$$

If $n+2 \leqslant j \leqslant k$, we have

$$\mu_{f,g_j}(x_0) \leqslant \operatorname{Min}(\mu_{f,g}(x_0) \mid g \in \mathfrak{G}_j) \leqslant \mu_{\mathfrak{G}_j}(x_0) . \tag{8.52}$$

Hence (8.51) and (8.52) imply

$$\sum_{j=1}^{k} \mu_{f,g_j}(x_0) \leqslant \nu_{F_n}(x_0) + \sum_{j=n+1}^{k} \mu_{\mathfrak{G}_j}(x_0) + \sum_{j=1}^{n+1} \mu_{f,g_j}^{(n)}(x_0)$$

$$\leqslant \nu_{F_n}(x_0) + \sum_{\mathfrak{G}\in\mathfrak{P}_{n+1}(\mathfrak{G})} \mu_{\mathfrak{G}}(x_0) + \sum_{j=1}^{k} \mu_{f,g_j}^{(n)}(x_0)$$

$$= \nu_{F_n}(x_0) + \mu_{\mathfrak{G}}(x_0) + \sum_{j=1}^{k} \mu_{f,g_j}^{(n)}(x_0) . \qquad \text{q.e.d.}$$

THEOREM 8.8. <u>Second Main Theorem (Maximal version for</u> $p = 0$). Assume that (B1) – (B8) hold. Assume that $\ell_g = 0$ for all $g \in \mathfrak{G}$. Assume that (f,g) is free for all $g \in \mathfrak{G}$. Assume that $\mathfrak{G}$ is in general position. Assume that $\#\mathfrak{G} = k \geqslant n+1$. For $0 < s < r$ abbreviate

$$Q(r,s) = 2n(n+1)k\varsigma(\log T_f(r,s) + \log Y(r) + \log^+ \operatorname{Ric}_\tau(r,s)) . \tag{8.53}$$

Take $\epsilon > 0$ and $s > 0$. Then we have the estimate

(8.54)

$$N_{F_n}(r,s) + \sum_{g\in\mathfrak{G}} m_{f,g}(r)$$

$$\underset{\bullet}{\leqslant} (n+1)T_f(r,s) + \frac{n(n+1)}{2} \operatorname{Ric}_\tau(r,s) + \left[nk - \frac{n(n+1)}{2}\right]\Gamma_{\mathfrak{G}}(r)$$

$$+ 2n(n+1)\varsigma \sum_{g\in\mathfrak{G}} \log^+ T_g(r,s) + Q(r,s) + \epsilon \log r$$

and

(8.55)

$$(k - n - 1)T_f(r,s)$$

$$\leq \sum_{g\in\mathfrak{G}} N_{f,g}^{(n)}(r,s) + \frac{n(n+1)}{2} \operatorname{Ric}_\tau(r,s)$$

$$+ nk\binom{k-1}{n} \sum_{g\in\mathfrak{G}} T_g(r,s) + Q(r,s) + \epsilon \log r .$$

PROOF. W.l.o.g. $0 < \epsilon < \frac{1}{3}$. Then Theorem 8.4 implies (8.54). Now the First Main Theorem (Corollary 3.3). Theorem 7.16 and Theorem 8.7 imply

$$(k - n - 1)T_f(r,s)$$

$$\leq \sum_{g\in\mathfrak{G}} N_{f,g}(r,s) - N_{F_n}(r,s) + \frac{n(n+1)}{2} \operatorname{Ric}_\tau(r,s)$$

$$+ \left(nk - \frac{n(n+1)}{2}\right)\Gamma_{\mathfrak{G}}(r)$$

$$+ 2n(n+1)c_1 \sum_{g\in\mathfrak{G}} \log^+ T_g(r,s) + Q(r,s) + \epsilon \log r + \sum_{g\in\mathfrak{G}} m_{f,g}(s)$$

$$\leq \sum_{g\in\mathfrak{G}} N_{f,g}^{(n)}(r,s) + \frac{n(n+1)}{2} \operatorname{Ric}_\tau(r,s)$$

$$+ \left(nk - \frac{n(n+1)}{2}\right)\Gamma_{\mathfrak{G}}(r) + N_{\mathfrak{G}}(r,s)$$

$$+ 2n(n+1)c_1 \sum_{g\in\mathfrak{G}} \log^+ T_g(r,s) + Q(r,s) + \epsilon \log r + \sum_{g\in\mathfrak{G}} m_{f,g}(s)$$

$$\leq \sum_{g\in\mathfrak{G}} N_{f,g}^{(n)}(r,s) + \frac{n(n+1)}{2} \operatorname{Ric}_\tau(r,s) + Q(r,s) + 2\epsilon \log r$$

$$+ \left(\left(nk - \frac{n(n+1)}{2}\right)\binom{k-1}{n} + 1\right) \sum_{g\in\mathfrak{G}} T_g(r,s)$$

$$+ \left[nk - \frac{n(n+1)}{2}\right] m_{\mathfrak{g}}(s)$$

$$\leqq \sum_{g\in\mathfrak{g}} N_{f,g}^{(n)}(r,s) + \frac{n(n+1)}{2} \operatorname{Ric}_{\tau}(r,s) + Q(r,s) + 3\epsilon \log r$$

$$+ nk\left[\frac{k}{n} - 1\right] \sum_{g\in\mathfrak{g}} T_{g}(r,s) .$$

Replacing ϵ by $\epsilon/3$ we obtain (8.55). q.e.d.

The assumptions of Theorem 8.8 are quite satisfying and $\ell_g = 0$ is possible while g is not constant. In Section 9, we will make use of this fact.

Thus in certain cases, the undesirable terms P_{pq}, R_{pq} are eliminated. However we may ask if it would be possible to find better remainder terms. For instance, we could split the integral

(8.56) $$\int_{M[t]} \Phi_{pq}^{\beta-1}(\Phi_{p+1,q}\mathbb{K}_p + \Phi_{p,q+1}\mathbb{H}_q + 2S^{p+1,\ q}_{\ p\ ,q+1}\mathbb{K}_{pq})$$

into the three integrals

(8.57) $$\int_{M[t]} \Phi_{pq}^{\beta-1}\Phi_{p+1,q}\mathbb{K}_p \qquad \int_{M[t]} \Phi_{pq}^{\beta-1}\Phi_{p,q+1}\mathbb{H}_q$$

(8.58) $$\int_{M[t]} \Phi_{pq}^{\beta-1}\, S^{p+1,\ q}_{\ p\ ,q+1}\mathbb{K}_{pq}$$

and try to incorporate (8.58) as a remainder term in the Second Main Theorem measuring the twist between f and g. Unfortunately, this idea does not work. From the Ahlfors estimates we know that the integral (8.56) exists. In Section 10, we will provide an example where all three integrals in (8.57) and (8.58) diverge while (8.56) exists as it must.

For the further discussion we will assume

(B9) Take $p \in \mathbb{Z}[0, \ell_{q}]$. Assume that

(8.59)
$$\frac{P_{pq}(r,g)}{T_f(r,s)} \longrightarrow 0 \quad \text{for} \quad r \longrightarrow \infty$$

(8.60)
$$\frac{R_{pq}(r,g)}{T_f(r,s)} \longrightarrow 0 \quad \text{for} \quad r \longrightarrow \infty .$$

(Observe that $T_f(r,s) \longrightarrow \infty$ by (B8).)

We define the customary defects (Stoll [28], [32]) by

(8.61)
$$R_f = \limsup_{r \to \infty} \frac{\mathrm{Ric}_\tau(r,s)}{T_f(r,s)}$$

(8.62)
$$Y_f = \limsup_{r \to \infty} \frac{\log Y(r)}{T_f(r,s)} .$$

R_f is called the <u>Ricci Defect</u> and Y_f is called the <u>Majorization Defect</u>. It is unknown, if $\mathrm{Ric}_\tau(r,s) \geqslant 0$. If $R_f < \infty$, then

(8.63)
$$\frac{\log^+ \mathrm{Ric}_\tau(r,s)}{T_f(r,s)} \longrightarrow 0 \quad \text{for} \quad r \longrightarrow \infty .$$

Since $Y(r) \geqslant 1$, we have $Y_f \geqslant 0$.

In the covering case, more can be said. Let (M,τ) be a covering parabolic space of $(\mathbb{C}^m, \tau_0)$. Then there is a proper surjective holomorphic map

(8.64)
$$\varphi = (\varphi_1, \ldots, \varphi_m) : M \longrightarrow \mathbb{C}^m$$

such that $\tau = \tau_0 \circ \varphi = \|\varphi\|^2$. The divisor of $d\varphi_1 \wedge \dots \wedge d\varphi_m \not\equiv 0$ is called the **branching divisor** of φ and denoted by ρ_φ. Then (Stoll [103])

$$\text{(8.65)} \qquad \mathrm{Ric}_\tau(r,s) = N_{\rho_\varphi}(r,s) \geq 0 .$$

Hence if f separates the fibers of φ, then $R_f < \infty$ by a theorem of Noguchi [18]. If there is a meromorphic function h on M which separates the fibers of φ such that $T_h(r,s)/T_f(r,s) \longrightarrow 0$ for $r \longrightarrow \infty$ then $R_f = 0$ by the same theorem of Noguchi.

Define $S = \mathrm{supp}\ \rho_\varphi$. Then $\varphi(S)$ is an analytic subset of $\mathbb{C}^m$. Let S_0 be the $(m-1)$-dimensional component of $\varphi(S)$. Observe that c is the sheet number of $\varphi : M \longrightarrow \mathbb{C}^m$. If S_0 is affine algebraic of degree d, then we have

$$\text{(8.66)} \qquad \mathrm{Ric}_\tau(r,s) = N_{\rho_\varphi}(r,s) \leq cd \log \frac{r}{s}$$

for $0 < s < r$. The map f is said to have **transcendental growth** if $A_f(\infty) = \infty$ which is equivalent to

$$\text{(8.67)} \qquad \frac{T_f(r,s)}{\log r} \longrightarrow \infty \qquad \text{for } r \longrightarrow \infty .$$

Thus if f has transcendental growth and if S_0 is affine algebraic, $R_f = 0$. If φ is biholomorphic, $\mathrm{Ric}_\tau(r,s) \equiv 0$ and $R_f = 0$.

Assume that $\hat{B}$ is a holomorphic form of bidegree $(m-1,0)$ on $\mathbb{C}^m$ that $\varphi^*(\hat{B}) = B$ and that $\hat{B}$ has polynomial coefficients of at most degree $n-1$, then there is a constant $c > 0$ such that

(8.68) $$Y(r) \leqslant 1 + cr^{2n-2} \qquad \text{for} \quad r \geqslant 1 .$$

Hence $Y_f = 0$ if f has transcendental growth.

PROPOSITION **8.9.** Assume that (B1) – (B5) are satisfied. Assume that f is not constant and that $\mathfrak{g}$ is in general position. Assume that

(8.69) $$\frac{T_g(r,s)}{T_f(r,s)} \longrightarrow 0 \qquad \text{for} \quad r \longrightarrow \infty$$

for every $g \in \mathfrak{g}$. Then

(8.70) $$\frac{\Gamma_{\mathfrak{g}}(r)}{T_f(r,s)} \longrightarrow 0 \qquad \text{for} \quad r \longrightarrow \infty .$$

The proposition follows from (7.77). If $p > 0$, the situation is not so satisfactory.

PROPOSITION **8.10.** Assume that (B1) – (B8) hold. Take $p \in \mathbb{Z}(0, \ell_{\mathfrak{g}}]$. Assume that $\mathfrak{g}_p$ is in general position. Define $k = \#\mathfrak{g}$. If $k \leqslant \binom{n}{p}$, define $c = 1$. If $k > \binom{n}{p}$, define $c = \left[\frac{k-1}{\binom{n}{p}}\right]$. Take $\epsilon > 0$ and $s > 0$. Then we have

(8.71)
$$\Gamma_{\mathfrak{g}_p}(r)$$
$$\leqslant c3^p \sum_{g \in} T_g(r,s) + \frac{kc}{2}(3^p - 1)(\varsigma \log Y(r) + \mathrm{Ric}_\tau(r,s) + \epsilon\varsigma \log r) .$$

If $Y_f = 0 = R_f$ and if (8.69) holds for all $g \in \mathfrak{g}$, then

(8.72) $$\liminf_{r\to\infty} \frac{\Gamma_{\mathfrak{g}_p}(r)}{T_f(r,s)} = 0 .$$

PROOF. Theorem 7.14 and 7.15 imply

$$\Gamma_{\mathfrak{g}_p}(r) \leqslant c \sum_{g \in \mathfrak{g}} T_{g_p}(r,s) + m_{\mathfrak{g}_p}(s) \tag{8.73}$$

for $0 < s < r \in \mathcal{L}_\tau$. Proposition 6.15 implies

(8.74)

$$T_{g_p}(r,s) \lessdot 3^p T_g(r,s) + \frac{1}{2}(3^p - 1)(c_\tau \log Y(r) + \mathrm{Ric}_\tau(r,s) + \epsilon c_\tau \log r) .$$

Now (8.73) and (8.74) imply (8.71) with any fixed $s > 0$, where $\epsilon c_\tau \log r$ absorbs the constants. If f has transcendental growth, (8.67), (8.69) and $Y_f = 0 = R_f$ imply (8.72). If f does not have rational growth, then $0 < A_f(\infty) < \infty$ and we obtain

$$\liminf_{r\to\infty} \frac{\Gamma_{\mathfrak{g}_p}(r)}{T_f(r,s)} \leqslant \frac{\epsilon}{A_f(\infty)} \tag{8.75}$$

for all $\epsilon > 0$. Hence $\epsilon \longrightarrow 0$ implies (8.72). q.e.d.

Assume that (B1) – (B8) hold. Take $p \in \mathbb{Z}[0, \ell_{\mathfrak{g}}]$. Assume that (f,g) is free of order $(0,p)$. Define the Nevanlinna defects

$$0 \leqslant \delta_f[g_p] = \liminf_{r\to\infty} \frac{m_{g_p \dot{L} f}(r)}{T_f(r,s)} \leqslant \infty \tag{8.76}$$

$$0 \leqslant \delta_f(g_p) = \liminf_{r\to\infty} \frac{m_{g_p \dot{L} f}(r)}{T_f(r,s) + T_{g_p}(r,s)} . \tag{8.77}$$

Then Theorem 3.2 implies

$$0 \leqslant \delta_f(g_p) = 1 - \limsup_{r\to\infty} \frac{N_{g_p \dot{L} f}(r) + T_{g_p L f}(r,s)}{T_f(r,s) + T_{g_p}(r,s)} \leqslant 1 . \tag{8.78}$$

Trivially we have

(8.79) $$0 \leqslant \delta_f(g_p) \leqslant \delta_f[g_p] \leqslant \infty .$$

If $T_{g_p}(r,s)/T_f(r,s) \longrightarrow 0$ for $r \longrightarrow \infty$, then $\delta_f(g_p) = \delta_f[g_p]$. Also we define

(8.80) $$0 \leqslant \lambda_f(g_p) = \limsup_{r\to\infty} \frac{T_{g_p}(r,s)}{T_f(r,s)} \leqslant \infty$$

(8.81) $$0 \leqslant \kappa_f(g_p) = \liminf_{r\to\infty} \frac{T_{g_p}(r,s)}{T_f(r,s)} \leqslant \lambda_f(g_p) \leqslant \infty .$$

If $\lambda_f(g_p) < \infty$, the definition of the Nevanlinna defects implies

(8.82) $$(1 + \kappa_f(g_p))\delta_f(g_p) \leqslant \delta_f[g_p] \leqslant (1 + \lambda_f(g_p))\delta_f(g_p)$$

If $Y_f = 0 = R_f$, then $\kappa_f(g_p) \leqslant 3^p\lambda_p(g)$ by (8.74). Hence if $Y_f = 0 = R_f$ and if $T_g(r,s)/T_f(r,s) \longrightarrow 0$ for $r \longrightarrow \infty$, then $\kappa_f(g_p) = 0$. If $\lambda_f(g) < \infty$, then

(8.83) $$\frac{\log^+ T_g(r,s)}{T_f(r,s)} \longrightarrow 0 \quad \text{for } r \longrightarrow \infty .$$

Now, Theorem 8.3 implies immediately:

<u>**THEOREM**</u> **8.11** <u>**Defect relation**</u>. Assume that (B1) – (B8) and (B9) hold with $p \in \mathbb{Z}[0,\ell_{\mathfrak{g}}]$. Assume that $R_f = 0 = Y_f$. Assume that $\lambda_f(g) < \infty$ for all $g \in \mathfrak{g}$. Assume that $\mathfrak{g}_p$ is in general position. Assume that

(8.84) $$\Gamma_{\mathfrak{g}_p}(r,s)/T_f(r,s) \longrightarrow 0 \quad \text{for } r \longrightarrow \infty .$$

Assume that (f,g) is free of order $(0,p)$, for all $g \in \mathfrak{g}$. Then we have

(8.85)
$$\sum_{g\in\mathfrak{G}} \delta_f[g_p] \leq \binom{n}{p+1} + \binom{n-1}{p} .$$

REMARK 1. If $T_{g_p}(r,s)/T_f(r,s) \longrightarrow 0$ for $r \longrightarrow \infty$ for all $g \in \mathfrak{G}$, then (8.84) is satisfied. Moreover we have $\delta_f[g_p] = \delta_f(g_p)$.

REMARK 2. We have $\delta_f(g_p) \leq \delta_f[g_p]$.

REMARK 3. If $p = 0$, then (8.85) reads

(8.86)
$$\sum_{g\in\mathfrak{G}} \delta_f[g_p] \leq n + 1 .$$

A different version of the defect relation can be obtained from Proposition 8.10. We combine (8.71) and Theorem 8.3 to obtain

(8.87)
$$\sum_{g\in\mathfrak{G}} m_{f\dot{L}g}(r) \underset{\cdot}{\leq} \left[\binom{n}{p+1} + \binom{n-1}{p}\right] T_f(r,s) + c_0 \operatorname{Ric}_\tau(r,s)$$

$$+ c_1 \log^+ \operatorname{Ric}_\tau(r,s)$$

$$+ \sum_{q=0}^{n-1} \sum_{g\in\mathfrak{G}} (P_{pq}(r,s) + kR_{pq}(r,g)) + \epsilon \log r$$

$$+ c_2(\log T_f(r,s) + \sum_{g\in\mathfrak{G}} T_g(r,s) + \log Y(r))$$

where c_0, c_1 and c_2 are constants. This form of the Second Main Theorem implies immediately:

THEOREM 8.12. Defect relation. Assume that (B1) -- (B8) and (B9) hold with $p \in \mathbb{Z}[0, \ell_{\mathfrak{G}}]$. Assume that $R_f = 0 = Y_f$. Assume that

(8.88) $\quad T_g(r,s)/T_f(r,s) \longrightarrow 0 \quad$ for $r \longrightarrow \infty$ for all $g \in \mathfrak{G}$.

Assume that $\mathfrak{G}_p$ is in general position. Assume that (f,g) is free of order $(0,p)$ for all $g \in \mathfrak{G}$. Then we have

(8.89) $$\sum_{g \in \mathfrak{G}} \delta_f(g_p) \leq \sum_{g \in \mathfrak{G}} \delta_f[g_p] \leq \binom{n}{p+1} + \binom{n-1}{p} .$$

Also Theorem 8.8 leads us to another version of the defect relation.

<u>THEOREM</u> 8.13. <u>Defect relation</u>. Assume that (B1) -- (B8) hold. Assume that $\ell_g = 0$ for all $g \in \mathfrak{G}$. Assume that (f,g) is free for all $g \in \mathfrak{G}$. Assume that $\mathfrak{G}$ is in general position. Assume that $R_f = 0 = Y_f$. Assume that (8.88) holds for all $g \in \mathfrak{G}$. Define

(8.90) $$\theta_f(g) = 1 - \limsup_{r \to \infty} \frac{N_{f,g}^{(n)}(r,s)}{T_f(r,s)}$$

for $g \in \mathfrak{G}$. Then $0 \leq \delta_f(g) \leq \theta_f(g) \leq 1$ and

(8.91) $$\sum_{g \in \mathfrak{G}} \delta_f(g) \leq \sum_{g \in \mathfrak{G}} \theta_f(g) \leq n + 1 .$$

§9. Value Distribution over a Function Field

Our theory can be used to prove a defect relation for hyperplanes defined over a field of meromorphic functions of rank $m - 1$. Our application will extend a defect relation to meromorphic maps established by Shiffman for meromorphic functions only.

Let M be a connected, complex manifold of dimension $m > 1$. Let $\mathfrak{K}_M$ be the field of meromorphic functions on M. If B is a meromorphic form on M, then

(9.1) $$\mathfrak{K}_M[B] = \{\varphi \in \mathfrak{K}_M \mid d\varphi \wedge B \equiv 0\}$$

is a subfield of $\mathfrak{K}_M$. Let $\varphi_1, \ldots, \varphi_k$ be meromorphic functions on M. Then $\varphi_1, \ldots, \varphi_k$ are said to be <u>analytically independent</u> if $d\varphi_1 \wedge \ldots \wedge d\varphi_k \not\equiv 0$ and <u>analytically dependent</u> if $d\varphi_1 \wedge \ldots \wedge d\varphi_k \equiv 0$. Let Φ be a set of meromorphic functions on M. A meromorphic function $\varphi \in \mathfrak{K}_M$ is said to be <u>analytically dependent</u> on Φ if either φ is constant or if there are analytically independent functions $\varphi_1, \ldots, \varphi_k$ in Φ such that $\varphi, \varphi_1, \ldots, \varphi_k$ are analytically dependent. Then

(9.2) $$\mathfrak{K}_M(\Phi) = \{\varphi \in \mathfrak{K}_M \mid \varphi \text{ analytically dependent on } \Phi\}$$

is a field containing Φ. The assignment $\Phi \longrightarrow \mathfrak{K}_M(\Phi)$ is a dependency relation in the sense of Van Der Waerden [34] p. 204 or Zariski-Samuel [41] p. 97 (See Andreotti-Stoll[2] Section 4). If $\varphi_1, \ldots, \varphi_k$ are analytically independent, then $k \leqslant m$.

Therefore if $\Phi \subseteq \mathfrak{K}_M$ but $\Phi \not\subseteq \mathbb{C}$, there are analytically independent $\varphi_1, \ldots, \varphi_k$ in $\mathfrak{K}_M(\Phi)$ such that

$$\mathfrak{K}_M(\Phi) = \mathfrak{K}_M(\varphi_1, \ldots, \varphi_k) . \tag{9.3}$$

Then $\varphi_1, \ldots, \varphi_k$ is called a <u>**base**</u> of $\mathfrak{K}_M(\Phi)$. The number k is independent of the choice of the base and is called the <u>**rank**</u> of Φ or also of $\mathfrak{K}_M(\Phi)$. There is a base of $\mathfrak{K}_M(\Phi)$ contained in Φ. If $\Phi \subseteq \mathbb{C}$, then $\mathfrak{K}_M(\Phi) = \mathbb{C}$ and Φ and $\mathfrak{K}_M(\Phi)$ are said to have <u>**rank**</u> 0. If $\varphi_1, \ldots, \varphi_k$ is a base of $\mathfrak{K}_M(\Phi)$, then

$$\mathfrak{K}_M(\Phi) = \mathfrak{K}_M[d\varphi_1 \wedge \ldots \wedge d\varphi_k] . \tag{9.4}$$

In general, a meromorphic form B is said to <u>**define**</u> $\mathfrak{K}_M(\Phi)$ if

$$\mathfrak{K}_M(\Phi) = \mathfrak{K}_M[B] . \tag{9.5}$$

If M is Stein, every meromorphic function on M is the quotient of holomorphic functions on M. If $\varphi_1, \ldots, \varphi_k$ is a base of $\mathfrak{K}_M(\Phi)$, then there are holomorphic fucntions $\psi_j \not\equiv 0$ and $\chi_j \not\equiv 0$ such that $\chi_j\varphi_j = \psi_j$. By (9.3) the holomorphic form

$$\chi_1^2 \ldots \chi_p^2 \; d\varphi_1 \wedge \ldots \wedge d\varphi_k = \bigwedge_{j=1}^{k} (\chi_j d\psi_j - \psi_j d\chi_j) \tag{9.6}$$

defines $\mathfrak{K}_M(\Phi)$. We have

$$\mathfrak{K}_M(\mathfrak{K}_M(\Phi)) = \mathfrak{K}_M(\Phi) . \tag{9.7}$$

A subset $\Phi \neq \emptyset$ of $\mathfrak{K}_M$ is said to be <u>**complete**</u> if $\mathfrak{K}_M(\Phi) = \Phi$.

Let V be a complex vector space of dimension $n + 1 > 1$. Let $f : M \longrightarrow \mathbb{P}(V)$ be a meromorphic map. A base $\alpha_0, \ldots, \alpha_n$ is said to be <u>**admissible**</u> for f if $\alpha_0^*, \ldots, \alpha_n^*$ is the dual base and if

$f(M) \not\subset \ddot{E}[\mathbb{P}(\mathfrak{a}_0^*)]$. If $\mathfrak{w} : U \longrightarrow V$ is any representation of f on an open, connected subset U of M then $\mathfrak{a}_0, \ldots, \mathfrak{a}_n$ is admissible, if and only if $\langle \mathfrak{w}, \mathfrak{a}_0^* \rangle \not\equiv 0$. Let $\mathfrak{a}_0, \ldots, \mathfrak{a}_n$ be an admissible base for f, then there are unique meromorphic functions $f_1, \ldots, f_n$ on M called the **coordinate function** of f in respect to the admissible base such that

$$(9.8) \qquad f_j | U = \frac{\langle \mathfrak{w}, \mathfrak{a}_j^* \rangle}{\langle \mathfrak{w}, \mathfrak{a}_0^* \rangle} \qquad \text{for } j = 1, \ldots, n$$

for any representation $\mathfrak{w} : U \longrightarrow V$. If $\tilde{\mathfrak{a}}_0, \ldots, \tilde{\mathfrak{a}}_n$ is any other admissible base for f of V, then there are complex numbers $c_{\mu\nu} \in \mathbb{C}$ such that

$$(9.9) \qquad \tilde{\mathfrak{a}}_\mu^* = \sum_{v\nu=0}^{n} c_{\mu\nu} \mathfrak{a}_\nu^* .$$

Then we have

$$(9.10) \qquad \tilde{f}_j = \frac{\sum_{\nu=0}^{n} c_{j\nu} f_\nu}{\sum_{\nu=0}^{n} c_{0\nu} f_\nu} .$$

Let $\Phi \neq \emptyset$ be a set of meromorphic functions on M. Then f is said to be **analytically dependent** on Φ and $\mathfrak{K}_M(\Phi)$ if and only if there is an admissible base $\mathfrak{a}_0, \ldots, \mathfrak{a}_n$ of V for f such that the coordinate functions $f_1, \ldots, f_n$ all belong to $\mathfrak{K}_M(\Phi)$. If f is analytically dependent on Φ, then $f_j \in \mathfrak{K}_M(\Phi)$ for $j = 1, \ldots, n$ for any choice of an admissible base $\mathfrak{a}_0, \ldots, \mathfrak{a}_n$ by the transition formula (9.10).

The meromorphic map f is said to be **analytically independent** of Φ and $\mathfrak{K}_M(\Phi)$ if and only if f is not analytically dependent on Φ. If $\mathfrak{a}_0, \ldots, \mathfrak{a}_n$ is a base of V admissible for f and if $f_1, \ldots, f_n$ are the coordinate functions, then f is analytically independent of Φ if anly on if there is an index $j \in \mathbb{N}[1,n]$ such that $f_j \notin \mathfrak{K}_M(\Phi)$.

PROPOSITION 9.1. Let $f : M \longrightarrow \mathbb{P}(V)$ be a meromorphic map. Let $\Phi \neq \emptyset$ be a set of meromorphic functions on M. Let B be a meromorphic form defining $\mathfrak{K}_M(\Phi)$. Let $\mathfrak{a}_0, \ldots, \mathfrak{a}_n$ be a base of V admissible for f and let $f_1, \ldots, f_n$ be the coordinate functions of f for this base. Take $0 \neq \mathfrak{b} \in V^*$ such that $f(M) \not\subseteq \ddot{E}(\mathbb{P}(\mathfrak{b}))$. Let $\mathfrak{v} : U \longrightarrow V$ be a representation of f. Then the following statements are equivalent:

(1) The map f depends analytically on $\mathfrak{K}_M(\Phi)$.

(2) $df_j \wedge B \equiv 0$ for $j = 1, \ldots, n$ on M.

(3) $d\frac{\mathfrak{v}}{<\mathfrak{v}, \mathfrak{b}>} \wedge B \equiv 0$ on U.

(4) $(\mathfrak{v} \wedge d\mathfrak{v}) \wedge B \equiv 0$ on U.

PROOF. (1) $\Leftrightarrow$ (2). Since $f_j \in \mathfrak{K}_M(\Phi)$ if and only if $df_j \wedge B \equiv 0$.

(2) $\Rightarrow$ (3). Let $\mathfrak{a}_0^*, \ldots, \mathfrak{a}_n^*$ be the dual base to $\mathfrak{a}_0, \ldots, \mathfrak{a}_n$. Then $b_\nu \in \mathbb{C}$ exist such that $\mathfrak{b} = \sum_{\nu=0}^{n} b_\nu \mathfrak{a}_\nu^*$. On U we have

$$\mathfrak{v} = \sum_{j=0}^{n} <\mathfrak{v}, \mathfrak{a}_j^*>\mathfrak{a}_j$$

$$<\mathfrak{v}, \mathfrak{b}> = \sum_{j=0}^{n} b_j <\mathfrak{v}, \mathfrak{a}_j^*>$$

$$g = \frac{<\mathfrak{v}, \mathfrak{b}>}{<\mathfrak{v}, \mathfrak{a}_0^*>} = b_0 + \sum_{j=1}^{n} b_j f_j \not\equiv 0$$

Hence $dg \wedge B = 0$. We have

$$d \frac{\mathfrak{w}}{\langle \mathfrak{w}, \mathfrak{b} \rangle} \wedge B = d\left[\frac{\mathfrak{w}}{\langle \mathfrak{w}, \mathfrak{a}_0^* \rangle} \frac{1}{g}\right] \wedge B$$

$$= d\left[\frac{1}{g} \mathfrak{a}_0 + \sum_{j=1}^{n} \frac{f_j}{g} \mathfrak{a}_j\right] \wedge B$$

$$= -\frac{dg \wedge B}{g^2} \mathfrak{a}_0 + \sum_{j=1}^{n} \frac{g\, df_j \wedge B - f_j\, dg \wedge B}{g^2} \mathfrak{a}_j = 0 .$$

(3) ⇒ (4). We have

$$(\mathfrak{w} \wedge d\mathfrak{w}) \wedge B = \langle \mathfrak{w}, \mathfrak{b} \rangle \mathfrak{w} \wedge \left[d \frac{\mathfrak{w}}{\langle \mathfrak{w}, \mathfrak{b} \rangle} \wedge B\right] = 0 .$$

(4) ⇒ (2). Let $\mathfrak{a}_0^*, \ldots, \mathfrak{a}_n^*$ be the dual base to $\mathfrak{a}_0, \ldots, \mathfrak{a}_n$. Then

$$df_j \wedge B = d \frac{\langle \mathfrak{w}, \mathfrak{a}_j^* \rangle}{\langle \mathfrak{w}, \mathfrak{a}_0^* \rangle} \wedge B$$

$$= \frac{\langle \mathfrak{w}, \mathfrak{a}_0^* \rangle \langle d\mathfrak{w}, \mathfrak{a}_j^* \rangle - \langle \mathfrak{w}, \mathfrak{a}_j^* \rangle \langle d\mathfrak{w}, \mathfrak{a}_0^* \rangle}{\langle \mathfrak{w}, \mathfrak{a}_0^* \rangle^2} \wedge B$$

$$= \frac{\langle (\mathfrak{w} \wedge d\mathfrak{w}) \wedge B, \mathfrak{a}_0^* \wedge \mathfrak{a}_j^* \rangle}{\langle \mathfrak{w}, \mathfrak{a}_0^* \rangle^2} = 0$$

on U. Hence $df_j \wedge B = 0$ on M. q.e.d.

The meromorphic map $f : M \longrightarrow \mathbb{P}(V)$ is said to be <u>free of</u> Φ <u>respectively</u> $\mathfrak{K}(\Phi)$ if and only if (f, g) is free for every meromorphic map $g : M \longrightarrow \mathbb{P}(V^*)$ which is analytically dependent on $\mathfrak{K}(\Phi)$.

<u>LEMMA</u> 9.2. If the meromorphic map $f : M \longrightarrow \mathbb{P}(V)$ is free of $\mathfrak{K}_M(\Phi)$, then f is analytically independent of $\mathfrak{K}_M(\Phi)$.

PROOF. Assume that f is analytically dependent on $\mathfrak{K}_M(\Phi)$. Let $\mathfrak{a}_0, \dots, \mathfrak{a}_n$ be a base of V admissible for f and let $f_1, \dots, f_n$ be the coordinate functions. Then $f_j \in \mathfrak{K}_M(\Phi)$ for $j = 1, \dots, n$. Let $\mathfrak{a}_0^*, \dots, \mathfrak{a}_n^*$ be the dual base of $\mathfrak{a}_0, \dots, \mathfrak{a}_n$ and define $a_j = \mathbb{P}(\mathfrak{a}_j^*)$ for $j = 0, \dots, n$. Let $\mathfrak{v} : U \longrightarrow V$ be an irreducible representation of f on the open, connected subset U of M. Then $<\mathfrak{v}, \mathfrak{a}_0^*> \not\equiv \emptyset$. Also

$$\mathfrak{w} = \mathfrak{a}_0^* \wedge \mathfrak{a}_1^* \llcorner \mathfrak{v} = <\mathfrak{v}, \mathfrak{a}_0^*>\mathfrak{a}_1^* - <\mathfrak{v}, \mathfrak{a}_1^*>\mathfrak{a}_0^* \not\equiv 0 .$$

Therefore $a_0 \wedge a_1 \llcorner f = g : M \longrightarrow \mathbb{P}(V^*)$ is defined as a meromorphic map where $\mathfrak{w}$ is a representation of g with $<\mathfrak{w}, \mathfrak{a}_1> = <\mathfrak{v}, \mathfrak{a}_0^*> \not\equiv 0$. Hence $\mathfrak{a}_1^*, \mathfrak{a}_0^*, \mathfrak{a}_2^*, \dots, \mathfrak{a}_n^*$ is an admissible base of g. Let $g_1, g_2, g_3, \dots, g_n$ be the coordinate functions of g for this base. Then

$$g_1 | U = - \frac{<\mathfrak{v}, \mathfrak{a}_1^*>}{<\mathfrak{v}, \mathfrak{a}_0^*>} = - f_1 | U .$$

Therefore $g_1 = - f_1 \in \mathfrak{K}_M(\Phi)$. Also $g_j \equiv 0$ for $j = 2, \dots, n$. Hence g is analytically dependent on $\mathfrak{K}_M(\Phi)$. We have

$$<\mathfrak{v}, \mathfrak{w}> = <\mathfrak{w}, \mathfrak{v}> = <(\mathfrak{a}_0^* \wedge \mathfrak{a}_1^*) \llcorner \mathfrak{v}, \mathfrak{v}>$$

$$= <\mathfrak{a}_0^* \wedge \mathfrak{a}_1^*, \mathfrak{v} \wedge \mathfrak{v}> \equiv 0 .$$

Therefore the pair (f,g) is not free. Contradiction! q.e.d.

LEMMA 9.3. Let f be a meromorphic function on M. Let $\Phi \neq \emptyset$ be a set of meromorphic functions on M. Then the following statements are equivalent:

(1) The map $f : M \longrightarrow \mathbb{P}_1$ is free of $\mathfrak{K}_M(\Phi)$.

(2) The map $f : M \longrightarrow \mathbb{P}_1$ is analytically independent of $\mathfrak{K}_M(\Phi)$.

(3) The meromorphic function f does not belong to $\mathfrak{K}_M(\Phi)$.

PROOF. (1) $\Rightarrow$ (2). by Lemma 8.2.

(2) $\longrightarrow$ (3). Let $\mathfrak{a}_0, \mathfrak{a}_1$ be the standard base of $\mathbb{C}^2$ and identify $\mathbb{P}(\mathbb{C}^2) = \mathbb{P}_1 = \mathbb{C} \cup \{\infty\}$ by $\mathbb{P}(\mathfrak{a}_0 + z\mathfrak{a}_1) = z$ for $z \in \mathbb{C}$. Then $\mathbb{P}(\mathfrak{a}_1) = \infty$. In respect to this base the meromorphic function f is the coordinate function of the meromorphic map f. Since f is analytically independent of $\mathfrak{K}_M(\Phi)$ we have $f \notin \mathfrak{K}_M(\Phi)$ as a function.

(3) $\longrightarrow$ (1). Assume that $f : M \longrightarrow \mathbb{P}_1$ is not free of $\mathfrak{K}_M(\Phi)$. Then there is a meromorphic map $g : M \longrightarrow \mathbb{P}_1^*$ which depends analytically on $\mathfrak{K}_M(\Phi)$ such that the pair (f,g) is not free. Let $\mathfrak{v} : U \longrightarrow V = \mathbb{C}^2$ and $\mathfrak{w} : U \longrightarrow V^*$ be reduced representations. Then $\langle \mathfrak{v}, \mathfrak{w} \rangle \equiv 0$ on U. Let $\mathfrak{a}_0^*, \mathfrak{a}_1^*$ be the dual base of $\mathfrak{a}_0, \mathfrak{a}_1$. We have

$$\mathfrak{v} = v_0\mathfrak{a}_0 + v_1\mathfrak{a}_1 \qquad v_0 \not\equiv 0 \qquad f\,|\,U = \frac{v_1}{v_0}$$

$$\mathfrak{w} = w_0\mathfrak{a}_0 + w_1\mathfrak{a}_1 \not\equiv 0$$

$$0 = \langle \mathfrak{v}, \mathfrak{w} \rangle = v_0 w_0 + v_1 w_1 .$$

If $v_1 \equiv 0$, then $f = 0 \in \mathfrak{K}_M(\Phi)$ which is wrong. Hence $v_1 \not\equiv 0$. Therefore $w_0 \not\equiv 0 \not\equiv w_1$. Hence

$$f\,|\,U = \frac{v_1}{v_0} = -\frac{w_0}{w_1} = -\tilde{g}\,|\,U$$

where $\tilde{g}$ is the meromorphic coordinate function for g in respect to the base $\mathfrak{a}_1^*, \mathfrak{a}_0^*$. Hence $f = -\tilde{g} \in \mathfrak{K}_M(\Phi)$. Contradiction! q.e.d.

Let B be a meromorphic form of bidegree $(m - 1, 0)$ on M. Let P be the set of poles of B such that $M - P = A$ in the largest open subset of M where B is holomorphic. Let $f : M \longrightarrow \mathbb{P}(V)$ be a meromorphic map. Then the concept of associated maps f_p for $p = 0,1, \ldots \ell_f = \ell_f(B)$ is defined on A and the associated maps extend to meromorphic maps $f_p : M \longrightarrow G_p(V)$. If $\mathfrak{v} : U \longrightarrow V$ is a reduced representation, then $\mathfrak{v}_{\underline{p}} : U \cap A \longrightarrow \tilde{G}_p(V)$ is a representation of f_p on $U \cap A$ and $\mathfrak{v}_{\underline{p}}$ extends to a meromorphic vector function on U. We say that f is <u>general for</u> B if $\ell_f(B) = n$.

<u>THEOREM</u> 9.4. Let $f : M \longrightarrow \mathbb{P}(V)$ be a meromorphic map. Let $\Phi \neq \emptyset$ be a set of meromorphic functions of rank $m - 1$. Let B be a meromorphic form of bidegree $(m - 1,0)$ on M. Assume that B defines $\mathfrak{K}_M(\Phi)$. Then f is free of $\mathfrak{K}_M(\Phi)$ if and only if f is general for B.

<u>PROOF</u>. a) <u>Assume that</u> f <u>is free of</u> $\mathfrak{K}_M(\Phi)$. Let P be the set of poles of B and define $A = M - P$. Then B is holomorphic on A. Let $(\alpha_\lambda, \mathfrak{v}_\lambda)_{\lambda \in \Lambda}$ be a representation atlas of f. Then $\alpha_\lambda : U_\lambda \longrightarrow U'_\lambda$ is a chart of M and $\mathfrak{v}_\lambda : U_\lambda \longrightarrow V$ is a reduced representation of f. Here $(U_\lambda)_{\lambda \in \Lambda}$ is an open covering of M and each U_λ is Stein and a Cousin II domain. The transition rules (4.10) and (4.11) hold. Abbreviate $p = \ell_f(B)$. We have to show that $p = n$.

<u>Assume that</u> $p < n$. For $(\lambda,\mu) \in \Lambda[1]$ define

$$(9.11) \qquad u_{\lambda\mu} = (v_{\lambda\mu})^{p+1}(\Delta_{\mu\lambda})^{\frac{p(p+1)}{2}} \quad \text{on } U_{\lambda\mu}$$

Then

$$(9.12) \qquad \mathfrak{v}_{\lambda\underline{p}} = u_{\lambda\mu}\mathfrak{v}_{\mu\underline{p}} \quad \text{on } U_{\lambda\mu}.$$

Here $\mathfrak{v}_{\lambda\underline{p}}$ is holomorphic on $U_\lambda \cap A \neq \emptyset$. Take $\lambda_0 \in \Lambda$ and $z_0 \in U_{\lambda_0} \cap A$ such that $\mathfrak{v}_{\lambda\underline{p}}(z_0) \neq 0$. Then there is a base $\mathfrak{n}_0, \ldots, \mathfrak{n}_n$ of V such that

(9.13) $$\mathfrak{w}_{\lambda_0 \underline{p}}(z_0) \wedge \mathfrak{n}_{p+1} \wedge \dots \wedge \mathfrak{n}_n \neq 0 .$$

For each $\lambda \in \Lambda$, define $\psi_\lambda = \mathfrak{w}_{\lambda\underline{p}} \wedge \mathfrak{n}_{p+1} \wedge \dots \wedge \mathfrak{n}_{n-1}$ on U_λ. Then ψ_λ is holomorphic on $U_\lambda \cap A$ and meromorphic on U_λ. Also

(9.14) $$\psi_\lambda = u_{\lambda\mu}\psi_\mu \quad \text{on } U_{\lambda\mu} \text{ if } (\lambda,\mu) \in \Lambda[1] .$$

Since $\psi_{\lambda_0} \not\equiv 0$, the connectivity of M implies, that $\psi_\lambda \not\equiv 0$ on U_λ for all $\lambda \in \Lambda$. Similar $\psi_\lambda \wedge \mathfrak{n}_n \not\equiv 0$ for all $\lambda \in \Lambda$.

Let $\mathfrak{n}_0^*, \dots ,\mathfrak{n}_n^*$ be the dual base to $\mathfrak{n}_0, \dots ,\mathfrak{n}_n$. A meromorphic vector function $\mathfrak{w}_\lambda : U_\lambda \longrightarrow V^*$ is defined by

(9.15) $$\mathfrak{w}_\lambda = \mathfrak{n}_0^* \wedge \dots \wedge \mathfrak{n}_n^* \llcorner \psi_\lambda .$$

Here $\mathfrak{w}_\lambda$ is holomorphic on $U_\lambda \cap A$. If $(\lambda,\mu) \in \Lambda[1]$, then

(9.16) $$\mathfrak{w}_\lambda = u_{\lambda\mu}\mathfrak{w}_\mu \quad \text{on } U_{\lambda\mu} .$$

We have

$$\langle \mathfrak{w}_\lambda, \mathfrak{n}_n \rangle = \langle \mathfrak{n}_0^* \wedge \mathfrak{n}_n^* \llcorner \psi_\lambda, \mathfrak{n}_n \rangle$$

$$= \langle \mathfrak{n}_0^* \wedge \dots \wedge \mathfrak{n}_n^*, \psi_\lambda \wedge \mathfrak{n}_n \rangle \not\equiv 0 \quad \text{on } U_\lambda .$$

Hence $\mathfrak{w}_\lambda \not\equiv 0$ on U_λ. Hence one and only one meromorphic map $g : M \longrightarrow \mathbb{P}(V^*)$ exists such that $\mathfrak{w}_\lambda | U_\lambda \cap A$ is a representation of g for each $\lambda \in \Lambda$. For $\lambda \in \Lambda$ we have

$$<\mathfrak{w}_\lambda, \nu\mathfrak{w}_\lambda> = <\nu\mathfrak{w}_\lambda, \mathfrak{w}_\lambda> = <\mathfrak{n}_0^* \wedge \ldots \wedge \mathfrak{n}_n^* L\mathfrak{y}_\lambda, \mathfrak{w}_\lambda>$$

$$= <\mathfrak{n}_0^* \wedge \ldots \wedge \mathfrak{n}_n^* L\mathfrak{y}_\lambda \wedge \mathfrak{w}_\lambda>$$

$$= <\mathfrak{n}_0^* \wedge \ldots \wedge \mathfrak{n}_n^*, \mathfrak{w}_\lambda \wedge \ldots \wedge \mathfrak{w}_\lambda^{(p)} \wedge \mathfrak{n}_{p+1} \wedge \ldots \wedge \mathfrak{n}_{n-1} \wedge \mathfrak{w}_\lambda> \equiv 0 .$$

Hence the pair (f,g) is not free.

Take $\lambda \in \Lambda$. On U_λ we have

$$\mathfrak{w}_\lambda \wedge \mathfrak{w}_\lambda' \wedge \ldots \wedge \mathfrak{w}_\lambda^{(p)} = \mathfrak{w}_{\lambda\underline{p}} \not\equiv 0$$

$$\mathfrak{w}_\lambda \wedge \mathfrak{w}_\lambda' \wedge \ldots \wedge \mathfrak{w}_\lambda^{(p)} \wedge \mathfrak{w}_\lambda^{(p+1)} = \mathfrak{w}_{\lambda\underline{p+1}} \equiv 0 .$$

An analytic subset S_λ of U_λ exists such that $S_\lambda \supseteq U_\lambda \cap P$ and such that $\mathfrak{w}_{\lambda\underline{p}}(z) \neq 0$ for all $z \in U_\lambda - S_\lambda \neq \emptyset$. Unique holomorphic functions $A_{\lambda\nu}$ exist on $U_\lambda - S_\lambda$ such that

$$\mathfrak{w}_\lambda^{(p+1)} = \sum_{\nu=0}^{p} A_{\lambda\nu}\mathfrak{w}_\lambda^{(\nu)} .$$

Then we have

$$(\mathfrak{w}_{\lambda\underline{p}})' = \mathfrak{w}_\lambda \wedge \mathfrak{w}_\lambda' \wedge \ldots \wedge \mathfrak{w}_\lambda^{(p-1)} \wedge \mathfrak{w}_\lambda^{(p+1)} = A_{\lambda p}\mathfrak{w}_{\lambda\underline{p}}$$

on $U_\lambda - S_\lambda$ which implies

$$\mathfrak{y}_\lambda' = (\mathfrak{w}_{\lambda\underline{p}})' \wedge \mathfrak{n}_{p+1} \wedge \ldots \wedge \mathfrak{n}_{n-1} = A_{\lambda p}\mathfrak{y}_\lambda \quad \text{on } U_\lambda - S_\lambda$$

$$\mathfrak{w}'_\lambda = \mathfrak{n}_0^* \wedge \dots \wedge \mathfrak{n}_n^* L\mathfrak{y}'_\lambda = A_{\lambda p}\mathfrak{n}_0^* \wedge \dots \wedge \mathfrak{n}_n^* L\mathfrak{y}_\lambda = A_{\lambda p}\mathfrak{w}_\lambda$$

on $U_\lambda - S_\lambda$. Hence $\mathfrak{w}_\lambda \wedge \mathfrak{w}'_\lambda = 0$ on $U_\lambda - S_\lambda$ and therefore on $U_\lambda \cap A$, which means

$$(\mathfrak{w}_\lambda \wedge d\mathfrak{w}_\lambda) \wedge B = \mathfrak{w}_\lambda \wedge \mathfrak{w}'_\lambda \mathfrak{z}_\lambda \equiv 0 \ .$$

By Proposition 9.1, the meromorphic map is analytically dependent on $\mathfrak{G}_m(\Phi)$. Since f is free of $\mathfrak{G}_M(\Phi)$, the pair (f,g) is free which contradicts an earlier observation. Hence $p = n$.

b) <u>Assume</u> <u>that</u> f <u>is</u> <u>general</u> <u>for</u> B. Let $G : M \longrightarrow \mathbb{P}(V^*)$ be a meromorphic map which depends analytically on $\mathfrak{G}_M(\Phi)$. Take reduced representations $\mathfrak{v} : U \longrightarrow V$ of f and $\mathfrak{w} : U \longrightarrow V^*$ of g and assume that $\mathfrak{z} : U \longrightarrow U'$ is a chart. Take vectors $\mathfrak{a} \in V^*$ and $\mathfrak{b} \in V$ such that $<\mathfrak{v},\mathfrak{a}> \not\equiv 0$ and $<\mathfrak{w},\mathfrak{b}> \not\equiv 0$ on U. In fact by taking U smaller we can assume that $<\mathfrak{v},\mathfrak{a}>$ and $<\mathfrak{w},\mathfrak{b}>$ are nowhere zero on U. Then $\tilde{\mathfrak{v}} = \mathfrak{v}/<\mathfrak{v},\mathfrak{a}>$ and $\tilde{\mathfrak{w}} = \mathfrak{w}/<\mathfrak{w},\mathfrak{b}>$ are reduced representations of f respectively g on U. By Proposition 9.1 we have

$$\tilde{\mathfrak{w}}'\mathfrak{z} = d\frac{\mathfrak{w}}{<\mathfrak{w},\mathfrak{b}>} \wedge B \equiv 0 \qquad \text{on } U .$$

Hence $\tilde{\mathfrak{w}}' \equiv 0$. Assume that the pair (f,g) is not free. Then $<\tilde{\mathfrak{v}},\tilde{\mathfrak{w}}> \equiv 0$ on U. By induction, we see that $<\tilde{\mathfrak{v}}^{(j)},\tilde{\mathfrak{w}}> \equiv 0$ for all $j = 1, \dots, n$. Here

$$\tilde{\mathfrak{v}} \wedge \tilde{\mathfrak{v}}' \wedge \dots \wedge \tilde{\mathfrak{v}}^{(n)} = \tilde{\mathfrak{v}}_{\underline{n}} \not\equiv 0$$

since f is general for B. Therefore $\tilde{\mathfrak{w}} \equiv 0$, which is impossible. q.e.d.

We may ask if the assumption in Theorem 9.4 that B has bidegree $(m-1,0)$ is a consequence of the other assumptions on Φ. In this repect we prove:

LEMMA 9.5. Let $\Phi \neq \emptyset$ be a set of meromorphic functions of rank $m-1$ on M. Assume that $\mathfrak{K}_M(\Phi) \neq \mathfrak{K}_M$. Let B be a meromorphic form on M defining $\mathfrak{K}_M(\Phi)$. Then B has bidegree $(m-1,0)$.

PROOF. Let $(p,0)$ be the bidegree of B. If $p = m$, then $d\varphi \wedge B \equiv 0$ for all $\varphi \in \mathfrak{K}_M$. Hence $\mathfrak{K}_M(\Phi) = \mathfrak{K}_M$ which is impossible. Hence $0 \leqslant p \leqslant m-1$. Assume that $p < m-1$. Let $g_1, \ldots, g_{m-1}$ be a base of $\mathfrak{K}_M(\Phi)$. Take $x_0 \in M$ such that g_j is holomorphic at x_0 and such that

$$dg_1(x_0) \wedge \ldots \wedge dg_{m-1}(x_0) \neq 0 .$$

An open, connected neighborhood U of x_0 and a holomorphic function g_m on U exist such that

$$g = (g_1, \ldots, g_m) : U \longrightarrow U'$$

is a chart. Let S be the set of all injective increasing maps $\mu : \mathbb{N}[1,p] \longrightarrow \mathbb{N}[1,m]$. There are meromorphic functions B_μ on U for $\mu \in S$ such that

$$B = \sum_{\mu \in S} B_\mu \, dg_{\mu(1)} \wedge \ldots \wedge dg_{\mu(p)} .$$

For $j \in \mathbb{N}[1,m-1]$ define $S_j = \{\mu \in S \mid j \notin \operatorname{Im} \mu\}$. Then we have

$$0 \equiv dg_j \wedge B = \sum_{\mu \in S_j} B_\mu \, dg_j \wedge dg_{\mu(1)} \wedge \ldots \wedge dg_{\mu(p)} .$$

Hence $B_\mu \equiv 0$ on U for all $\mu \in S_j$ if $j \in \mathbb{N}[1.m-1]$. Take any $\mu \in S$. Then $\#\operatorname{Im} \mu = p < m-1$. Hence $j \in \mathbb{N}[1,m-1]$

exists such that $\mu \in S_j$. Therefore $B_\mu \equiv 0$. Thus $B|U \equiv 0$ which implies $B \equiv 0$. Hence $\mathfrak{K}_M(\Phi) = \mathfrak{K}_M[B] = \mathfrak{K}_M$ which is impossible. Hence $p = m - 1$. q.e.d.

Let M and N be connected complex manifolds of dimension m. Let $f : M \longrightarrow N$ be a meromorphic map of rank m and with indeterminancy I_f . Then $\dim I_f \leqslant m - 2$. Then there exists a non-negative divisor $\rho = \rho_f$ on M called the **ramification** divisor of f (or also the **branching divisor**) such that we have the following. Take $x_0 \in M - I_f$. Let $\mathfrak{z} : U \longrightarrow U'$ be a chart of $f(x_0)$ and let W be an open neighborhood of x_0 with $W \subseteq M - I_f$ such that $f(W) \subseteq U$. Then $\mathfrak{z} \circ f|W = (f_1, \dots ,f_m)$ where $f_1, \dots ,f_m$ are holomorphic functions on W. Since f has rank m, we know that $df_1 \wedge \dots \wedge df_m \not\equiv 0$. Then $\rho|W$ is the zero divisor of $df_1 \wedge \dots \wedge df_m$ on W. This defines ρ on $M - I_f$. Since $\dim I_f \subseteq m - 2$, the divisor ρ extends uniquely to a divisor on M.

LEMMA 9.6. Let $g_1, \dots ,g_m$ be analytically independent meromorphic functions on the connected complex manifold M of dimension m. A meromorphic map g of rank m is defined by

$$g = (g_1, \dots ,g_m) : M \longrightarrow \mathbb{P}_1^m . \tag{9.17}$$

Let ρ_g be the ramification divisor of g. Assume that there are holomorphic functions $v_j \not\equiv 0 \not\equiv w_j$ such that $w_j g_j = v_j$ on M for $j = 1, \dots ,m$. Let ν_j be the zero divisor of the representation (v_j,w_j) of g_j for $j = 1, \dots ,m$. Define the holomorphic form

$$\gamma = w_1^2 \dots w_m^2 dg_1 \wedge \dots \wedge dg_m = \bigwedge_{j=1}^{m} (w_j dv_j - v_j dw_j) . \tag{9.18}$$

Then $\gamma \not\equiv 0$. let μ_γ be the zero divisor of γ. Then we have

(9.19) $$\mu_\gamma = \rho_g + 2 \sum_{j=1}^{m} \nu_j .$$

<u>PROOF</u>. It suffices to prove (9.19) at every point $x_0 \in M - I_g$. Then either g_j or $1/g_j$ or both are holomorphic at x_0. By a change of enumeration, we can assume without less of generality, that $g_1, \dots, g_n$ are holomorphic at x_0 and that $g_{n+1}, \dots, g_m$ have poles at x_0. If $1 \leqslant j \leqslant n$, then $(g_j, 1)$ is a reduced reprensentation of g_j at x_0 and we have $(v_j, w_j) = (g_j, 1)w_j$. Hence $\nu_j(x_0) = \mu^0_{w_j}(x_0)$. If $n + 1 \leqslant j \leqslant m$, then $(1, 1/g_j)$ is a reduced representation of g_j at x_0 and we have $(v_j, w_j) = (1/1/g_j)v_j$. Hence $\nu_j(x) = \mu^0_{v_j}(x_0)$. Also ρ_g is the divisor of

$$\beta = dg_1 \wedge \dots \wedge dg_n \wedge d(1/g_{n+1}) \wedge \dots \wedge d(1/g_m)$$

at x_0 where

$$\gamma = w_1^2 \dots w_n^2 v_{n+1}^2 \dots v_m^2 \beta .$$

Therefore

$$\mu_\gamma(x_0) = 2 \sum_{j=1}^{n} \mu^0_{w_j}(x_0) + 2 \sum_{j=n+1}^{m} \mu^0_{v_j}(x_0) + \rho_g(x_0)$$

$$= 2 \sum_{j=1}^{m} \nu_j(x_0) + \rho_g(x_0) . \qquad \text{q.e.d.}$$

LEMMA 9.7. Let $f, g_1, \dots, g_{m-1}$ be analytically independent meromorphic functions on the connected, m-dimensional, complex manifold M. Let $v_j \not\equiv 0 \not\equiv w_j$ be holomorphic functions on M such that $w_j g_j = v_j$ for $j = 1, \dots, m-1$. Let ν_j be the divisor of the representation (v_j, w_j) of g_j for $j = 1, \dots, m-1$. A holomorphic form B of bidegree $(m-1, 0)$ is defined by

$$B = w_1^2 \dots w_{m-1}^2 dg_1 \wedge \dots \wedge dg_{m-1} = \bigwedge_{j=1}^{m-1} (w_j dv_j - v_j dw_j) . \tag{9.20}$$

Consider $f : M \longrightarrow \mathbb{P}_1$ as a meromorphic map. Let μ_{F_1} be the divisor of the 1st (and last) associated section F_1 of f in respect to B. Let ρ be the ramification divisor of the meromorphic map

$$(f, g_1, \dots, g_{m-1}) : M \longrightarrow \mathbb{P}_1^m . \tag{9.21}$$

Then we have

$$\mu_{F_1} = \rho + 2 \sum_{j=1}^{m-1} \nu_j . \tag{9.22}$$

PROOF. Let $\mathfrak{z} : U \longrightarrow U'$ be a chart. Let $\mathfrak{w} = (v_0, w_0)$ be a reduced representation of f on U. Then $w_0 f = v_0$. Also $\mathfrak{w}' = (v_0', w_0')$. Define $\mathfrak{n}_0 = (1,0)$ and $\mathfrak{n}_1 = (0,1)$. Then

$$\mathfrak{w} \wedge \mathfrak{w}' = \begin{vmatrix} v_0 & w_0 \\ v_0' & w_0' \end{vmatrix} \mathfrak{n}_0 \wedge \mathfrak{n}_1 = (v_0 w_0' - w_0 v_0') \mathfrak{n}_0 \wedge \mathfrak{n}_1$$

$$(\mathfrak{w} \wedge \mathfrak{w}')\mathfrak{z} = (\mathfrak{w} \wedge d\mathfrak{w}) \wedge B = \bigwedge_{j=0}^{m-1} (v_j dw_j - w_j dv_j) \mathfrak{n}_0 \wedge \mathfrak{n}_1 .$$

Since v_0, w_0 are coprime at every point of U, Lemma 9.6 implies

$$\mu_{F_2}|U = \mu_{\mathfrak{w}\wedge\mathfrak{w}'} = \rho|U + 2\sum_{j=1}^{m-1}\nu_j|U. \qquad \text{q.e.d.}$$

If f is holomorphic on U and if $w_j = 1$ on U, then

(9.23) $$\mathfrak{w}\wedge\mathfrak{w}' = f'\mathfrak{n}_0\wedge\mathfrak{n}_1 \quad\text{and}\quad f'\mathfrak{z} = df\wedge dg_1\wedge\ldots\wedge dg_{m-1}\,.$$

We will make the following **general assumptions**:

(C1) Let M be a connected, complex manifold of dimension m.

(C2) Let τ be a parabolic exhaustion of M.

(C3) Let $\Phi \neq \emptyset$ be a set of meromorphic functions of rank $m-1$ on M.

(C4) Let V be a hermitian vector space of dimension $n+1 > 1$.

(C5) Let $f : M \longrightarrow \mathbb{P}(V)$ be a meromorphic map.

(C6) Assume that f is free of $\mathfrak{K}_M(\Phi)$.

(C7) Let $\mathfrak{g} \neq \emptyset$ be a finite set of meromorphic maps $g : M \longrightarrow \mathbb{P}(V^*)$ with $k = \#\mathfrak{g} \geq n+1$.

(C8) Assume that every $g \in \mathfrak{g}$ is analytically dependent on $\mathfrak{K}_M(\Phi)$.

(C9) Assume that $\mathfrak{g}$ is in general position.

(C10) Assume that B is a holomorphic form of bidegree $(m-1,0)$ on M which defines $\mathfrak{K}_M(\Phi)$.

(C11) Assume that τ majorizes B with majorant Y.

According to Theorem 9.4, the meromorphic map f is general for B, that is $\ell_f = n$ in respect to B. By Proposition 9.1 (4) we have

(9.23) $$\ell_g = \ell_{\mathfrak{g}} = 0 \qquad \text{for all } g \in \mathfrak{g}\,.$$

Assumptions (C1) to (C11) and Theorem 9.4 imply (B1) – (B8), where $\ell_g = 0$ for all $g \in \mathfrak{g}$. Therefore Theorem 8.8 implies

THEOREM 9.8. **Second Main Theorem for functions fields.** Assume that (C1) – (C11) hold. Take $\epsilon > 0$ and $s > 0$. Then we have

(9.24)

$$N_{F_n}(r,s) + \sum_{g \in \mathfrak{G}} m_{f,g}(r)$$

$$\leq (n+1)T_f(r,s) + \frac{n(n+1)}{2} \operatorname{Ric}_\tau(r,s) + \left[nk - \frac{n(n+1)}{2}\right]\Gamma_{\mathfrak{G}}(r)$$

$$+ 2n(n+1)C_0(k \log T_f(r,s) + \sum_{g \in \mathfrak{G}} \log^+ \Gamma_g(r,s) + k \log \operatorname{Ric}_\tau^+(r,s))$$

$$+ 2n(n+1)C_0 k \log Y(r) + \epsilon \log r .$$

Also we have

(9.25)

$$(k - n - 1)T_f(r,s)$$

$$\leq \sum_{g \in \mathfrak{G}} N^{(n)}_{f,g}(r,s) + \frac{n(n+1)}{2} \operatorname{Ric}_\tau(r,s) + nk\left[\frac{k}{p} - 1\right] \sum_{g \in \mathfrak{G}} T_g(r,s)$$

$$+ 2n(n+1)C_0 k(\log T_f(r,s) + \log Y(r) + \log^+ \operatorname{Ric}_\tau(r,s)) + \epsilon \log r .$$

In addition, we assume

(C12) $\qquad T_g(r,s)/T_f(r,s) \longrightarrow 0$ for $r \longrightarrow \infty$ for each $g \in \mathfrak{G}$

(C13) $$\limsup_{r\to\infty} \frac{\operatorname{Ric}_\tau(r,s)}{T_f(r,s)} \leq 0$$

(C14) $$\frac{\log Y(r)}{T_f(r,s)} \longrightarrow 0 \text{ for } r \longrightarrow \infty$$

Suppose that (C1) – (C14) are satisfied. Then we define the Nevanlinna defects

(9.26) $$\delta_f(g) = \delta_f[g] = \liminf_{r\to\infty} \frac{m_{f,g}(r)}{T_f(r,s)}$$

$$0 \leq \delta_f(g) = 1 - \limsup_{r\to\infty} \frac{N_{f,g}(r,s)}{T_f(r,s)} \leq 1 \tag{9.27}$$

$$\theta_f(g) = 1 - \limsup_{r\to\infty} \frac{N_{f,g}^{(n)}(r,s)}{T_f(r,s)} . \tag{9.28}$$

Since $N_{f,g}^{(n)}(r,s) \leq N_{f,g}(r,s)$, we have

$$0 \leq \delta_f(g) \leq \theta_f(g) \leq 1 . \tag{9.29}$$

The Second Main Theorem 9.8 implies the defect relation.

THEOREM 9.9. Defect relation. Assume that (C1) – (C14) hold. Then

$$\boxed{\sum_{g\in\mathfrak{G}} \delta_f(g) \leq \sum_{g\in\mathfrak{G}} \theta_f(g) \leq n+1} \tag{9.30}$$

Our statement (9.53) does not contain any term depending on B. However, the assumptions (C10), (C11) and (C14) depend on B. We are in a situation deplored by Royden. An assumption which has seemingly nothing to do when the statement is made for the sake of the proof only. In the case $M = \mathbb{C}^m$ we will be able to eliminate these assumptions on B by constructing a form B satisfying (C10), (C11) and (C14) and to some extent the result can be established even if (M,τ) is a covering parabolic space of $\mathbb{C}^m$.

First we consider the situation on $\mathbb{C}^m$ where the exhaustion is given by $\tau(z) = \|z\|^2 = |z_1|^2 + \ldots + |z_m|^2$. Then $\upsilon > 0$ and τ majorizes any holomorphic form of bidegree $(m-1,0)$.

For any holomorphic vector function $g : \mathbb{C}^m \longrightarrow V$ define

$$M(r,g) = \operatorname{Max}\{\|g(\mathfrak{z})\| \mid \|\mathfrak{z}\| \leq r\} . \tag{9.31}$$

LEMMA 9.10. Let B be a holomorphic form of bidegree $(m-1,0)$ on $\mathbb{C}^m$. For $\mu \in \mathbb{N}[1,m]$ define

$$\zeta_\mu = (-1)^{\mu-1} dz_1 \wedge \dots \wedge dz_{\mu-1} \wedge dz_{\mu+1} \wedge \dots \wedge dz_m . \tag{9.32}$$

Then

$$B = \sum_{\mu=1}^{m} B_\mu \zeta_\mu . \tag{9.33}$$

Define the holomorphic vector function $\mathfrak{B} = (B_1, \dots, B_m) : \mathbb{C}^m \longrightarrow \mathbb{C}^m$. Let Y be the majorant of B in respect to τ. Then we have

$$Y(r) = \operatorname{Max}(1, mM(r,\mathfrak{B})^2) \tag{9.34}$$

$$\log Y(r) \leqslant 2 \log^+ M(r,\mathfrak{B}) + \log m . \tag{9.35}$$

PROOF. We have

$$m i_{m-1} B \wedge \bar{B} = m i_{m-1} \sum_{\mu,\nu=1}^{m} B_\mu \bar{B}_\nu \zeta_\mu \wedge \bar{\zeta}_\nu$$

$$\upsilon^{m-1} = i_{m-1} \sum_{\mu=1}^{m} \zeta_\mu \wedge \bar{\zeta}_\mu .$$

Take $r > 0$. Take $c > 0$, then $m i_{m-1} B \wedge \bar{B} \leqslant c\upsilon^{m-1}$ on $\mathbb{C}^m[r]$ if and only if

$$m \sum_{\mu,\nu=1}^{m} B_\mu(\mathfrak{z}) \bar{B}_\nu(\mathfrak{z}) \bar{x}_\mu x_\nu \leqslant c \sum_{\mu=1}^{m} |x_\mu|^2 = c\|\mathfrak{x}\|^2 \tag{9.36}$$

for all $\mathfrak{z} \in \mathbb{C}^m[r]$ and $\mathfrak{x} \in \mathbb{C}^m$. Let $Y_0(r)$ be the infimum of all those constants c. Then

$$m\,|(\mathfrak{G}(\mathfrak{z})\,|\,\mathfrak{x})|^2 \leqslant Y_0(r)\,\|\mathfrak{x}\|^2 .$$

Take $\mathfrak{x} = \mathfrak{G}(\mathfrak{z})$, then $m\|\mathfrak{G}(\mathfrak{z})\|^4 \leqslant Y_0(r)\|\mathfrak{z})\|^2$ or

$$m\|\mathfrak{G}(\mathfrak{z})\|^2 \leqslant Y_0(r)$$

which implies $mM(r,\mathfrak{G})^2 \leqslant Y_0(r)$. Therefore

$$\text{(9.37)} \qquad \mathrm{Max}(1, mM(r,\mathfrak{G})^2) \leqslant Y(r) .$$

If $mM(r,\mathfrak{G})^2 \leqslant 1$, then $m\,|(\mathfrak{G}(\mathfrak{z})\,|\,\mathfrak{x})| \leqslant \|\mathfrak{x}\|^2$ and $Y_0(r) \leqslant 1$. Hence $Y(r) = 1$ and we have equality in (9.37). If $mM(r,\mathfrak{G})^2 > 1$ then $Y_0(r) \geqslant 1$ and $Y_0(r) = Y(r)$. Also

$$m\,|(\mathfrak{G}(\mathfrak{z})\,|\,\mathfrak{x})|^2 \leqslant mM(r,\mathfrak{G})^2\|\mathfrak{x}\|^2$$

which implies $Y_0(r) \leqslant mM(r,\mathfrak{G})^2$. Hence

$$Y_0(r) = \mathrm{Max}(1, mM(r,\mathfrak{G})^2) .$$

In particular, $Y(r)$ is continuous in r which proves (9.34) and (9.35). q.e.d.

<u>LEMMA</u> 9.11. Let f be a holomorphic function on $\mathbb{C}^m$. Take $j \in \mathbb{N}[1,m]$, $r > 0$ and $\epsilon > 0$. Then

$$\text{(9.38)} \qquad M\left[r, \frac{\partial f}{\partial z_j}\right] \leqslant \frac{1}{\epsilon} M(r + \epsilon, f) .$$

<u>PROOF</u>. Take $\mathfrak{z} \in \mathbb{C}^m[r]$ such that $|f_{z_j}(\mathfrak{z})| = M(r, f_{z_j})$. Let $\mathfrak{n}_1, \ldots, \mathfrak{n}_m$ be the standard base of $\mathbb{C}^m$. A holomorphic function $h : \mathbb{C} \longrightarrow \mathbb{C}$ is defined by $h(\zeta) = f(\mathfrak{z} + \zeta\mathfrak{n}_j)$. Then $h'(0) = f_{z_j}(\mathfrak{z})$. If

$\zeta \in \mathbb{C}[\epsilon]$, then $\|\mathfrak{z} + \zeta \mathfrak{n}_j\| \leq \|\mathfrak{z}\| + |\zeta| \leq v + \epsilon$. Hence $|h(\zeta)| \leq M(r + \epsilon, f)$ for all $\zeta \in \mathbb{C}[\epsilon]$. The Cauchy estimates give

$$M\left(r, \frac{\partial f}{\partial z_j}\right) = |f_{z_j}(\mathfrak{z})| = |h'(0)| \leq \frac{1}{\epsilon} M(r + \epsilon, f) . \qquad \text{q.e.d.}$$

We need the following result of H. Skoda [85] Theorem 3 on $\mathbb{C}^m$.

THEOREM 9.12. Let f be a non-constant meromorphic function on $\mathbb{C}^m$. Take $\epsilon > 0$.

a) There are entire functions $g \not\equiv 0$ and $h \not\equiv 0$ on $\mathbb{C}^m$ such that $hf = g$ and such that for each $s > 0$ there is a constant $c(\epsilon, s) > 0$ such that

$$\text{(9.39)} \qquad \operatorname{Max}(\log^+ M(r,g), \log^+ M(r,h)) \leq c(\epsilon,s)(1 + r)^{4m-1} T_f(r + \epsilon, s)$$

for each $r > s$.

b) There are entire functions $g \not\equiv 0$ and $h \not\equiv 0$ on $\mathbb{C}^m$ such that $hf = g$ and such that for each $s > 0$ there is a constant $c(\epsilon, s) > 0$ such that

$$\text{(9.40)} \quad \operatorname{Max}(\log^+ M(r,g), \log^+ M(r,h)) \leq c(\epsilon,s)(1 + (\log(1 + r^2))^2) T_f(r + \epsilon r, s)$$

for each $r > s$.

REMARK. The pair (g,h) may have a common divisor, may depend on ϵ but may not depend on s, and may be different in b) from the pair chosen in a). The case a) is good for rapid growth, the case b) is good for slow growth.

PROOF. Skoda proved the theorem if $f(0) = 1$ and $s = 0$. If $s > 0$ and $0 \neq a \in \mathbb{C}$. Then there is a constant $c_0(a)$ such that

$$T_{af}(r,s) \leq T_f(r,s) + c_0(a) .$$

Also we have

$$T_f(r + \epsilon,s) + T_f(s,0) = T_f(r + \epsilon,0) .$$

Hence the theorem extends to the case $s > 0$ where f is holomorphic at 0 and $f(0) \neq 0$.

Assume that either f is holomorphic at 0 with $f(0) = 0$ or that f is not holomorphic at zero. Take $\mathfrak{a} \in \mathbb{C}^m$ with $\|\mathfrak{a}\| < \epsilon$ such that f is holomorphic at $\mathfrak{a}$ and $f(\mathfrak{a}) \neq 0$. Define a meromorphic function $\tilde{f}$ on M by $\tilde{f}(\mathfrak{z}) = f(\mathfrak{z} + \mathfrak{a})$ for $\mathfrak{z} \in \mathbb{C}^m$.
The function $\tilde{f}$ is holomorphic at 0 with $\tilde{f}(0) \neq 0$. Let $\tilde{g}, \tilde{h}$ be the solution for a) respectively b). Define entire function g and h by

$$g(\mathfrak{z}) = \tilde{g}(\mathfrak{z} - \mathfrak{a}) \qquad h(\mathfrak{z}) = \tilde{h}(\mathfrak{z} - \mathfrak{a}) .$$

Then we have $hf = g \not\equiv 0$ and

$$M(r,g) \leq M(r + \epsilon,\tilde{g}) \qquad M(r,h) \leq M(r + \epsilon,\tilde{h})$$

for all $r > 0$. Define $\mathbb{C}^m[r,\mathfrak{a}] = \{\mathfrak{z} \in \mathbb{C}^m \mid \|\mathfrak{z} - \mathfrak{a}\| \leq r\}$. For $r > 0$, we have $\mathbb{C}^m[r,\mathfrak{a}] \subseteq \mathbb{C}^m[r + \epsilon]$. Hence

$$A_{\tilde{f}}(r) = r^{2-2m} \int_{\mathbb{C}^m[r,\mathfrak{a}]} f^*(\Omega) \wedge \upsilon^{m-1}$$

$$\leq r^{2-2m} \int_{\mathbb{C}^m[r+\epsilon]} f^*(\Omega) \wedge \upsilon^{m-1}$$

$$\leq \left(1 + \frac{\epsilon}{r}\right)^{2m-2} A_f(r + \epsilon)$$

and

$$\int_s^r A_{\hat{f}}(t)\,\frac{dt}{t} \leqslant \int_s^r \left[1 + \frac{\epsilon}{t}\right]^{2m-1} A_f(t + \epsilon)\,\frac{dt}{t + \epsilon}$$

$$T_{\hat{f}}(r,s) \leqslant \left[1 + \frac{\epsilon}{s}\right]^{2m-1} T_f(r + \epsilon, s + \epsilon)$$

$$\leqslant \left[1 + \frac{\epsilon}{s}\right]^{2m-1} T_f(r + \epsilon, s)\ .$$

In case a) we obtain

$$\operatorname{Max}(\log^+ M(r,g), \log^+ M(r,h))$$

$$\leqslant \operatorname{Max}(\log^+ M(r + \epsilon, \hat{g}), \log^+ M(r + \epsilon, \tilde{h}))$$

$$\leqslant c(\epsilon,s)(1 + r + \epsilon)^{4m-1} T_{\hat{f}}(r + 2\epsilon, s)$$

$$\leqslant c(\epsilon,s)\left[1 + \frac{\epsilon}{s}\right]^{2m-1}(1 + r + \epsilon)^{4m-1} T_f(r + 3\epsilon, s)$$

$$\leqslant c(\epsilon,s)\left[1 + \frac{\epsilon}{s}\right]^{2m-1}(1 + \epsilon)^{4m-1}(1 + r)^{4m-1} T_f(r + 3\epsilon, s)\ .$$

Thus if we replace ϵ by $\epsilon/3$ and redefine the constant $c(\epsilon,s)$ we obtain (9.39).

Now consider the case b):

$$\operatorname{Max}(\log^+ M(r,g), \log^+ M(r,h))$$

$$\leqslant \operatorname{Max}(\log^+ M(r + \epsilon, \hat{g}), \log^+ M(r + \epsilon, \tilde{h}))$$

$$\leqslant c(\epsilon,s)(1 + (\log(1 + (r + \epsilon)^2))^2 T_{\hat{f}}(r + \epsilon r, s)\ .$$

Assume that $r > 1 + s$, then we continue

$$\leqslant c(\epsilon,s)(1 + (\log(1 + r^2) + \log(1 + \epsilon)^2)^2) T_f(r + 2\epsilon r, s)$$

$$\leqslant c_1(\epsilon,s)(1 + (\log(1 + r)^2)^2) T_f(r + 2\epsilon r, s)\ .$$

Since $T_f(r + 2\epsilon r,s)$ is bounded above and below for $s \leqslant r \leqslant s + 1$, there is a constant $c_2(\epsilon,s)$ such that

$$\mathrm{Max}(\log^+ M(r,g),\log^+ M(r,h))$$

$$\leqslant c_2(\epsilon,s)(1 + (\log(1 + r^2))^2)T_f(r + 2\epsilon r,s) .$$

Replace ϵ by $\epsilon/2$ and redefine the constant. Then we have (9.40). q.e.d.

<u>LEMMA</u> **9.13.** Let $v_1, \ldots ,v_{m-1}$ and $w_1, \ldots ,w_{m-1}$ be holomorphic functions on $\mathbb{C}^m$. Define

$$B = \bigwedge_{j=1}^{m-1} (w_j dv_j - v_j dw_j) . \tag{9.41}$$

Let Y be the majorant of B in respect to τ. Take $\epsilon > 0$ and $r > 0$. Then we have

(9.42)

$$\log Y(r) \leqslant 2 \sum_{j=1}^{m} \log^+ M(r + \epsilon,w_j) + 2 \sum_{j=1}^{m} \log^+ M(r + \epsilon,v_j)$$

$$+ 2(m - 1)\log^+ \frac{2m}{\epsilon} + \log m .$$

<u>PROOF</u>. Let $\mathfrak{n}_1, \ldots ,\mathfrak{n}_m$ be the standard base of $\mathbb{C}^m$. Define $\hat{\mathfrak{n}}_j$ by

$$\hat{\mathfrak{n}}_j = (-1)^{j-1}\mathfrak{n}_1 \wedge \ldots \wedge \mathfrak{n}_{j-1} \wedge \mathfrak{n}_{j+1} \wedge \ldots \wedge \mathfrak{n}_m . \tag{9.43}$$

Let $z_1, \ldots ,z_m$ be the coordinates of $\mathbb{C}^m$ and define $\frac{\partial v_j}{\partial z_\mu} = v_{j\mu}$ and

$$\zeta_j = (-1)^{j-1}dz_1 \wedge \ldots \wedge dz_{j-1} \wedge dz_{j+1} \wedge \ldots \wedge dz_m . \tag{9.44}$$

Then we have

$$(9.45) \qquad B = \sum_{j=1}^{m} B_j \zeta_j = \bigwedge_{j=1}^{m-1} \sum_{\mu=1}^{m} (w_j v_{j\mu} - v_j w_{j\mu}) dz_\mu .$$

Define $y_{j\mu} = w_j v_{j\mu} - v_j w_{j\mu}$ and $\varphi_j = y_{j1}\eta_1 + \ldots + y_{jm}\eta_m$. Define $\mathfrak{G} = \sum_{\mu=1}^{m} B_\mu \eta_\mu$. Then we have

$$\sum_{\mu=1}^{m} B_\mu \hat{\eta}_\mu = \varphi_1 \wedge \ldots \wedge \varphi_{m-1}$$

$$\|\mathfrak{G}\| = \|\varphi_1 \wedge \ldots \wedge \varphi_{m-1}\| \leqslant \|\varphi_1\| \ldots \| \varphi_{m-1}\|$$

$$\|\mathfrak{G}\|^2 \leqslant \prod_{j=1}^{m-1} \sum_{\mu=1}^{m} |u_{j\mu}|^2 \leqslant \prod_{j=1}^{m-1} \sum_{\mu=1}^{m} (|w_j v_{j\mu}| + |v_j w_{j\mu}|)^2$$

or

$$\|\mathfrak{G}\| \leqslant \prod_{j=1}^{m-1} \sum_{\mu=1}^{m} (|w_j| \ |v_{j\mu}| + |v_j| \ |w_{j\mu}|) .$$

Hence we have

$$M(r,\mathfrak{G}) \leqslant \prod_{j=1}^{m-1} \sum_{\mu=1}^{m} (M(r,w_j)M(r,v_{j\mu}) + M(r,v_j)M(r,w_{j\mu}))$$

$$\leqslant \left[\frac{2m}{\epsilon}\right]^{m-1} \prod_{j=1}^{m-1} M(r + \epsilon, w_j) M(r + \epsilon, v_j) .$$

The estimate (9.58) implies

$$\log Y(r) \leqslant 2 \log^+ M(r,\mathfrak{G}) + \log m$$

$$\leqslant 2 \sum_{j=1}^{m} (\log^+ M(r + \epsilon, w_j) + \log^+ M(r + \epsilon, v_j))$$

$$+ 2(m - 1)\log^+ \frac{2m}{\epsilon} + \log m . \qquad \text{q.e.d.}$$

THEOREM 9.14. Let $\varphi_1, \ldots, \varphi_{m-1}$ be analytically independent meromorphic functions on $\mathbb{C}^m$. Take $\epsilon > 0$.

a) There there is a holomorphic form B of bidegree $(m-1,0)$ on $\mathbb{C}^m$ which defines $\mathfrak{K}_{\mathbb{C}^m}(\varphi_1, \ldots, \varphi_{m-1})$ such that for each $s > 0$ there is a constant $c(\epsilon,s) > 0$ such that the majorant Y of B for τ can be estimated by

$$\text{(9.46)} \qquad \log Y(r) \leqslant c(\epsilon,s)(1+r)^{4m-1} \sum_{j=1}^{m-1} T_{\varphi_j}(r+\epsilon,s)$$

for all $r > s$.

b) There is a holomorphic form B of bidegree $(m-1,0)$ on $\mathbb{C}^m$ which defines $\mathfrak{K}_{\mathbb{C}^m}(\varphi_1, \ldots, \varphi_{m-1})$ such that for each $s > 0$ there is a constant $c(\epsilon,s) > 0$ such that the majorant Y of B for τ can be estimated by

$$\text{(9.47)} \qquad \log Y(r) \leqslant c(\epsilon,s)(1+(\log(1+r^2))^2) \sum_{j=1}^{m-1} T_{\varphi_j}(r+\epsilon r,s)$$

for $r > s$.

c) If $\varphi_1, \ldots, \varphi_{m-1}$ are rational functions, then there is a holomorphic form B of bidegree $(m-1,0)$ on $\mathbb{C}^m$ which defines $\mathfrak{K}_{\mathbb{C}^m}(\varphi_1, \ldots, \varphi_{m-1})$ and a constant $c > 0$ such that the majorant Y of B for τ can be estimated by

$$\text{(9.48)} \qquad \log Y(r) \leqslant c \log r \qquad \text{for } r \geqslant 2\,.$$

REMARK. In all there cases B can be obtained in the form (9.44) where $v_j \not\equiv 0 \not\equiv w_j$ are holomorphic functions on $\mathbb{C}^m$ with $w_j\varphi_j = v_j$ for $j = 1, \ldots, m-1$. In a) and b), the form B depends on ϵ but not on s.

PROOF. If B is defined as indicated in the Remark, then B defines $\mathfrak{G}_{\mathbb{C}^m}(\varphi_1, \dots, \varphi_{m-1})$ by (9.6). Hence it remains to choose v_j, w_j for $j = 1, \dots, m-1$. If $\varphi_1, \dots, \varphi_{m-1}$ are rational, then v_j, w_j can be taken as polynomials and (9.48) follows from (9.42). This settles the case c).

a) Each φ_j is not constant. By Theorem 9.12, we can choose v_j, w_j to satisfy Thereom 9.12 a) with $w_j\varphi_j = v_j \not\equiv 0$. We can assume that $c(\epsilon,s)$ does not depend on j. The estimates (9.39) and (9.42) imply

$$\log Y(r) \leqslant 4c(\epsilon,s)(1 + r + \epsilon)^{4m-1} \sum_{j=1}^{m-1} T_{\varphi_j}(r + 2\epsilon,s) + c_0(\epsilon)$$

where $c_0(\epsilon)$ is a constant. Here $1 + r + \epsilon \leqslant (1 + r)(1 + \epsilon)$. Also $T_{\varphi_j}(r,s) \longrightarrow \infty$ for $r \longrightarrow \infty$. Hence increasing $c(\epsilon,s)$ to a sufficiently large constant $c_1(\epsilon,s)$ we obtain

$$\log Y(r) \leqslant 4c_1(\epsilon,s)(1 + r)^{4m-1} \sum_{j=1}^{m-1} T_{\varphi_j}(r + 2\epsilon,s) .$$

If we replace ϵ by $\epsilon/2$ and redefine the constant, we obtain (9.46).

b) By (9.42) and Theorem 9.12 b) we obtain

$$\log Y(r) \leqslant 4c(\epsilon,s)(1 + (\log(1 + (r + \epsilon)^2))^2) \sum_{j=1}^{m-1} T_{g_j}((r + \epsilon)(1 + \epsilon)) + c_0(\epsilon)$$

where the constant $c_0(\epsilon)$ can be absorpt again by increasing $c(\epsilon,s)$. Without loss of generality we can assume that $0 < \epsilon < \mathrm{Min}(1,s)$ and we can replace ϵ by $\epsilon^2/3$. Then $\epsilon < s < r$ and

$$\left(r + \frac{\epsilon^2}{3}\right)\left(1 + \frac{\epsilon^2}{3}\right) \leqslant r\left(1 + \frac{\epsilon}{3}\right)^2 = r\left(1 + \frac{2\epsilon}{3} + \frac{\epsilon^2}{9}\right) < r\,(1 + \epsilon)$$

$$1 + \left[\log\left[1 + \left[r + \frac{\epsilon^2}{3}\right]\right]\right]^2 \leqslant 1 + \left[\log\left[1 + r^2\left[1 + \frac{\epsilon}{3}\right]^2\right]\right]^2$$

$$\leqslant 1 + (\log(1 + 4r^2))^2$$

$$\leqslant 1 + (2 \log 2 + \log(1 + r^2))^2$$

$$\leqslant 1 + (2 + \log(1 + r^2))^2$$

$$\leqslant 1 + 8 + 2(\log(1 + r^2))^2$$

$$\leqslant 9(1 + (\log(1 + r^2))^2) .$$

Hence redefining the constant $c(\epsilon,s)$ yields (9.47). q.e.d.

Let M be a connected, complex manifold of dimension $m > 1$. Let $\pi : M \longrightarrow \mathbb{C}^m$ be a proper, surjective holomorphic map. Then $\tau = \|\pi\|^2$ is a parabolic exhaustion of M and (M,τ) is called a <u>parabolic</u> <u>covering</u> <u>space</u> of $\mathbb{C}^m$. The generic sheet number of π is given by ς as defined by (2.14). Let ρ be the branching divisor of π. Then

(9.49) $$\mathrm{Ric}_\tau(r,s) = N_\rho(r,s)$$

(Stoll [103]). A divisor $\nu \geqslant 0$ on (M,τ) has <u>rational</u> <u>growth</u> if $n_\nu(\infty) < \infty$, that is, if $N_\nu(r,s) = O(\log r)$ which is the case if and only if the $(m-1)$-dimensional component of $\pi(\mathrm{supp}\ \nu)$ is affine algebraic. A meromorphic map

(9.50) $$f : M \longrightarrow \mathbb{P}(V)$$

has <u>rational</u> <u>growth</u> if and only if

(9.51) $$\lim_{r\to\infty} \frac{T_f(r,s)}{\log r} = A_f(\infty) < \infty$$

and **transcendental growth** if $A_f(\infty) = \infty$. The **order of** f and the **lower order** of f are given by

(9.52)
$$\limsup_{r\to\infty} \frac{\log T_f(r,s)}{\log r} = \mathrm{Ord}\, f \qquad \liminf_{r\to\infty} \frac{\log T_f(r,s)}{\log r} = \underline{\mathrm{Ord}}\, f .$$

If there is a meromorphic map $h : \mathbb{C}^m \longrightarrow \mathbb{P}(V)$ such that $f = h \circ \pi$, then

(9.53)
$$T_f(r,s) = T_h(r,s) .$$

Hence f has rational growth if and only if h has rational growth which is the case if and only if h is rational. Also f has finite order if and only if h has finite order with $\mathrm{Ord}\, f = \mathrm{Ord}\, h$.

THEOREM 9.15. **The Second Main Theorem for function fields over** $\mathbb{C}^m$. Let V be a hermitian vector space of dimension $n + 1 > 1$. Let M be a connected, complex manifold of dimension $m > 1$. Let $\pi : M \longrightarrow \mathbb{C}^m$ be a surjective, proper, holomorphic map of sheet number $\mathfrak{c}$ and with branching divisor ρ. Define $\tau = \|\pi\|^2$. Then τ is a parabolic exhaustion of M. Let $\varphi_1, \dots, \varphi_{m-1}$ analytically independent meromorphic functions on $\mathbb{C}^m$. Define $\psi_j = \varphi_j \circ \pi$. Then $\psi_1, \dots, \psi_{m-1}$ are analytically independent meromorphic functions on M. Define $\mathfrak{K} = \mathfrak{K}_M(\psi_1, \dots, \psi_{m-1})$. Let $\mathfrak{G}$ be a finite set of meromorphic maps $g : M \longrightarrow \mathbb{P}(V^*)$ with $k = \#\mathfrak{G} > n + 1$. Assume that $\mathfrak{G}$ is in general position. Assume that each $g \in \mathfrak{G}$ is analytically dependent on $\mathfrak{K}$. Let $f : M \longrightarrow \mathbb{P}(V)$ be a meromorphic map which is free of $\mathfrak{K}$. Then we have

(9.54)
$$(k - n - 1)T_f(r,s)$$
$$\overset{\cdot}{\leq} \sum_{g\in\mathfrak{G}} N^{(n)}_{f,g}(r,s) + nk\begin{pmatrix} k-1 \\ n \end{pmatrix} \sum_{g\in\mathfrak{G}} T_g(r,s) + \frac{n(n+1)}{2} N_\rho(r,s)$$
$$+\ 2n(n+1)\mathfrak{c}k(\log T_f(r,s) + \log^+ N_\rho(r,s)) + S(r)$$

where the following estimates of the remainder terms $S(r)$ are available.
(A) Take $\epsilon > 0$ and $s > 0$. Then there is a constant $c_0(\epsilon,s) > 0$ such that

(9.55) $$S(r) \leqslant c_0(s,\epsilon)(1 + r)^{4m-1} \sum_{j=1}^{m-1} T_{\psi_j}(r + \epsilon,s) \quad \text{for all } r > s .$$

(B) Take $\epsilon > 0$ and $s > 0$. Then there is a constant $c_1(\epsilon,s) > 0$ such that

(9.56)
$$S(r) \leqslant c_1(\epsilon,s)(1 + (\log(1 + r^2))^2) \sum_{j=1}^{m-1} T_{\psi_j}(r + \epsilon r,s) \quad \text{for all } r > s .$$

(C) Assume that each ψ_j has finite order and that $\mathrm{Ord}\,\psi_j < \lambda$ for $j = 1, \ldots ,m - 1$. Then there is a constant $c_2(\lambda) > 0$ such that

(9.57) $$S(r) < c_2(\lambda)r^{\lambda} \quad \text{for } r > 1 .$$

(D) Assume that $\varphi_1, \ldots ,\varphi_{m-1}$ are rational functions. Then there is a constant $c_3 > 0$ such that

(9.58) $$S(r) \leqslant c_3 \log r \quad \text{for } r \geqslant 2 .$$

<u>PROOF</u>. Clearly (C1) – (C9) are satisfied with $\Phi = \mathfrak{F}$. There are holomorphic functions v_j, w_j on $\mathbb{C}^m$ such that $w_j\varphi_j = v_j \not\equiv 0$ and such that

(9.59) $$B_0 = \bigwedge_{j=1}^{m-1} (w_j dv_j - v_j dw_j) = w d\varphi_1 \wedge \ldots \wedge d\varphi_{m-1}$$

(9.60) $$w_0 = w_1^2 \ldots w_{m-1}^2 \not\equiv 0$$

is a holomorphic form of bidegree $(m - 1,0)$ which defines $\mathfrak{F}_0 = \mathfrak{F}_{\mathbb{C}^m}(\varphi_1, \ldots ,\varphi_{m-1})$ and which is majorized by τ_0 on $\mathbb{C}^m$ with majorant Y satisfying the estimates (9.46) or (9.47) or (9.48) in Theorem 9.14. Then $\tau = \tau_0 \circ \pi$ majorizes the holomorphic form B of bidegree $(m-1,0)$ on M with the same majorant Y. Also $w = w_0 \circ \pi \not\equiv 0$ is a holomorphic function on M with

$$B = w d\psi_1 \wedge \dots \wedge d\psi_{m-1} \not\equiv 0 \ . \tag{9.61}$$

Hence B defines $\mathfrak{F} = \mathfrak{F}_M(\psi_1, \dots, \psi_{m-1})$. The assumptions (C10) and (C11) are satisfied and (9.25) holds. If we assume without loss of generality, that $0 < \epsilon \leqslant 1$ and if we define

$$S(r) = 2kn(n+1)\varsigma \log Y(r) + \log^+ r \tag{9.62}$$

then (9.25) implies (9.54). Observe that $T_{\psi_j}(r,s) \longrightarrow \infty$ for $s \longrightarrow \infty$ since ψ_j is not constant, and that $T_{\varphi_j} = T_{\psi_j}$.

In the case of Theorem 9.14 a), the constant $c(\epsilon,s)$ can be increased to a constant $c_0(\epsilon,s)$ as to absorb $2kn(n+1)\varsigma$ and $\log^+ r$ and we obtain (A). In the case of Theorem 9.1 b), the constant $c(\epsilon,s)$ can be increased to a constant $c_1(\epsilon,s)$ as to imply (B).

If $\varphi_1, \dots, \varphi_{m-1}$ are rational, Theorem 9.14 c) and (9.62) imply (C).

Assume that $\psi_1, \dots, \psi_{m-1}$ have finite order and that $\operatorname{Ord} \varphi_j < \ell < \infty$ for $j = 1, \dots, m-1$. Take $\eta > 0$ such that $\operatorname{Ord} \varphi_j < \lambda - \eta$ for $j = 1, \dots, m-1$. Then there are constants $c_4 > 0$ and $c_5 > 0$ such that

$$T_{\psi_j}(r,s) \leqslant c_4 r^{\lambda-\eta} \text{for} \quad r \geqslant 2$$

$$1 + (\log(1+r^2))^2 \leqslant c_5 r^{\eta} \quad \text{for} \quad r \geqslant 1 \ .$$

Applying (9.56) with $s = \epsilon = 1$. Then we have

$$S(r) \leq c_1(1,1)(m - 1)c_4c_5 2^{\lambda-\eta} r^{\lambda-\eta} r^{\eta} = c_2(\lambda) r^{\lambda} \quad \text{for} \quad r > 1 .$$

q.e.d.

<u>THEOREM</u> 9.16. <u>Defect relation for function fields over</u> $\mathbb{C}^m$. Let V be a hermitian vector space of dimension $n + 1 > 1$. Let M be a connected, complex manifold of dimension $m > 1$. Let $\pi : M \longrightarrow \mathbb{C}^m$ be a surjective, proper, holomorphic map of sheet number ς and with branching divisor ρ. Let $\varphi_1, \ldots, \varphi_{m-1}$ be analytically independent meromorphic functions on $\mathbb{C}^m$. Define $\psi_j = \varphi_j \circ \pi$. Then $\psi_1, \ldots, \psi_{m-1}$ are analytically independent functions on M. Define $\mathfrak{f} = \mathfrak{K}_M(\psi_1, \ldots, \psi_{m-1})$. Let $\tau = \|\pi\|^2$. Then τ is a parabolic exhaustion of M. Let $\mathfrak{G}$ be a finite set of meromorphic maps $g : M \longrightarrow \mathbb{P}(V^*)$. Assume that each $g \in \mathfrak{G}$ is analytically dependent on $\mathfrak{f}$. Let $f : M \longrightarrow \mathbb{P}(V)$ be a meromorphic map which is free of $\mathfrak{f}$. Assume that

(9.63) $$\frac{T_g(r,s)}{T_f(r,s)} \longrightarrow 0 \quad \text{for} \quad r \longrightarrow \infty \quad \text{for all} \quad g \in \mathfrak{G}$$

(9.64) $$\frac{N_\rho(r,s)}{T_f(r,s)} \longrightarrow 0 \quad \text{for} \quad r \longrightarrow \infty .$$

Assume that at least one of the following assumptions (A) or (B) or (C) or (D) is satisfied.

(A) There is a number $\epsilon > 0$ such that

(9.65) $$\frac{(1 + r)^{4m-1} T_{\psi_j}(r,s)}{T_f(r,s)} \longrightarrow 0 \quad \text{for} \quad r \longrightarrow \infty$$

for $j = 1, \ldots, m - 1$.

(B) There is a number $\epsilon > 0$ such that

(9.66)

$$\frac{(1 + (\log(1 + r^2))^2 T_{\psi_j}(r + \epsilon r, s)}{T_f(r,s)} \longrightarrow 0 \quad \text{for} \quad r \longrightarrow \infty$$

for $j = 1, \ldots, m-1$.

(C) The functions $\psi_1, \ldots, \psi_{m-1}$ have finite order with

(9.67)
$$\operatorname{Ord} \psi_j < \underline{\operatorname{Ord}}\, f \leqslant \infty \quad \text{for} \quad j = 1, \ldots, m-1 .$$

(D) The functions $\varphi_1, \ldots, \varphi_{m-1}$ are rational and f has transcendental growth.

Then we have

(9.68)
$$\boxed{\sum_{g \in \mathfrak{G}} \delta_f(g) \leqslant \sum_{g \in \mathfrak{G}} \theta_f(g) \leqslant n + 1}$$

PROOF. The defect relation (9.68) immediately follows from the Second Main Theorem 9.15. In Case (C) take λ with $\operatorname{Ord} \psi_j < \lambda < \underline{\operatorname{Ord}}\, f$. Then $T_f(r,s) r^{-\lambda} \longrightarrow \infty$ for $r \longrightarrow \infty$ and

(9.69)
$$\frac{S(r)}{T_f(r,s)} \longrightarrow 0 \quad \text{for} \quad r \longrightarrow \infty$$

by (9.62). If $\varphi_1, \ldots, \varphi_{m-1}$ are rational and if f has trancendental growth, then $T_f(r,s)/\log r \longrightarrow \infty$ for $r \longrightarrow 0$ and (9.69) follows from (9.63). q.e.d.

Each of the conditions (A), (B), (C), and (D) implies, that f has transcendental growth. Thus if the $(m-1)$-dimensional component of $\pi(\operatorname{supp} \rho)$ is affine algebraic, (9.64) is automatically satisfied. If f separates the fibers of π, then $\limsup_{r \to \infty} N_\rho(r,s)/T_f(r,s) = R_f < \infty$

by a theorem of Noguchi [70] and the Ricci defect R_f appears in (9.68). If $g \in \mathfrak{G}$ exists such that g separates the fibers of π, then

$$\limsup_{r\to\infty} \frac{N_\rho(r,s)}{T_g(r,s)} \leqslant 2(\varsigma - 1)$$

by the theorem of Noguchi [70]. Then (9.63) implies (9.64) and (9.68) holds.

Naturally, the Second Main Theorem and the Defect Relation simplify, if we take $M = \mathbb{C}^m$ and π as the identity. Then $\psi_j = \varphi_j$ and $N_\rho(r,s) \equiv 0$. Also $\varsigma = 1$. We leave it to the reader to formulate this special case explicitly.

If $n = 1$, the assumptions again become simplier. In this case, $M = \mathbb{C}^m$ with $m > 1$ and $n = 1$, B. Shiffman already obtained the two theorems under some what weaker results. Instead of the assumptions (A) or (B) or (C) or (D) he needs only

$$\text{(9.70)} \qquad \frac{T_{\varphi_j}(r,s)}{T_f(r,s)} \longrightarrow 0 \qquad \text{for } r \longrightarrow \infty$$

for $j = 1, \ldots, m-1$. For comparison we state Shiffman's results [83], [84].

THEOREM 9.17. **Shiffman [83], [84].** Let $f, g_1, \ldots, g_q$ be distinct meromorphic functions on $\mathbb{C}^m$ with $m > 1$. Assume that

$$\text{(9.71)} \qquad \operatorname{rank}(f, g_1, \ldots, g_q) = \operatorname{rank}(g_1, \ldots, g_q) + 1 .$$

Then there is a constant $c_1 > 0$ such that

$$\text{(9.72)} \qquad (q-2)T_f(r,s)$$

$$\leqslant \sum_{j=1}^{q} N^{(1)}_{f,g_j}(r,s) + c_1\left[\sum_{j=1}^{q} T_{g_j}(r,s) + \log T_f(r,s) + \log^+ r\right] .$$

If $T_{g_j}(r,s)/T_f(r,s) \longrightarrow 0$ for $r \longrightarrow \infty$ and $j = 1, \ldots, q$, then we have

$$(9.73) \qquad \sum_{j=1}^{q} \delta_f(g_j) \leqslant \sum_{j=1}^{q} \theta_f(g_j) \leqslant 2 \ .$$

B. Shiffman shows that Theorem 9.17 follows from:

<u>**THEOREM**</u> **9.18.** <u>**Shiffman**</u> **[83], [84].** Let $f, g_1, \ldots, g_q$ be distinct V meromorphic functions on $\mathbb{C}^m$ with $q \geqslant m - 1 \geqslant 0$. Assume that

$$(9.74) \qquad \operatorname{rank}(g_1, \ldots, g_q) = m - 1 < \operatorname{rank}(f, g_1, \ldots, g_{m-1})$$

Let ρ be the ramification divisor of the meromorphic map

$$(9.75) \qquad (f, g_1, \ldots, g_{m-1}) : \mathbb{C}^m \longrightarrow \mathbb{P}^m$$

Then there is a constant $c > 0$ such that

$$(9.76) \qquad N_\rho(r,s) + (q - 2)T_f(r,s)$$

$$\leqslant \sum_{j=1}^{q} N_{f,g_j}(r,s) + c\left[\sum_{j=1}^{q} T_{g_j}(r,s) + \log T_f(r,s) + \log^+ r\right]$$

Our general theory forced us to make maximum modulus estimates for B, that is for the functions v_j, w_j such that $w_j\varphi_j = v_j$. Thus applying Skoda's results we have to make assumptions (A), (B), (C), or (D) while Shiffman needs (9.70) only. Perhaps a modification of the B-method will yield the same results. Except for this small deviation Theorems 9.15 and 9.16 can be considered as an extension of Shiffman's Theorems 9.17 and 9.18. Hopefully, the assumptions that $\psi_1, \ldots, \psi_{m-1}$ are lifted from $\mathbb{C}^m$ can be eliminated by further research.

The B-method was invented almost 30 years ago, and has been used almost exclusively by the author (Stoll [93], [100], [108]).

The B-method is justified again by the results of this section.

Mori [63] obtains a defect relation for meromorphic maps $f : \mathbb{C}^m \longrightarrow \mathbb{P}(V)$ and for moving targets $g : \mathbb{C}^m \longrightarrow \mathbb{P}(V)$. We will discuss and extend his results in Section 11.

§10. An Example

We will give an example where the integral in (6.19) cannot be split into three convergent integrals according to (6.8).

We take $m = 1$, $M = \mathbb{C}$, $B = 1$, $n = 2$, $V = \mathbb{C}^3$, $\mathbb{P}(V) = \mathbb{P}_2$.

Let $\mathfrak{n}_0, \mathfrak{n}_1, \mathfrak{n}_2$ be the standard base of $\mathbb{C}^3$ and let $\mathfrak{n}_0^*, \mathfrak{n}_1^*, \mathfrak{n}_2^*,$ be the dual base. Define a holomorphic map $f : \mathbb{C} \longrightarrow \mathbb{P}_2$ be the reduced representation

$$(10.1) \qquad \mathfrak{w}(t) = (1,t,t^2) = \mathfrak{n}_0 + \mathfrak{n}_1 t + \mathfrak{n}_2 t^2 .$$

Then

$$(10.2) \qquad \mathfrak{w}'(t) = (0,1,2t) = \mathfrak{n}_1 + 2t\mathfrak{n}_2$$

$$(10.3) \qquad \mathfrak{w}''(t) = (0,0,2) = 2\mathfrak{n}_2$$

$$(10.4) \qquad \mathfrak{w}_{\underline{1}}(t) = \mathfrak{w}(t) \wedge \mathfrak{w}'(t) = \mathfrak{n}_0 \wedge \mathfrak{n}_1 + 2t\mathfrak{n}_0 \wedge \mathfrak{n}_2 + t^2\mathfrak{n}_1 \wedge \mathfrak{n}_2$$

$$(10.5) \qquad \mathfrak{w}_{\underline{2}}(t) = \mathfrak{w}(t) \wedge \mathfrak{w}'(t) \wedge \mathfrak{w}'' = 2\mathfrak{n}_0 \wedge \mathfrak{n}_1 \wedge \mathfrak{n}_2 .$$

We abbreviate

$$\textbf{(10.6)} \qquad A = 1 + |t|^2 + |t|^4 \qquad\qquad B = 1 + 4|t|^2 + |t|^4$$

$$\textbf{(10.7)} \qquad C = 2 + |t|^2 \qquad\qquad B = A + 3|t|^2 = C^2 - 3\,.$$

Then we have

$$\textbf{(10.8)} \qquad \|\mathfrak{w}\|^2 = A \qquad \|\mathfrak{w}_{\underline{1}}\|^2 = B \quad \|\mathfrak{w}_{\underline{2}}\|^2 = 4$$

$$\textbf{(10.9)} \qquad (\mathfrak{w}_{\underline{1}} L^* \mathfrak{w}) = -\bar{t}(1 + 2|t|^2)\mathfrak{n}_0 + (1 - |t|^4)\mathfrak{n}_1 + t(2 + |t|^2)\mathfrak{n}_2$$

$$\textbf{(10.10)} \qquad (\mathfrak{w}_{\underline{2}} L^* \mathfrak{w}_1) = 2\bar{t}^2\mathfrak{n}_0 - 4\bar{t}\mathfrak{n}_1 + 2\mathfrak{n}_2\,.$$

A meromorphic map $g : \mathbb{C} \longrightarrow \mathbb{P}_2^*$ is defined by the reduced representation

$$\textbf{(10.11)} \qquad \mathfrak{w}(t) = (-t, 1, t^2) = -t\mathfrak{n}_0^* + \mathfrak{n}_1^* + t^2\mathfrak{n}_2^*$$

$$\textbf{(10.12)} \qquad \mathfrak{w}'(t) = (-1, 0, 2t) = -\mathfrak{n}_0^* + 2t\mathfrak{n}_2^*$$

$$\textbf{(10.13)} \qquad \mathfrak{w}''(t) = (0, 0, 2) = 2\mathfrak{n}_2^*$$

$$\textbf{(10.14)} \qquad \mathfrak{w}_{\underline{1}}(t) = \mathfrak{w}(t) \wedge \mathfrak{w}'(t) = \mathfrak{n}_0^* \wedge \mathfrak{n}_1^* - t^2\mathfrak{n}_0^* \wedge \mathfrak{n}_2^* + 2t\mathfrak{n}_1^* \wedge \mathfrak{n}_2^*$$

$$\textbf{(10.15)} \qquad \mathfrak{w}_{\underline{2}}(t) = \mathfrak{w}(t) \wedge \mathfrak{w}'(t) \wedge \mathfrak{w}''(t) = 2\mathfrak{n}_0^* \wedge \mathfrak{n}_1^* \wedge \mathfrak{n}_2^*\,.$$

Then we have

(10.17) $$\|\mathfrak{v}_{\underline{2}}\|^2 = 4 = \|\mathfrak{w}_{\underline{2}}\|^2$$

(10.18) $$(\mathfrak{v}_{\underline{1}} L^* \mathfrak{v}) = -(1 - |t|^4)\mathfrak{n}_0^* - \bar{t}(1 + 2|t|^2)\mathfrak{n}_1^* + t(2 + |t|^2)\mathfrak{n}_2^*$$

(10.19) $$(\mathfrak{v}_{\underline{2}} L^* \mathfrak{v}_1) = 4\bar{t}\mathfrak{n}_0^* + 2\bar{t}^2\mathfrak{n}_1^* + 2\mathfrak{n}_2^* .$$

These maps are related by the identities

(10.20) $$\langle \mathfrak{w}, \mathfrak{v} \rangle = t^4$$

(10.21) $$\langle \mathfrak{w}_{\underline{1}} L^* \mathfrak{w}, \mathfrak{v} \rangle = A + t^3 C$$

(10.22) $$\langle \mathfrak{w}_{\underline{2}} L^* \mathfrak{w}_{\underline{1}}, \mathfrak{v} \rangle = 2(t^2 - \bar{t}C)$$

(10.23) $$\langle \mathfrak{w}, \mathfrak{v}_{\underline{1}} L^* \mathfrak{v} \rangle = t^3 C - A$$

(10.24) $$\langle \mathfrak{w}, \mathfrak{v}_{\underline{2}} L^* \mathfrak{v}_{\underline{1}} \rangle = 2(t^2 + \bar{t}C)$$

(10.25) $$\langle \mathfrak{w}_1 L^* \mathfrak{v}, \mathfrak{v}_{\underline{1}} L^* \mathfrak{v} \rangle = t^2 C^2$$

(10.26) $$\langle \mathfrak{w}_{\underline{2}} L^* \mathfrak{w}_{\underline{1}}, \mathfrak{v}_{\underline{1}} L^* \mathfrak{v} \rangle = 2(\bar{t}^2 B + tC)$$

(10.27) $$\langle \mathfrak{w}_{\underline{1}} L^* \mathfrak{w}, \mathfrak{v}_{\underline{2}} L^* \mathfrak{v}_{\underline{1}} \rangle = 2(tC - \bar{t}^2 B)$$

(10.28) $$\langle \mathfrak{w}_{\underline{2}} L^* \mathfrak{w}_{\underline{1}}, \mathfrak{v}_{\underline{2}} L^* \mathfrak{v}_1 \rangle = 4 .$$

Recall

(10.29)
$$\Lambda_{pq} = \frac{\langle \omega_{\underline{q}} L^* \omega_{\underline{q-1}}, \tilde{\omega}_{\underline{p}} L^* \tilde{\omega}_{\underline{p-1}} \rangle}{\| \omega_{\underline{q}} \| \, \| \omega_{\underline{q-1}} \| \, \| \tilde{\omega}_{\underline{p}} \| \, \| \omega_{\underline{p-1}} \|} .$$

Hence

(10.30)
$$\Lambda_{00} = \frac{t^4}{A}, \quad \Lambda_{01} = \frac{t^3 C + A}{A\sqrt{B}}, \quad \Lambda_{02} = \frac{t^2 - C\bar{t}}{\sqrt{A}\,\sqrt{B}}$$

(10.31)
$$\Lambda_{10} = \frac{t^3 C - A}{A\sqrt{B}}, \quad \Lambda_{11} = \frac{t^2 C^2}{AB}, \quad \Lambda_{12} = \frac{tC + \bar{t}^2 B}{B\sqrt{A}}$$

(10.32)
$$\Lambda_{20} = \frac{t^2 + tC}{\sqrt{A}\,\sqrt{B}}, \quad \Lambda_{21} = \frac{tC - \bar{t}B}{B\sqrt{A}}, \quad \Lambda_{22} = \frac{1}{B}$$

where the matrix (Λ_{jk}) is unitary. We have the volume elements

(10.33)
$$h_0 = \frac{\| \omega_{\underline{1}} \|}{\| \omega \|^2} = \frac{\sqrt{B}}{A} = \frac{\| \tilde{\omega}_{\underline{1}} \|}{\| \tilde{\omega} \|^2} = k_0$$

(10.34)
$$h_1 = \frac{\| \omega \| \, \| \omega_{\underline{2}} \|}{\| \omega_{\underline{1}} \|^2} = \frac{2\sqrt{A}}{B} = \frac{\| \tilde{\omega} \| \, \| \tilde{\omega}_{\underline{2}} \|}{\| \tilde{\omega}_{\underline{1}} \|^2} = k_1 .$$

Therefore we have

(10.35)
$$\begin{aligned}
\xi_{00} &= \Phi_{10} k_0^2 + \Phi_{01} h_0^2 + 2 S_{01}^{10} \, k_0 h_0 \\
&= (2|\Lambda_{00}|^2 + |\Lambda_{10}|^2 + |\Lambda_{01}|^2 + \Lambda_{10}\bar{\Lambda}_{01} + \bar{\Lambda}_{10}\Lambda_{01}) \frac{B}{A^2} \\
&= (2|\Lambda_{00}|^2 + |\Lambda_{10} + \Lambda_{01}|^2) \frac{B}{A^2} \\
&= \left[\frac{2|t|^8}{A^2} + \frac{2|t|^6 C^2}{A^2 C} \right] \frac{B}{A^2}
\end{aligned}$$

(10.36) $$\xi_{00} = \frac{2|t|^6}{A^4}(2C^2 + |t|^2B)$$

The Φ matrix computes as

(10.37) $$\Phi_{00} = \frac{|t|^8}{A^2},\quad \Phi_{01} = \frac{|t|^8}{A^2} + \frac{|t^3C + A|^2}{A^2B},\quad \Phi_{02} = 1$$

(10.38) $$\Phi_{10} = \frac{|t|^8}{A^2} + \frac{|t^3C - A|^2}{A^2B},\quad \Phi_{11} = 1 + \frac{1}{B^2},\quad \Phi_{12} = 2$$

(10.39) $$\Phi_{20} = 1,\ \Phi_{21} = 2,\ \Phi_{22} = 3\ .$$

Also we have

(10.40) $$S^{01}_{10} = \frac{|t|^6C^2 - A^2}{A^2B}$$

(10.41)

$$S^{01}_{10} = \frac{(|t| - 1)}{A^2B}\left[\sum_{\nu=0}^{q} t^\nu + |t|^2(|t|^2 + 1)(|t| + 1)(2 + 3|t|^2)\right]$$

Hence

(10.42) $$S^{01}_{10}\begin{cases} < 0 & \text{if } |t| < 1 \\ = 0 & \text{if } |t| = 1 \\ > 0 & \text{if } |t| > 1 \end{cases} \qquad S^{01}_{10} = -1 \text{ if } t = 0$$

We have $\Phi_{00}(t) > 0$ if $t \neq 0$ and $\Phi_{01} > 0$, $\Phi_{10} > 0$ and $\Phi_{11} > 0$ everywhere. Take $0 < \beta < \frac{3}{4}$. The the integrals

(10.43) $$\int_{\mathbb{C}(r)} \frac{\Phi_{01}h_0^2}{\Phi_{00}^{1-\beta}}\,\upsilon \qquad \int_{\mathbb{C}(r)} \frac{\Phi_{01}k_0^2}{\Phi_{00}^{1-\beta}}\,\upsilon \qquad \int_{\mathbb{C}(r)} \frac{S^{01}_{10}h_0k_0}{\Phi_{00}^{1-\beta}}\,\upsilon$$

do not exist since the integrands become infinite of order $8 - 8\beta > 2$ at 0, but the integral

(10.44) $$\int_{\mathbb{C}(r)} \frac{\xi_{00}}{\Phi_{00}^{1-\beta}}\,\upsilon = \int_{\mathbb{C}(r)} \left[\frac{\Phi_{01}h_0^2}{\Phi_{00}^{1-\beta}} + \frac{\Phi_{10}k_0^2}{\Phi_{00}^{1-\beta}} + \frac{2S_{10}^{01}\,h_0k_0}{\Phi_{00}^{1-\beta}}\right]\upsilon$$

exists since the integrand becomes infinite of order $1 - 8\beta < 2$ at 0 and is of class C^∞ elsewhere. Since $h_0 = k_0$ and since $h_0/h_1 = 2\sqrt{A}/\sqrt{B}$ is bounded we have

(10.45) $$R_{00}(r) = \int_{\mathbb{C}\langle r\rangle} \log\left(1 + \frac{k_0}{h_0}\right)\sigma = \log 2$$

(10.46) $$R_{01}(r) = \int_{\mathbb{C}\langle r\rangle} \log\left(1 + \frac{k_0}{h_1}\right)\sigma \leqslant c < \infty$$

for some constant $c > 0$. Since $S_{0,1}^{1,0} \geqslant 0$ for $|t| > 1$, we have

(10.47) $$P_{00}(r) = \int_{\mathbb{C}\langle r\rangle} \log\left(\frac{\Phi_{01}h_0^2}{\xi_{00}}\right)\sigma \leqslant c_0$$

for some constant $c_0 > 0$. In order to compute $P_{01}(r)$ we calculate

(10.48) $$S_{02}^{11} = \frac{|t|^4c^2 - \frac{1}{2}c^3(t^3 + \bar{t}^3)}{AB\sqrt{AB}}$$

(10.49) $$\xi_{01} = \Phi_{02}h_1^2 + \Phi_{11}k_0^2 + 2S_{02}^{11}h_1k_0$$

$$= \frac{4A}{B^2} + \left(1 + \frac{1}{B^2}\right)\frac{B}{A^2} + \frac{(2|t|^4c^2 - c^3(t^3 + \bar{t}^3))2}{A^2B^2}$$

$$= \frac{4A^3 + (B^2 + 1)B + 2|t|^4c^2 - 2c^3(t^3 + \bar{t}^3)}{A^2B^2}$$

$$\frac{\xi_{01}}{\Phi_{02} h_1^2} = 1 + \frac{B^3 + B}{4A^3} + \frac{|t|^4 C^2}{A^3} - \frac{C^3(t^3 + \bar{t}^3)}{2A^3} . \tag{10.50}$$

Hence

$$\frac{\xi_{01}}{\Phi_{02} h_1^2} \longrightarrow \frac{5}{4} \quad \text{for } t \longrightarrow \infty .$$

A constant $c_1 > 0$ exists such that

$$P_{01}(r) = \int_{\mathbb{C}\langle r\rangle} \log \frac{\Phi_{02} h_1^2}{\xi_{01}} \sigma \leqslant c_1 . \tag{10.51}$$

Thus the remainder terms $R_{00}, R_{01}, P_{00}, P_{01}$ remain unbounded.

§11. The Theorem of Nevanlinna–Mori

Let f be a non–constant, meromorphic function on $\mathbb{C}$. Let $\mathfrak{G}$ be a finite set of meromorphic functions on $\mathbb{C}$. Assume that $f \not\equiv g$ for all $g \in \mathfrak{G}$ and that

$$T_g(r,s)/T_f(r,s) \longrightarrow 0 \quad \text{for} \quad r \longrightarrow \infty . \tag{11.1}$$

Then Nevanlinna [66] wondered if his defect relation

$$\sum_{g \in \mathfrak{G}} \delta_f(g) \leqslant 2 \tag{11.2}$$

remains true. He proved this conjecture if $\mathfrak{G}$ consists of three elements g_1, g_2, g_3. He defines

$$h = \frac{f - g_1}{f - g_2} \frac{g_2 - g_3}{g_2 - g_3} \tag{11.3}$$

and shows that $\delta_f(g_j) = \delta_j(a_j)$ for $j = 1,2.3$ where $a_1 = 0$, $a_2 = 1$ and $a_3 = \infty$.

Mori [63] extends this result to a meromorphic map $f : \mathbb{C}^m \longrightarrow \mathbb{P}_n$ and a set $\mathfrak{G}$ of meromorphic maps $g : \mathbb{C}^m \longrightarrow \mathbb{P}^*_n$ where $\#\mathfrak{G} = n + 2$ and where (11.1) holds for all $g \in \mathfrak{G}$. He assumes that $\mathfrak{G}$ is in general position and that f satisfies a certain non–degeneracy condition. Then

$$\sum_{g \in \mathfrak{G}} \delta_f(g) \leqslant n + 2 . \tag{11.4}$$

Mori enumerates $\mathfrak{G} = \{g_1, \ldots, g_{n+2}\}$ and he constructs a meromorphic map $h : \mathbb{C}^m \longrightarrow \mathbb{P}_n$ and hyperplanes $a_j \in \mathbb{P}_n^*$ in general position such that $\delta_f(g_j) = \delta_h(a_j)$ for $j = 1, \ldots, n + 2$.

Mori's theorem will be extended to maps defined on a parabolic manifold. Although the basic idea is the same, the modifications are considerable, but provide better results and deeper insights.

The following <u>General</u> <u>Assumption</u> <u>E</u> will be made

(E1) Let (M, τ) be a parabolic manifold of dimension m.
(E2) Let V be a hermitian vector space of dimension $n + 1 > 1$.
(E3) Let $f : M \longrightarrow \mathbb{P}(V)$ be a non-constant, meromorphic map.
(E4) Let $\mathfrak{G} \neq \emptyset$ be a set of meromorphic maps $g : M \longrightarrow \mathbb{P}(V^*)$.
(E5) Assume that $\mathfrak{G}$ is in general position with $\#\mathfrak{G} = n + 2$.
(E6) Let (f, g) be free for each $g \in \mathfrak{G}$. Enumerate $\mathfrak{G} = \{g_0, \ldots, g_{n+1}\}$.
(E7) Let $\alpha_0, \ldots, \alpha_n$ be an orthonormal base of V^*.
Define $\alpha_{n+1} = - \alpha_0 - \ldots - \alpha_n$. Define $a_j = \mathbb{P}(\alpha_j)$ for $j = 0, 1, \ldots, n + 1$.
(E8) If $j \in \mathbb{Z}[0, n + 1]$, then $T_{g_j}(r, s)./T_f(r, s) \longrightarrow 0$ for $r \longrightarrow \infty$.

<u>REMARK</u> 1. These assumptions suffice for the construction of the Nevanlinna–Mori map and the comparison of the defects. The defect relation requires additional assumptions.

<u>REMARK</u> 2. It would be suffice to assume that $a_0, \ldots, a_{n+1}$ are in general position, since the theory does not depend on the choice of the hermitian metric on V.

<u>THEOREM</u> 11.1. (Mori [63].) Assume that (E1) – (E7) hold. Then there exists a meromorphic map $h : M \longrightarrow \mathbb{P}(V)$ such that (h, a_j) are free for $j = 0, 1, \ldots, n + 1$ and such that for each $s > 0$ there are constants $c_p(s)$ for $p = 1, \ldots, 4$ such that for all $r > s$ we have the estimates

$$(11.5)\qquad T_h(r,s) \leqslant T_f(r,s) + \sum_{j=0}^{n+1} T_{g_j}(r,s) + c_1(s)$$

$$(11.6)\qquad T_f(r,s) \leqslant T_h(r,s) + n \sum_{j=0}^{n+1} T_{g_j}(r,s) + c_2(s)$$

$$(11.7)\qquad N_{h,a_j}(r,s) \leqslant N_{f,g_j}(r,s) + \sum_{\substack{k=0\\k\neq j}}^{n+1} T_{g_k}(r,s) + c_3(s)$$

$$(11.8)\qquad N_{f,g_j}(r,s) \leqslant N_{h,a_j}(r,s) + (n+1) \sum_{k=0}^{n+1} T_{g_k}(r,s) + c_4(s) .$$

If in addition we assume that also (E8) holds, then we have

$$(11.9)\qquad T_h(r,s)/T_f(r,s) \longrightarrow 1 \qquad \text{for } r \longrightarrow \infty$$

$$(11.10)\qquad \delta_f(g_j) = \delta_f(a_j) \qquad \text{for } j = 0,1, \dots ,n+1 .$$

Basically the theorem is due to Mori, but does not give such explicit estimates. The proof is obtained in several steps.

a) A special operation. Let $\mathfrak{n}_0, \dots ,\mathfrak{n}_n$ be the dual base of $\mathfrak{a}_0, \dots ,\mathfrak{a}_n$. The base $\mathfrak{n}_0, \dots ,\mathfrak{n}_n$ is orthonormal. Abbreviate

$$(11.11)\qquad \mathfrak{n} = \mathfrak{n}_0 \wedge \dots \wedge \mathfrak{n}_n \qquad \mathfrak{a} = \mathfrak{a}_0 \wedge \dots \wedge \mathfrak{a}_n$$

If $\mathfrak{z}_0, \dots ,\mathfrak{z}_{n+1}$ are vectors in V^*, define

$$(11.12)\qquad \hat{\mathfrak{z}}_j = (-1)^j \mathfrak{z}_0 \wedge \dots \wedge \mathfrak{z}_{j-1} \wedge \mathfrak{z}_{j+1} \wedge \dots \wedge \mathfrak{z}_{n+1}$$

for $j \in \mathbb{Z}[0,n+1]$. If $j \in \mathbb{Z}[0,n]$, define

$$(11.13)\qquad \check{\mathfrak{z}}_j = (-1)^j \mathfrak{z}_0 \wedge \dots \wedge \mathfrak{z}_{j-1} \wedge \mathfrak{z}_{j+1} \wedge \dots \wedge \mathfrak{z}_n$$

Then $\hat{\mathfrak{z}}_j = \check{\mathfrak{z}}_j \wedge \mathfrak{z}_{n+1}$. A homogeneous, projective operation

(11.14)
$$\nabla : V \times (V^*)^{n+2} \longrightarrow V$$

is defined for $\mathfrak{x} \in V$, $\mathfrak{z}_j \in V^*$ with $j = 0, \ldots, n+1$ by

(11.15)
$$\mathfrak{x} \nabla \mathfrak{z}_0 \nabla \cdots \nabla \mathfrak{z}_{n+1} = \frac{1}{\sqrt{n+1}} \sum_{j=0}^{n} \langle \mathfrak{x}, \mathfrak{z}_j \rangle \langle \hat{\mathfrak{z}}_j, \mathfrak{n} \rangle \mathfrak{n}_j .$$

In fact, the operation is (n+3)–linear.

<u>LEMMA</u> 11.2. If $\mathfrak{x} \in V$ and $\mathfrak{z}_j \in V^*$ for $j = 0, \ldots, n+1$, then

(11.6)
$$\| \mathfrak{x} \nabla \mathfrak{z}_0 \cdots \nabla \mathfrak{z}_{n+1} \| \leqslant \| \mathfrak{x} \| \, \| \mathfrak{z}_0 \| \cdots \| \mathfrak{z}_{n+1} \| .$$

<u>PROOF</u>. If $j \in \mathbb{Z}[0,n]$, we have

$$| \langle \mathfrak{x}, \mathfrak{z}_j \rangle | \; | \langle \hat{\mathfrak{z}}_j, \mathfrak{n} \rangle | \leqslant \| \mathfrak{x} \| \, \| \mathfrak{z}_j \| \prod_{\substack{k=0 \\ k \neq j}}^{n} \| \mathfrak{z}_k \| = \| \mathfrak{x} \| \, \| \mathfrak{z}_0 \| \cdots \| \mathfrak{z}_{n+1} \|$$

Therefore we obtain

$$\| \mathfrak{x} \nabla \mathfrak{z}_0 \nabla \cdots \nabla \mathfrak{z}_{n+1} \| = \frac{1}{\sqrt{n+1}} \left[\sum_{j=0}^{n} | \langle \mathfrak{x}, \mathfrak{z}_j \rangle |^2 \; | \langle \hat{\mathfrak{z}}_j, \mathfrak{n} \rangle |^2 \right]^{1/2}$$

$$\leqslant \| \mathfrak{x} \| \, \| \mathfrak{z}_0 \| \cdots \| \mathfrak{z}_{n+1} \| . \qquad \text{q.e.d.}$$

<u>COROLLARY</u> 11.3. If $x \in \mathbb{P}(V)$ and $z_j \in \mathbb{P}(V^*)$ for $j = 0, \ldots, n+2$, then

(11.17)
$$\square \, x \, \dot{\nabla} \, z_0 \, \dot{\nabla} \, x_1 \, \dot{\nabla} \cdots \dot{\nabla} \, z_{n+1} \, \square \leqslant 1$$

<u>LEMMA</u> 11.4. If $x \in \mathbb{P}(V)$ and $z_j \in \mathbb{P}(V^*)$ for $j = 0, 1, \ldots, n+1$, then

$$(11.18)\qquad \prod_{j=0}^{n+1} \Box\, z_0 \,\dot\wedge\, \dots \,\dot\wedge\, z_{j-1} \,\dot\wedge\, z_{j+1} \,\dot\wedge\, \dots \,\dot\wedge\, z_{n+1}\, \Box$$

$$\leqslant (n+1)\, \Box\, x \,\dot\nabla\, z_0 \,\dot\nabla\, \dots \,\dot\nabla\, z_{n+1}\, \Box\, .$$

PROOF. Take $\mathfrak{x} \in V$ with $\|\mathfrak{x}\| = 1$ and $x = \mathbb{P}(\mathfrak{x})$. Take $\mathfrak{z}_j \in V^*$ with $\|\mathfrak{z}_j\| = 1$ and $z_j = \mathbb{P}(\mathfrak{z}_j)$ for $j = 0,1, \dots ,n+1$. Then

$$(11.19)\qquad \Box\, z_0 \,\dot\wedge\, \dots \,\dot\wedge\, z_{j-1} \,\dot\wedge\, z_{j+1} \,\dot\wedge\, \dots \,\dot\wedge\, \mathfrak{z}_{n+1}\, \Box = \|\hat{\mathfrak{z}}_j\| = |\langle \hat{\mathfrak{z}}_j, \mathfrak{n}\rangle|$$

$$(11.20)\qquad \Box\, x \,\dot\nabla\, z_0 \,\dot\nabla\, \dots \,\dot\nabla\, z_{n+1}\, \Box = \|\mathfrak{x} \,\dot\nabla\, \mathfrak{z}_0 \,\dot\nabla\, \dots \,\dot\nabla\, \mathfrak{z}_{n+1}\| \,.$$

We have

$$(11.21)\qquad \sum_{j=0}^{n} |\langle \mathfrak{x}, \mathfrak{a}_j\rangle|^2 = \Big\| \sum_{j=0}^{n} \langle \mathfrak{x}, \mathfrak{a}_j\rangle \mathfrak{n}_j \Big\|^2 = \|\mathfrak{x}\|^2 = 1$$

$$(11.22)\qquad \mathfrak{a} \llcorner \mathfrak{x} = (\mathfrak{a}_0 \wedge \dots \wedge \mathfrak{a}_n) \llcorner \mathfrak{x} = \sum_{j=0}^{n} \langle \mathfrak{x}, \mathfrak{a}_j\rangle \hat{\mathfrak{a}}_j \,.$$

Since $\hat{\mathfrak{a}}_0, \dots ,\hat{\mathfrak{a}}_n$ is an orthonormal base of $\bigwedge_n V^*$, we obtain

$$(11.23)\qquad \|\mathfrak{a} \llcorner \mathfrak{x}\| = \left[\sum_{j=0}^{n} |\langle \mathfrak{x}, \mathfrak{a}_j\rangle|^2 \right]^{1/2} = 1 \,.$$

If $j \in \mathbb{Z}[0,n+1]$, we have $\|\hat{\mathfrak{z}}_j\| \leqslant 1$. If $j \in \mathbb{Z}[0,n]$, we have $\|\hat{\mathfrak{z}}_j\| = \|\check{\mathfrak{z}}_j \wedge \mathfrak{z}_{n+1}\| \leqslant \|\check{\mathfrak{z}}_j\| \leqslant 1$. We obtain

$$\|\hat{\mathfrak{z}}_{n+1}\| = |\langle \hat{\mathfrak{z}}_{n+1}, \mathfrak{n}\rangle|\ \|\mathfrak{a} \llcorner \mathfrak{x}\|$$

$$= \|\langle \hat{\mathfrak{z}}_{n+1}, \mathfrak{n}\rangle \mathfrak{a} \llcorner \mathfrak{x}\|$$

$$= \|\hat{\mathfrak{z}}_{n+1} \llcorner \mathfrak{x}\|$$

$$= \| \sum_{j=0}^{n} \langle \mathfrak{z}, \mathfrak{a}_j \rangle \mathfrak{y}_j \|$$

$$\leq \sqrt{n+1} \left[\sum_{j=0}^{n} | \langle \mathfrak{z}, \mathfrak{a}_j \rangle |^2 \right]^{1/2}$$

$$\prod_{j=0}^{n+1} \|\hat{\mathfrak{a}}_j\| \leq \sqrt{n+1} \left[\sum_{j=0}^{n} | \langle \mathfrak{z}, \mathfrak{a}_j \rangle |^2 \, | \langle \hat{\mathfrak{a}}_j, \mathfrak{n} \rangle |^2 \right]^{1/2}$$

$$= (n+1) \| \mathfrak{z} \nabla \mathfrak{a}_0 \nabla \mathfrak{a}_1 \nabla \ldots \nabla \mathfrak{a}_{b+1} \| . \qquad \text{q.e.d.}$$

<u>**COROLLARY**</u> **11.5.** Take $x \in \mathbb{P}(V)$ and $z_j \in \mathbb{P}(V^*)$ for $j = 0,1, \ldots ,n+1$. Assume that $z_0, \ldots ,z_{n+1}$ are in general position. Then

$$\square\, x \,\dot{\nabla}\, z_0 \,\dot{\nabla} \ldots \dot{\nabla}\, z_{n+1} \,\square > 0 \tag{11.24}$$

and $x \nabla z_0 \nabla \ldots \nabla z_{n+1}$ exists.

<u>**PROOF**</u>. Since $z_0, \ldots ,z_{n+1}$ are in general position,

$$\square\, z_0 \,\dot{\wedge} \ldots \dot{\wedge}\, z_{j-1} \,\dot{\wedge}\, z_{j+1} \,\dot{\wedge} \ldots \dot{\wedge}\, z_{n+1} \,\square > 0$$

for all $j \in \mathbb{Z}[0,n+1]$. Thus (11.18) implies (11.24). q.e.d.

b) <u>**The Mori map**</u>. We assume that (E1) – (E7) hold. Take a point $x \in M$ such that $f, g_0, \ldots ,g_{n+1}$ are holomorphic at x and such that $g_0(x), \ldots ,g_{n+1}(x)$ is in general position. Then $f(x) \nabla g_0 \nabla \ldots \nabla g_{n+1}(x)$ exists. Hence $f, g_0, \ldots ,g_{n+1}$ is free for ∇. A meromorphic map

$$h = f \nabla g_0 \nabla \ldots \nabla g_{n+1} : M \longrightarrow \mathbb{P}(V) \tag{11.25}$$

called the <u>**Mori map**</u>, is defined. We obtain the <u>**operation divisor**</u>

(11.26) $$\eta = \mu_{f \dot{\nabla} g_0 \dot{\nabla} \ldots \dot{\nabla} g_{n+1}}$$

with <u>valence</u> <u>function</u> N_η and the distance

(11.27) $$\gamma = \Box f \dot{\nabla} g_0 \dot{\nabla} \ldots \dot{\nabla} g_{n+1} \Box \leqslant 1$$

with the <u>compensation</u> <u>function</u>

(11.28) $$m(r) = \int_{M\langle r\rangle} \log \frac{1}{\gamma}\, \sigma > 0 \qquad \text{for} \quad r \in \mathcal{L}_\tau$$

which extends to all $r > 0$ as a continuous function. Theorem 3.4 implies:

<u>THEOREM</u> 11.6 (The First Main Theorem for the Mori map). We assume that (E1) – (E7) hold. If $0 < s < r$, then

(11.29) $$T_f(r,s) + \sum_{j=0}^{n+1} T_{g_j}(r,s) = N_\eta(r,s) + m(r) - m(s) + T_h(r,s) \, .$$

<u>COROLLARY</u> 11.7. If $0 < s < r$, then

(11.30) $$T_h(r,s) \leqslant T_f(r,s) + \sum_{j=0}^{n+1} T_{g_j}(r,s) + m(s) \, .$$

Thus (11.5) is proved.

c) <u>Estimation</u> <u>of</u> **m(r)** <u>and</u> $N_\eta(r,s)$. For $j \in \mathbb{Z}[0,n+1]$, the function $g_0, \ldots, g_{j-1}, g_{j+1}, \ldots, g_{n+1}$ are free for $\wedge$. Hence the divisor

(11.31) $$\rho_j = \mu_{g_0 \dot{\wedge} \ldots \dot{\wedge} g_{j-1} \dot{\wedge} g_{j+1} \dot{\wedge} \ldots \dot{\wedge} g_{n+1}} \geqslant 0$$

exists with valence funtion $N_{\rho_j} \geqslant 0$. Also the distance

(11.32) $$\gamma_j = \Box g_0 \dot{\wedge} \ldots \dot{\wedge} g_{j-1} \dot{\wedge} g_{j+1} \dot{\wedge} \ldots \dot{\wedge} g_{n+1} \Box \leqslant 1$$

does not vanish identically. The compensation function

$$(11.33)\qquad m_j(r) = \int_{M\langle r\rangle} \log \frac{1}{\gamma_j}\,\sigma \geq 0$$

is defined for all $r \in \mathcal{E}_\tau$ and extends to all $r > 0$ as a continuous function. Abbreviate

$$(11.34)\qquad \hat{T}_j(r,s) = \sum_{\substack{k=0\\k\neq j}}^{n+1} T_k(r,s)\ .$$

The First Main Theorem for the exterior product (3.36) respectively (7.60) implies:

<u>LEMMA</u> **11.8.** If $0 < s < r$ and if $j \in \mathbb{Z}[0,n+1]$, then we have

$$(11.35)\qquad \hat{T}_j(r,s) = N_{\rho_j}(r,s) + m_j(r) - m_j(s)\ .$$

<u>LEMMA</u> **11.9.** For $r > 0$ we have

$$(11.36)\qquad m(r) \leq \sum_{j=0}^{n+1} m_j(r) + \log(n+1)$$

<u>PROOF</u>. Lemma 11.4, (11.7) and (11.32) imply $\gamma_0 \cdots \gamma_{n+1} \leq (n+1)\gamma$. Hence (11.28) and (11.33) imply (11.36). q.e.d.

<u>LEMMA</u> **11.10.** Let $U \neq \emptyset$ be an open, connected subset of M. Assume that there are reduced representations $\mathfrak{v} : U \longrightarrow V$ of f, $\mathfrak{w}_j : U \longrightarrow V^*$ of g_j for $j = 0,1,\ldots,n+1$ and $\mathfrak{f} : U \longrightarrow V$ of h. Then there is a holomorphic function $Y \not\equiv 0$ on U such that

$$(11.37)\qquad \mathfrak{y} = \sqrt{n+1}\ \mathfrak{v} \wedge \mathfrak{w}_0 \wedge \ldots \wedge \mathfrak{w}_{n+1} = Y\mathfrak{f}\ .$$

Moreover, if $j \in \mathbb{Z}[0,n+1]$, then

(11.38) $$\langle \mathfrak{w}, \mathfrak{v}_j\rangle\langle \hat{\mathfrak{v}}_j, \mathfrak{n}\rangle = Y\langle \mathfrak{f}, \mathfrak{a}_j\rangle .$$

PROOF. Since $\mathfrak{g} : U \longrightarrow V$ is a representation of h on U, a holomorphic function $Y \not\equiv 0$ on U exists such that $\mathfrak{g} = Y\mathfrak{f}$. We have

(11.39) $$\mathfrak{g} = \sum_{j=0}^{n} \langle \mathfrak{w}, \mathfrak{v}_j\rangle\langle \hat{\mathfrak{v}}_j, \mathfrak{n}\rangle\mathfrak{n}_j .$$

If $j \in \mathbb{Z}[0,n]$, we obtain

(11.40) $$\langle \mathfrak{w}, \mathfrak{v}_j\rangle\langle \hat{\mathfrak{v}}_j, \mathfrak{n}\rangle = \langle \mathfrak{g}, \mathfrak{a}_j\rangle = Y\langle \mathfrak{f}, \mathfrak{a}_j\rangle .$$

which proves (11.38) for $j \in \mathbb{Z}[0,n]$.

There are meromorphic functions $X_j \not\equiv 0$ on U such that

(11.41) $$\mathfrak{v}_{n+1} = \sum_{j=0}^{n} X_j\mathfrak{v}_j .$$

For $j \in \mathbb{Z}[0,n]$, we obtain

(11.42) $$\hat{\mathfrak{v}}_j = \check{\mathfrak{v}}_j \wedge \mathfrak{v}_{n+1} = X_j\, \check{\mathfrak{v}}_j \wedge \mathfrak{v}_j = -X_j\, \hat{\mathfrak{v}}_{n+1}$$

$$Y\langle \mathfrak{f}, \mathfrak{a}_{n+1}\rangle = \langle \mathfrak{g}, \mathfrak{a}_{n+1}\rangle = -\sum_{j=0}^{n} \langle \mathfrak{g}, \mathfrak{a}_j\rangle$$

$$= -\sum_{j=0}^{n} \langle \mathfrak{w}, \mathfrak{v}_j\rangle\langle \hat{\mathfrak{v}}_j, \mathfrak{n}\rangle = \sum_{j=0}^{n} \langle \mathfrak{w}, \mathfrak{v}_j\rangle X_j\langle \hat{\mathfrak{v}}_{n+1}, \mathfrak{n}\rangle$$

$$= \langle \mathfrak{w}, \mathfrak{v}_{n+1}\rangle\langle \hat{\mathfrak{v}}_{n+1}, \mathfrak{n}\rangle$$

which proves (11.38) for $j = n + 1$. q.e.d.

LEMMA 11.11. If $j \in \mathbb{Z}[0, n+1]$, then we have

$$\mu_{f,g_j} + \mu_{\rho_j} = \mu_\eta + \mu_{h,a_j} \tag{11.43}$$

$$\mu_\eta \leqslant \sum_{j=0}^{n+1} \mu_{\rho_j} \tag{11.44}$$

$$N_\eta(r,s) \leqslant \sum_{j=0}^{n+1} N_{\rho_j}(r,s) \qquad \text{for } 0 < s < r\,. \tag{11.45}$$

PROOF. It suffices to prove (11.43) and (11.44) on open, connected subsets $U \neq 0$ of M where we have representations $\mathfrak{w}$, $\mathfrak{v}_j$, $\mathfrak{y}$ and $\mathfrak{f}$ as described in Lemma 11.10. We have

$$\mu_{f,g_j}|U = \mu^0_{\langle \mathfrak{w}, \mathfrak{v}_j \rangle} \qquad \mu_{h,a_j}|U = \mu^0_{\langle \mathfrak{f}, \mathfrak{a}_j \rangle} \tag{11.46}$$

$$\mu_{\rho_j}|U = \mu^0_{\langle \hat{\mathfrak{v}}_j, \mathfrak{n} \rangle} \qquad \mu_\eta|U = \mu^0_Y\,. \tag{11.47}$$

Thus (11.38) implies (11.43). Also we have

$$\langle \hat{\mathfrak{v}}_{n+1}, \mathfrak{n} \rangle \mathfrak{a} \llcorner \mathfrak{w} = \hat{\mathfrak{v}}_{n+1} \llcorner \mathfrak{w} = (-1)^{n+1} \sum_{j=0}^{n} \langle \mathfrak{w}, \mathfrak{v}_j \rangle \check{\mathfrak{v}}_j\,. \tag{11.48}$$

Since $\mu_{\mathfrak{a} \llcorner \mathfrak{w}} = 0$, we obtain

$$\begin{aligned}
\mu_{\rho_{n+1}}|U &= \mu^0_{\langle \hat{\mathfrak{v}}_{n+1}, \mathfrak{n} \rangle} \\
&\geqslant \operatorname{Min}\{\mu^0_{\langle \mathfrak{w}, \mathfrak{v}_j \rangle} + \mu_{\hat{\mathfrak{v}}_j} \mid j \in \mathbb{Z}[0,n]\} \\
&\geqslant \operatorname{Min}\{\mu^0_{\langle \mathfrak{w}, \mathfrak{v}_j \rangle} \mid j \in \mathbb{Z}[0,n]\}
\end{aligned}$$

$$= \operatorname{Min}\{\mu_\eta | U + \mu_{h,a_j} | U - \mu_{\rho_j} | U \mid j \in \mathbb{Z}[0,n]\}$$

$$\geqslant \mu_\eta | U - \operatorname{Max}\{\mu_{\rho_j} | U \mid j \in \mathbb{Z}[0,n]\}$$

$$\geqslant \mu_\eta | U - \sum_{j=0}^{n} \mu_{\rho_j} | U$$

which proves (11.44). Trivially (11.44) implies (11.45). q.e.d.

<u>COROLLARY</u> **11.12.** If $0 < s < r$, then

(11.49) $$T_f(r,s) \leqslant T_h(r,s) + n \sum_{j=0}^{n+1} T_{g_j}(r,s) + \sum_{j=0}^{n+1} m_j(s) + \log(n+1)$$

<u>PROOF</u>. According to (11.29), (11.45), (11.36), (11.35) and (11.34) we have

$$T_f(r,s) + \sum_{j=0}^{n+1} T_{g_j}(r,s)$$

$$\leqslant T_h(r,s) + N_\eta(r,s) + m(r) - m(s)$$

$$\leqslant T_h(r,s) + \sum_{j=0}^{n+1} (N_{\rho_j}(r,s) + m_j(r)) + \log(n+1)$$

$$= T_h(r,s) + \sum_{j=0}^{n+1} \hat{T}_j(r,s) + \sum_{j=0}^{n+1} m_j(s) + \log(n+1)$$

$$= T_h(r,s) + (n+1) \sum_{j=0}^{n+1} T_{g_j}(r,s) + \sum_{j=0}^{n+1} m_j(s) + \log(n+1)$$

which implies (11.49). q.e.d.

Thus Corollary 11.12 proves (11.6).

d) <u>**The estimation of the valence functions**</u>. (11.43) and (11.35) imply

$$N_{h,a_j}(r,s) \leqslant N_{f,g_j}(r,s) + N_{\rho_j}(r,s) \leqslant N_{f,g_j}(r,s) + \hat{T}_j(r,s) + m_j(s)$$

which is

$$(11.50)\qquad N_{h,a_j}(r,s) \leqslant N_{f,g_j}(r,s) + \sum_{\substack{k=0\\k\neq j}}^{n+1} T_{g_k}(r,s) + m_j(s) .$$

Thus (11.7) is proved. (11.43), (11.45) and (11.35) imply

$$N_{f,g_j}(r,s) \leqslant N_{h,a_j}(r,s) + N_{\eta}(r,s)$$

$$\leqslant N_{h,a_j}(r,s) + \sum_{k=0}^{n+1} N_{\rho_k}(r,s)$$

$$\leqslant N_{h,a_j}(r,s) + \sum_{k=0}^{n+1} \hat{T}_k(r,s) + \sum_{k=0}^{n+1} m_k(s)$$

which is

$$(11.51)\qquad N_{f,g_j}(r,s) \leqslant N_{h,a_j}(r,s) + (n+1) \sum_{k=0}^{n+1} T_{g_k}(r,s) + \sum_{k=0}^{n+1} m_k(s) .$$

Thus (11.8) is proved. If we assume that (E8) holds in addition, then (11.5) and (11.6) prove (11.9) trivially. Also (11.7) and (11.8) proved (11.10) immediately. **Theorem 11.1 is proved.**

e) **Linear non–degeneracy.** In order to obtain a defect relation we must find conditions on f which assure that h is linearly non–degenerate.

Take $\mathfrak{b} = (b_0, \ldots, b_{n+2}) \in \mathbb{C}^{n+2}$. Let $U \neq \emptyset$ be a connected, open subset of M. Let $\mathfrak{w}_j : U \longrightarrow V^*$ be a reduced representation of g_j on U for $j \in \mathbb{Z}[0, n+1]$. A holomorphic vector function

$$(11.52)\qquad \mathfrak{y}_{\mathfrak{b}} = \sum_{j=0}^{n+1} b_j \langle \hat{\mathfrak{w}}_j, \mathfrak{n} \rangle \mathfrak{w}_j : U \longrightarrow V^*$$

is defined. The map $\mathfrak{b} \longrightarrow \mathfrak{y}_{\mathfrak{b}}$ is linear. Define the diagonal

(11.53) $$D = \{(b, \ldots ,b) \in \mathbb{C}^{n+2} \mid b \in \mathbb{C}\} .$$

LEMMA 11.13. $\mathfrak{y}_{\mathfrak{b}} \equiv 0$ if and only if $\mathfrak{b} \in D$.

PROOF. Define the meromorphic functions X_j by (11.41). Since $g_0, \ldots ,f_{n+1}$ are in general position, $X_j \not\equiv 0$. With (11.42) we have

(11.54) $$\mathfrak{y}_{\mathfrak{b}} = \sum_{j=0}^{n} (b_j\langle\widehat{\mathfrak{w}}_j,\mathfrak{n}\rangle + b_{n+1}X_j\langle\widehat{\mathfrak{w}}_{n+1},\mathfrak{n}\rangle)\mathfrak{w}_j$$

$$= \sum_{j=0}^{n} (b_j - b_{n+1})\langle\widehat{\mathfrak{w}}_j,\mathfrak{n}\rangle\mathfrak{w}_j .$$

Since $\langle\widehat{\mathfrak{w}}_j,\mathfrak{n}\rangle \not\equiv 0$ and $\mathfrak{w}_k, \ldots ,\mathfrak{w}_n$ are linearly independent, $\mathfrak{y}_{\mathfrak{b}} \equiv 0$ if and only if $b_j = b_{n+1}$ for all $j \in \mathbb{Z}[0,n]$. q.e.d.

Since $\mathfrak{y}_{\mathfrak{b}}$ is linear in $\mathfrak{b}$, we have $\mathfrak{y}_{\mathfrak{b}} = \mathfrak{y}_{\mathfrak{r}}$ if and only if $\mathfrak{b} - \mathfrak{r} \in D$.

Let $E = \mathbb{C}^{n+2}/D$ be the quotient vector space. Let $\rho : \mathbb{C}^{n+1} \longrightarrow E$ be the residual map. If $\mathfrak{r} \in E$ with $\rho(\mathfrak{b}) = \mathfrak{r}$, then $\mathfrak{y}_{\mathfrak{r}} = \mathfrak{y}_{\mathfrak{b}}$ is well-defined and the map $\mathfrak{r} \longrightarrow \mathfrak{y}_{\mathfrak{r}}$ is linear and injective.

If $U^0 \neq \emptyset$ is an open, connected subset of M with $U \cap U^0 \neq \emptyset$, and if $\mathfrak{w}_j^0 : U^0 \longrightarrow V^*$ is a reduced representation of g_j on U^0 for $j \in \mathbb{Z}[0,n + 1]$, then there are zero free holomorphic functions w_j on $U \cap U^0$ such that $\mathfrak{w}_j^0 = w_j\mathfrak{w}_j$ on $U \cap U^0$. If we define $w = w_0 \cdots w_{n+1}$ and

(11.55) $$\mathfrak{y}^0_{\mathfrak{b}} = \sum_{j=0}^{n+1} b_j\langle \widehat{\mathfrak{w}}_j, \mathfrak{n}\rangle \mathfrak{w}^0_j : U^0 \longrightarrow V^*$$

then $\mathfrak{y}^0_{\mathfrak{b}} = w\mathfrak{y}_{\mathfrak{b}}$ on $U \cap U^0$. Hence, take $c \in \mathbb{P}(E)$. Then there is one and only one meromorphic map $g_c : M \longrightarrow \mathbb{P}(V^*)$ defined by the property:

"Take $\mathfrak{b} \in \mathbb{C}^{n+2}$ with $c = \mathbb{P}(\rho(\mathfrak{b}))$. Then $\mathfrak{b} = (b_0, \ldots ,b_{n+1}) \neq 0$. let $\mathfrak{w}_j : U \longrightarrow V^*$ be a reduced representation of g_j for $j = 0,1, \ldots ,n + 1$. Then $\mathfrak{y}_{\mathfrak{b}}$ defined by (11.61) is a representation of g_c on U."

A set of meromorphic maps $\mathfrak{L} = \{g_c \mid c \in \mathbb{P}(E)\}$ is defined.

__LEMMA__ **11.14.** The map $\mathbb{P}(E) \longrightarrow \mathfrak{L}$ defined by $c \longrightarrow g_c$ is bijective.

__PROOF__. By definition, the map is surjective. Assume that $g_{c_1} = g_{c_2}$ for $c_i \in \mathbb{P}(E)$. Then $c_j = \mathbb{P}(\rho(\mathfrak{b}_j))$ where $\mathfrak{b}_i = (b^i_0, \ldots ,b^i_{n+1})$. Let $\mathfrak{w}_j : U \longrightarrow V^*$ be reduced representations of g_j for $j = 0,1, \ldots ,n+1$. Then there is a meromorphic function $G \not\equiv 0$ on U such that $\mathfrak{y}_{\mathfrak{b}_1} = G\mathfrak{y}_{\mathfrak{b}_2}$ on U. Hence (11.54) implies

(11.56) $$(b^1_j - b^1_{n+1}) = G(b^2_j - b^2_{n+1}) \qquad \text{for all } j \in \mathbb{Z}[0,n] .$$

Since $\mathfrak{b} \notin D$, we have $b^1_j - b^1_{n+1} \not\equiv 0$ for some $j \in \mathbb{Z}[0,n]$. Hence G is a constant and $\rho(b_1) = G\rho(\mathfrak{b}_2)$ which implies $c_1 = c_2$.
q.e.d.

Thus $\mathfrak{L}$ can be regarded as a projective space of dimension n.

The meromorphic map $f : M \longrightarrow \mathbb{P}(V)$ is said to be <u>linearly non-degenerate over</u> $\mathfrak{g}$ (or $g_0, \ldots, g_{n+1}$) if and only if (f, g_c) is free for every $c \in \mathbb{P}(E)$.

THEOREM 11.15. The meromorphic map $f : M \longrightarrow \mathbb{P}(V)$ is linearly non-degenerate over $\mathfrak{g}$ if and only if the Mori map $h : M \longrightarrow \mathbb{P}(V)$ is linearly non-degenerate.

PROOF. a) <u>Assume that h is linearly non-degenerate.</u> Take any $c \in \mathbb{P}(E)$. Take $\mathfrak{b} \in \mathbb{C}^{n+1} - D$ with $c = \mathbb{P}(\rho(\mathfrak{b}))$. Let $U \neq \emptyset$ be an open connected subset of M such that there are reduced representations $\mathfrak{v} : U \longrightarrow V$ of f and $\mathfrak{w}_j : U \longrightarrow V^*$ of g_j for $j = 0, 1, \ldots, n+1$ and $\mathfrak{f} : U \longrightarrow V$ of the Mori map h. Define $\mathfrak{y}$ and Y as in Lemma 11.10 and $\mathfrak{y}_{\mathfrak{b}}$ by (11.52). Then (11.54) implies

$$\langle \mathfrak{v}, \mathfrak{y}_{\mathfrak{b}} \rangle = \langle \mathfrak{v}, \sum_{j=0}^{n} (b_j - b_{n+1}) \langle \hat{\mathfrak{w}}_j, \mathfrak{n} \rangle \mathfrak{w}_j \rangle$$

$$= \sum_{j=0}^{n} (b_j - b_{n+1}) \langle \hat{\mathfrak{w}}_j, \mathfrak{n} \rangle \langle \mathfrak{v}, \mathfrak{w}_j \rangle$$

$$= \langle \mathfrak{y}, \sum_{j=0}^{n} (b_j - b_{n+1}) \alpha_j \rangle$$

$$= Y \langle \mathfrak{f}, \sum_{j=0}^{n} (b_j - b_{n+1}) \alpha_j \rangle \not\equiv 0 \, .$$

Hence (f, g_c) is free. The map f is linearly non-degenerate over $\mathfrak{g}$.

b) <u>Assume that f is linearly non-degenerate over</u> $\mathfrak{g}$. Take $b \in \mathbb{P}(V^*)$. Then $b = \mathbb{P}(\mathfrak{a})$ with $0 \neq \mathfrak{a} \in V^*$. Then $\mathfrak{a} = b_0 \alpha_0 + \ldots + b_n \alpha_n$. Put $b_{n+1} = 0$ and $\mathfrak{b} = (b_0, \ldots, b_{n+1}) \in \mathbb{C}^{n+1}$. Then $\mathfrak{b} \notin D$ and $c = \mathbb{P}(\rho(\mathfrak{b})) \in \mathbb{P}(E)$ exists. Take U, $\mathfrak{v}$, $\mathfrak{w}_j$, Y, $\mathfrak{y}$, $\mathfrak{f}$ as in a) and define $\mathfrak{y}_{\mathfrak{b}}$ by (11.52). By assumption $\langle \mathfrak{v}, \mathfrak{y}_{\mathfrak{b}} \rangle \not\equiv 0$. As in a) we have

$$\langle \mathfrak{f}, \mathfrak{r} \rangle = \frac{1}{Y} \langle \mathfrak{w}, \sum_{j=0}^{n} (b_j - b_{n+1}) \mathfrak{a}_j \rangle = \frac{1}{Y} \langle \mathfrak{w}, \mathfrak{g}_b \rangle \not\equiv 0 .$$

Therefore (h,b) is free. The map h is linearly non-degenerate. q.e.d.

If $g_0, \dots, g_{n+1}$ are constant, then f is linearly non-degenerate over $\mathfrak{G}$, if and only if f is linearly non-degenerate.

We shall restrict ourself to the case of a covering parabolic space, since the condition "h is general for B" does not easily translate into a condition on f. Thus we assume in addition

(E9) There is a proper, surjective, holomorphic map $\pi : M \longrightarrow \mathbb{C}^m$ such that $\tau = \|\pi\|^2$.

(E10) Let ρ be the branching divisor of π. Assume that

$$R_f = \limsup_{r\to\infty} \frac{N_\rho(r,s)}{T_f(r,s)} < \infty . \tag{11.57}$$

(E11) Assume that f is linearly non-generate over $\mathfrak{G}$.

(E12) At least one map g_j is not constant.

REMARK 1. Since the case of constant targets is already solved, it is reasonable to assume that at least one g_j is not constant. Then

$$\frac{T_{g_j}(r,s)}{\log r} \longrightarrow A_{g_j}(\infty) > 0 \qquad \text{for} \quad r \longrightarrow \infty \tag{11.58}$$

for this j. Hence we obtain

$$\frac{T_f(r,s)}{\log r} = \frac{T_f(r,s)}{T_{g_j}(r,s)} \frac{T_{g_j}(r,s)}{\log r} \longrightarrow \infty \qquad \text{for} \quad r \longrightarrow \infty . \tag{11.59}$$

Therefore f has transcendental growth. By (11.9) also the Mori map h has trancendental growth.

REMARK 2. By (E11) and Theorem 11.11 the Mori map h is linearly non-degenerate.

REMARK 3. By (11.9) $R_h = R_f < \infty$.

REMARK 4. If π is biholomorphic, $\rho = 0$ and $R_f = 0$. If π is not biholomorphic, but if the $(m-1)$-dimensional component of $\pi(\mathrm{supp}\ \rho)$ is affine algebraic, then $N_\rho(r,s) = O(\log r)$. By (11.58), $R_f = 0$. By a Theorem of Noguchi [70], $R_f < \infty$ if f separates the fibers of π. If $j \in \mathbb{Z}[0,n+1]$ exists such that g_j separates the fibers of π, then $R_{g_j} < \infty$ by Noguchi's theorem. Then (E8) implies $R_f = 0$.

The First Main Theorem for fixed targets (A.176) implies

THEOREM 11.16. Assume that (E1) – (E12) hold. Then

(11.60)
$$\boxed{\sum_{j=0}^{n+1} \delta_f(g_j) \leqq n + 1 + \frac{n(n+1)}{2} R_f}$$

If $M = \mathbb{C}^m$ and $\tau(\mathfrak{z}) = \|\mathfrak{z}\|^2$, the theorem was proved by Mori [63]. (He remarks that his original proof is incorrect and he is publishing a correction.) Our proof shows the power of the theory developed here, provides better estimates and applies to a more general situation.

§12. References

[1] L. Ahlfors, The theory of meromorphic curves, Acta Soc. Sci. Fenn, Nova Ser. A 3(4)(1941), 171-183.

[2] A. Andreotti and W. Stoll, Analytic and algebraic dependence of meromorphic functions. Lecture Notes in Mathematics, 234(1971), 390 pp. Springer-Verlag.

[3] A. Baerenstein, Proof of Edrei's spread conjecture. Proc. London Math. Soc., 26(1973), 418-434.

[4] A. Biancofiore, A hypersurface defect relation for a class of meromorphic maps, Trans. Amer. Math. Soc., 270(1982), 47-60.

[5] A. Biancofiore and W. Stoll, Another proof of the lemma of the logarithmic derivative in several complex variables. In "Recent developments in several complex variables," Annals of Math. Studies, 100, Princeton University Press, Princeton NJ, (1981), 29-45.

[6] L. Bieberbach, Beispiel zweier ganzer Funktionen zweier komplexer Variablen, welche ein schlichte volum treue Abbildung des $\mathbb{R}_4$ auf einen Teil seiner selbst vermitteln. Sitz. Ber. preuss. Akad. Wiss, 1933.

[7] E. Borel, Sur les zéros des fonctions entières, Acta Math., 20(1897), 357-396.

[8] R. Bott and S. S. Chern, Hermitian vector bundles and the equidistribution of the zeros of their holomorphic sections, Acta Math., 114(1965), 71-112.

[9] D. Burns, Curvature of the Monge-Ampère foliations and parabolic manifolds, Ann. of Math., 115(1982), 349-373.

[10] J. Carlson, A result on the value distribution of holomorphic maps $f : \mathbb{C}^n \longrightarrow \mathbb{C}^n$, Proc. Symp. in pure Math., 30 part 2(1977), 225-228.

[11] J. Carlson and Ph. Griffiths, Defect relation for equidimensional holomorphic mappings between algebraic varieties, Ann. of Math., 95(1972), 557-584.

[12] H. Cartan, Sur les zéros des combinaisons linéaires de p fonctions holomorphes données, Mathematica (cluj), 7(1933), 80-103.

[13] S. S. Chern, Complex analytic mappings of Riemann surfaces I, Amer. J. Math., 82(1960), 323-337.

[14] S. S. Chern, The integrated form of the first main theorem for complex analytic mappings in several variables, Ann. of Math., (2)71(1960), 536-551.

[15] C. T. Chuang, Une généralisation d'une inégalité de Nevanlinna, Scientia Sinica, 13(1964), 887-895.

[16] C. T. Chuang, On the distribution of values of meromorphic functions, Chinese Ann. Math., 1(1980), 91-114.

[17] E. F. Collingwood, Sur quelques théorèmes de M. Nevanlinna, CR Acad. Sci. Paris, 179(1924), 955-957.

[18] M. Cornalba and Ph. Griffiths, Analytic cycles and vector bundles on non-compact algebraic varieties, Invent. Math., 28(1975), 1-106.

[19] M. Cornalba and B. Shiffman, A counter example to the "Transcendental Bezout Problem," Ann. of Math., 96(1972), 402-406.

[20] M. Cowen, Hermitian vector bundles and value distribution for Schubert cycles, Trans. Amer. Math. Soc., 180(1973), 189-228.

[21] M. Cowen and Ph. Griffiths, Holomorphic curves and metrics of non-negative curvature, J. Analyse Math., 29(1976), 93-153.

[22] L. Dektjarev, The general first fundamental theorem of value distribution, Dokl. Akad. Nauk. SSR, 193(1970), (Soviet Math. Dokl., 11(1970), 961-963.)

[23] A. Dinghas, Wertverteilung meromorpher Funktionen in ein-und mehrfach zusammenhängenden Gebieten, Lecture Notes in Mathematics, 783(1980), 145 pp., Springer-Verlag.

[24] D. Drasin, The inverse problem of Nevanlinna theory, Acta Math., 138(1977), 83-151.

[25] J. Dusfresnoy, Sur les valeurs exceptionelles des fonctions méromorphes voisines d'une fonction méromorphe donnée, CR Acad. Sci. Parris, 208(1939), 255-257.

[26] A. Edrei, Solution of the deficiency problem for functions of small lower order, Proc. Lond Math. Soc., 26(1973), 435-445.

[27] P. Fatou, Sur les fonctions méromorphes de deux variable, CR Acad. Sci. Paris, 175(1922), 862-865, 1030-1033.

[28] W. H. J. Fuchs, The development of the theory of deficient values since Nevanlinna, Ann. Acad. Scie. Fennicae Ser. A. I. Math., 7(1982), 33-84.

[29] H. Fujimoto, Remarks to the uniqueness problem of meromorphic maps. I, II, III, IV Nagoya Math. J., 71(1978), 13-24, 25-41, ibid, 75(1979), 71-85, ibid 83(1981), 153-181.

[30] H. Fujimoto, On the defect relation for the derived curves of a holomorphic curve in $\mathbb{P}^n(\mathbb{C})$, Tôkoku Math. J., 34(1982), 141-160.

[31] H. Fujimoto, Non-integrated defect relation for meromorphic maps into $\mathbb{P}^{N_1}(\mathbb{C}) \times \ldots \times \mathbb{P}^{N_k}(\mathbb{C})$, (1983), pp. 44, preprint.

[32] P. M. Gauthier and W. Hengartner, The value distribution of most functions of one or several complex variables, Ann. of Math., 96(1972), 31-52.

[33] M. Green, Holomorphic maps into complex projective space omitting hyperplanes, Trans. Amer. Math. Soc., 169(1972), 89-103.

[34] M. Green, Some Picard theorems for holomorphic maps to algebraic varieties, Amer. J. Math., 97(1975), 43-75.

[35] Ph. Griffiths, Entire holomorphic mappings in one and several complex variables, Annals of Math. Studies, 85(1976), 99 pp., Princeton Univ. Press, Princeton, NY.

[36] Ph. Griffiths and J. King, Nevanlinna theory and holomorphic mappings between algebraic varieties, Acta Math., 130(1973), 145-220.

[37] F. Gross, Factorization of meromorphic functions, Math. Research Center. Naval Research Laboratory, Washington, D. C. (1972), pp 258.

[38] G. af Hällström, Über meromorphe Funktionen mit mehrfach zusammenhängenden Existenzgebieten, Acta Acad. Abo. Math. Phys., 12, 8, (1939), pp 100.

[39] W. K. Hayman, Meromorphic functions, Oxford Math. Monographs, (1964), pp 191.

[40] W. K. Hayman, Some achievements of Nevanlinna theory, Ann. Acad. Scie. Fennicae. Ser. AI Math., 7(1982), 65-71.

[41] W. K. Hayman, and P. B. Kennedy, Subharmonic functions, London Math. Soc. Monographs 9 Academic Press, London - New York - San Francisco (1976), pp. 281.

[42] W. Hengartner, Famille des traces sur les droites complexes d'une fonction plurisous harmonic du entiere dans $\mathbb{C}^n$, Comment. Math. Helv., 43(1968), 358-377.

[43] G. M. Henkin, Solutions with estimates of the H. Lewy and Poincare-Lelong equations. Constructions of functions of the Nevanlinna class with prescribed zeros in strictly pseudoconvex domains, Dokl. Akad. Nauk. SSSR 210(1975), 771-774. (Soviet Math. Dokl., 16(1975), 1310-1314.)

[44] C. W. Henson and L. A. Rubel, Some applications of Nevanlinna theory to mathematical logic: Identities of exponential functions, Trans. Amer. Math. Soc., 282(1984), 1-32.

[45] J. Hirschfelder, The first main theorem of value distribution in several variables, Invent. Math., 8(1969), 1-33.

[46] H. Kneser, Zur Theorie der gebrochenen Funktionen mehrerer Veränderlichen. Jber. Deutsch. Math. Verein., 18(1938), 1-38.

[47] R. O. Kujala, Functions of finite λ-type in several complex variables, Trans. Amer. Math. Soc., 161(1970), 327-358.

[48] O. Lehto, On the birth of Nevanlinna theory, Ann. Acad. Sci. Fennicae Ser. A, I. Math., 7(1982), 5-23.

[49] P. Lelong, Sur l'extension aux fonctions entières de n variables, d'ordre fini, d'un development canonique de Weierstrass, CR Acad. Sci. Paris, 237(1953), 865-867.

[50] P. Lelong, Intégration sur une ensemble analytique complexe, Bull. Soc. Math. France, 85(1975), 328-370.

[51] P. Lelong, Fonctions entières (n-variables) et fonctions plurisous-harmoniques d'ordre fini dans $\mathbb{C}^n$, J. Analyse Math., 12(1964), 365-407.

[52] L. Lempert, Boundary behavior of meromorphic functions of several variables, Acta Math., 144(1980), 1-25.

[53] B. Ja. Levin, Distribution of zeros of entire functions, Transl. of Math. Monographs 5, Americ. Math. Soc., (1964), pp 493.

[54] H. Levine, A theorem on holomorphic mappings into complex projective space, Ann. of Math., (2) 71(1960), 529-535.

[55] J. Miles, Quotient representation of meromorphic functions, J. Analyse Math., 25(1972), 371-388.

[56] R. E. Molzon, Sets omitted by equidimensional holomorphic mappings, Amer. J. Math., 101(1979), 1271-1283.

[57] R. E. Molzon, Degeneracy theorems for holomorphic mappings between algebraic varieties, Trans. Amer. Math. Soc., 270(1982), 183-192.

[58] R. E. Molzon, Some examples in value distribution theory, Lecture Notes in Mathematics, 981(1983), 90-101. Springer-Verlag.

[59] R. E. Molzon, B. Shiffman and N. Sibony, Average growth estimates for hyperplane sections of entire analytic sets, Math. Ann., 257(1981), 43-59.

[60] S. Mori, On the deficiencies of meromorphic maps of $\mathbb{C}^m$ into $\mathbb{P}^n(\mathbb{C})$, Nagoya Math. J., 67(1977), 165-176.

[61] S. Mori, The deficiencies and the order of holomorphic mappings of $\mathbb{P}^n$ into a compact complex manifold, Tôhoku Math. J., 31(1979), 285-291.

[62] S. Mori, Holomorphic curves with maximal deficiency sum, Kodai Math., J., 2(1979), 116-122.

[63] S. Mori, Remarks on holomorphic mappings, Contempory Math., 25(1983), 101-114.

[64] J. Murray, A second main theorem of value distribution theory on Stein manifolds with pseudoconvex exhaustion, Thesis, Notre Dame (1974), pp 1-69.

[65] R. Nevanlinna, Einige Eindentigkeitssätze in der Theorie der meromorphen Funktionen, Acta Math., 48(1926), 367-391.

[66] R. Nevanlinna, Le Théorème de Picard-Borel et la Théorie des Fonctions Meromorphes, Gauthiers-Villars, Paris (1929), reprint Chelsea-Publ. Co., New York (1974), pp 171.

[67] R. Nevanlinna, Eindeutige analytische Funktionen 2nd ed. Die Grundl. d. Math. Wiss., 46(1953), pp 379. Springer-Verlag.

[68] D. J. Newman, Problem 84-6*, The Math. Intelligencer, 6(2) (1984), 39.

[69] J. Noguchi, A relation between order and defects of meromorphic mappings of $\mathbb{C}^n$ into $\mathbb{P}^N(\mathbb{C})$, Nagoya Math. J., 59(1975), 97-106.

[70] J. Noguchi, Meromorphic mappings of a covering space over $\mathbb{C}^m$ into a projective variety and defect relations, Hiroshima Math. J., 6(1976), 265-280.

[71] J. Noguchi, Holomorphic curves in algebraic varieties, Hiroshima Math. J., 7(1977), 833-853. Supplement: Hiroshima Math. J., 10(1980), 229-231.

[72] J. Noguchi, On value distribution of meromorphic mappings of covering spaces over $\mathbb{C}^m$ into algebraic varieties, pp 35, preprint.

[73] G. Patrizio, Boundary behavior of meromorphic maps, Math. Ann., 261(1982), 111-132.

[74] E. Picard, Sur une propriété des fonctions entiéres, CR Acad. Sci. Paris, 88(1879), 1024-1027.

[75] J. L. Potier, Fibrés vectoriels de rang 1 d'ordre fini, Bull. Soc. Math. de France, 104(1976), 349-367.

[76] S. Rickmann, Value distribution of quasimeromorphic mappings, Ann. Acad. Sci. Fenn. A I. Math 7(1981), 81-85.

[77] L.I. Ronkin, Introduction to the theory of entire functions of several variables. 44 Transl. of Math Monog. (1974) pp273.

[78] L. Sario and K. Noshiro, Value distribution theory, Van Nostrand, Princeton, NJ, (1966), pp 236.

[79] B. Shiffman, Nevanlinna defect relations for singular divisors, Invent. math., 31(1975), 155-182.

[80] B. Shiffman, Holomorphic curves in algebraic manifolds, Bull. Amer. Math. Soc., 83(1977), 553-568.

[81] B. Shiffman, On holomorphic curves and meromorphic maps in projective spaces, Indiana Univ. Math. J., 28(1979), 627-641.

[82] B. Shiffmann, Introduction to Carlson-Griffiths equidistribution theory, Lecture Notes in Mathematics, 981(1983), 44-89. Springer-Verlag.

[83] B. Shiffman, New defect relations for meromorphic functions on $\mathbb{C}^m$, Bull Amer. Math. Soc. (New Series), 7(1982), 599-601.

[84] B. Shiffman, A general second main theorem for meromorphic functions on $\mathbb{C}^n$, Amer. J. Math., 106(1984), 509-531.

[85] H. Skoda, Croissance des fonctions entières s'annulant sur une hypersurface donnée de $\mathbb{C}^n$, Seminair P. Lelong 1970-71, Lecture Notes in Mathematics, 275(1972), 82-105. Springer-Verlag.

[86] H. Skoda, Valeurs au bord les solutions de l'opérateur d", et caracterisation des zeros des fonctions de la classe Nevanlinna, Bull. Soc. Math. France, 104(1976), 225-299.

[87] L. Smiley, Dependence theorems for meromorphic maps, Thesis, Notre Dame (1979), pp 57.

[88] L. Smiley, Geometric conditions for unicity of holomorphic curves, Contemp. Math., 25(1983), 149-154.

[89] J. Spellecy, A defect relation on polydiscs, Thesis, Notre Dame, pp 63.

[90] W. Stoll, Mehrfache Integrale auf komplexen Mannigfaltigkeiten, Math. Zeitschr., 57(1952), 116-154.

[91] W. Stoll, Ganze Funktionen endlicher Ordnung mit gegebenen Nullstellenflächen, Math. Zeitschr., 57(1953), 211-237.

[92] W. Stoll, Konstruktion Jacobischer und mehrfach periodischer Funktionen zu gegebenen Nullstellenflächen, Math. Zeitschr., 126(1953), 31-43.

[93] W. Stoll, Die beiden Hauptsätze der Wertverteilungstheorie bei Funktionen mehrerer komplexen Veränderlichen, I., Acta Math., 90(1953), 1-115, II Acta Math., 92(1954), 55-169.

[94] W. Stoll, The growth of the area of a transcendental analytic set I, II Math. Ann., 156(1964), 47-78, 144-170.

[95] W. Stoll, Normal families of non-negative divisors, Math. Zeitschr., 84(1964), 154-218.

[96] W. Stoll, A general first main theorem of value distribution, Acta Math., 118(1967), 111-191.

[97] W. Stoll, About the value distribution of holomorphic maps into projective space, Acta Math., 123(1969), 83-114.

[98] W. Stoll, Value distribution of holomorphic maps into compact, complex manifolds, LecAure Notes in Mathematics, 135(1970), pp. 267. Springer-Verlag.

[99] W. Stoll, Value distribution of holomorphic maps. Several Complex Variables I, Lecture Notes in Mathematics, 155(1970), 165-190. Springer-Verlag.

[100] W. Stoll, Deficit and Bezout estimates. Value Distribution Theory. Part B. (edited by R. O. Kujala and A. L. Vitter III), Pure and Appl. Math., 25 Marcell Dekker, New York (1973), pp 271.

[101] W. Stoll, Holomorphic functions of finite order in several complex variables, CBMS Regional Conference Series in Mathematics 21 Amer. Math. Soc., Providence, RI, (1974), pp 83.

[102] W. Stoll, Aspects of value distribution theory in several complex variables, Bull. Amer. Math. Soc., 83(1977), 166-183.

[103] W. Stoll, Value distribution on parabolic spaces, Lecture Notes in Mathematics, 600(1977), pp 216. Springer-Verlag.

[104] W. Stoll, A Casorati-Weierstrass theorem for Schubert zeros of semi-ample, holomorphic vector bundles, Atti Acad. Naz. Lincei. Mem. C1. Sci. Fis. Mat. Natur. Ser. VIIIm 15(1978), 63-90.

[105] W. Stoll, The characterization of strictly parabolic manifolds, Ann. Scuola. Norm. Sup. Pisa, 7(1980), 87-154.

[106] W. Stoll, The characterization of strictly parabolic spaces, Compositio Mathematics, 44(1981), 305-373.

[107] W. Stoll, Introduction to value distribution theory of meromorphic maps, Lecture Notes in Mathematics, 950(1982), 210-359. Springer-Verlag.

[108] W. Stoll, The Ahlfors-Weyl theory of meromorphic maps on parabolic manifolds, Lecture Notes in Mathematics, 981(1983), 101-219. Springer-Verlag.

[109] W. Stoll, Value distribution and the lemma of the logarithmic derivative on polydiscs, Internat. J. Math. Sci., 6(1983), no. 4, 617-669.

[110] P. Thie, The Lelong number of a point of a complex analytic set, Math. Ann., 172(1967), 269-312.

[111] M. Tsuji, Potential theory in modern function theory, Chelsea Publ. Co., New York, NY, 1975, pp 590.

[112] Ch. Tung, The first main theorem of value distribution on complex spaces, Atti della Acc. Naz d. Lincei Serie VIII,15(1979), 93-262.

[113] G. Valiron, Lectures on general theory of integral functions, Chelsea Publ. Co., New York, NY, 1949, pp 208.

[114] B. L. Van Der Waerden, Moderne Algebra I, 1 ed. Die Grundl. d. Math. Wiss., 33(1930), pp 243. Springer-Verlag.

[115] A. Vitter, The lemma of the logarithmic derivative in several complex variables, Duke Math. J., 44(1977), 89-104.

[116] K. T. W. Weierstrass, Theorie der eindeutigen analytischen Funktionen, Abhandl. Kön. Preuss. Akad. Wiss. Berlin, (1876), 11-60.

[117] A. Weitsman, A theorem on Nevanlinna deficiencies, Acta. Math., 128(1972), 41-52.

[118] H. Weyl and J. Weyl, Meromorphic curves, Ann. of Math., 39(1938), 516-538.

[119] H. Weyl and J. Weyl, Meromorphic functions and analytic curves, Annals of Mathematics Studies, 12, Princeton University Press, Princeton, NY, (1943), pp 269.

[120] W. Wirtinger, Ein Integral satz über analytische Gebilde im Gebiete von mehreren komplexen Veränderlichen, Monatshefte Math. Phys., 45(1937), 418-431.

[121] H. Wittich, Neuere Unterschungen über eindeutige Funktionen, Erg. d. Math. und ihrer Grenzgeb, 2 ed. (1968). Springer-Verlag.

[122] H. Wittich, Anwendungen der Wertverteilungslehre auf gewöhnliche Differentialgleichungen, Ann. Acad. Scie. Fennicae Ser. AI Math., 7(1982), 89-97.

[123] P. M. Wong, Defect relations for maps on parabolic spaces and Kobayashi metrics on projective spaces omitting hyperplanes, Thesis, Notre Dame, (1976), pp 231.

[124] P. M. Wong, Geometry of the homogeneous complex Monge-Ampere equation, Invent. Math., 67(1982), 261-274.

[125] H. Wu, Remarks on the first main theorem in equidistribution theory, I, II, III, IV, J. Differential Geometry, 2(1968), 197-202, 369-384, ibid 3(1969), 83-94, 433-446.

[126] H. Wu, The equidistribution theory of holomorphic curves, Annals of Mathematics Studies, 64, Princeton Univ. Press, Princeton, NJ, (1970), pp 219.

[127] Lo. Yang, Deficient functions of meromorphic functions, Scientia Sinica, 24(1981), 1179-1189.

[128] O. Zariski and P. Samuel, Commutative Algebraic I, D. Van Nostrand Co., Princeton, NJ, (1958), pp 329.

[129] H. J. W. Ziegler, Vector valued Nevanlinna Theory, Pitman Advanced Publ. Program. Research Notes in Math., 73(1982), pp 201.

REMARK. When this manuscript was being completed for publication, Professor Shiffman sent me a preprint of the paper:

[130] Charles F. Osgood, Sometimes effect Thue-Siegel-Roth-Schmidt-Nevanlinna Bounds, or better, to appear in Journal of Number Theory, pp 51.

Charles Osgood asserts that he proved the Nevanlinna Conjecture (11.2) under the assumption (11.1) by means of number theory. The paper is difficult to understand and is still under investigation. The result was announced in:

[131] Charles F. Osgood, A fully general Nevanlinna N-small function theorem and a sometimes effective Thue-Siegel-Roth-Schmidt Theorem for solutions to linear differential equation, Contemp. Math., 25(1983), 129-130.

Although the announcement seems to indicate, that the Nevanlinna conjecture was proved, it lacks a clear cut formulation of the result and it is not easy to understand.

Index